T0180047

Molecules of Death

Second Edition

Editors

R H Waring
University of Birmingham, UK

G B Steventon
King's College London, UK

S C Mitchell
Imperial College London, UK

Imperial College Press

ICP

Published by

Imperial College Press
57 Shelton Street
Covent Garden
London WC2H 9HE

Distributed by

World Scientific Publishing Co. Pte. Ltd.
5 Toh Tuck Link, Singapore 596224
USA office: 27 Warren Street, Suite 401-402, Hackensack, NJ 07601
UK office: 57 Shelton Street, Covent Garden, London WC2H 9HE

British Library Cataloguing-in-Publication Data
A catalogue record for this book is available from the British Library.

MOLECULES OF DEATH (2nd Edition)

ISBN-13 978-1-86094-814-5
ISBN-10 1-86094-814-6
ISBN-13 978-1-86094-815-2 (pbk)
ISBN-10 1-86094-815-4 (pbk)

Printed in Singapore.

Molecules
of Death

Second Edition

"And Darkness and Decay and the Red Death held illimitable dominion over all."

The Masque of the Red Death
Edgar Allan Poe 1809–1849

INTRODUCTION

"The 'Red Death' had long devastated the country. No pestilence had ever been so fatal, or so hideous. Blood was its Avatar and its seal — the redness and the horror of blood. There were sharp pains, and sudden dizziness, and then profuse bleeding at the pores, with dissolution. The scarlet stains upon the body and especially upon the face of the victim, were the pest ban which shut him out from the aid and from the sympathy of his fellow-men. And the whole seizure, progress, and termination of the disease, were the incidents of half-an-hour".

The Masque of the Red Death
Edgar Allan Poe 1809–1849

Simply a terrifying Gothic tale conjured from the deep recesses of a tormented mind? No, not necessarily. Such diseases exist, albeit thankfully rare, within the group of haemorrhagic fevers caused by viral infection. In the more subdued language of modern science, "they are characterised by an insidious onset of influenza-like symptoms followed, in severe cases, by a generalised bleeding diathesis (spontaneous bleeding), encephalopathy and death" (Cummins D, 1991, *Blood Rev*, 5, 129–137). The time-scale may be several days not half-an-hour but, with poetic licence, the visible oozing of blood from mucous membranes, the feverish state of mind and the general disintegration of the body induced by massive internal haemorrhage and its sequelae are correct. The latter aspect of bodily destruction can be exemplified by another disease, necrotising fasciolitis, an as yet uncommon soft-tissue infection usually caused by toxin-producing virulent bacteria such as ß-haemolytic Streptococci, and which is characterised by wide-spread cellular necrosis. "It starts typically with a purple lesion of the skin, followed by necrotising fascial infection with secondary necrosis of the overlying skin (and underlying muscle) and rapid progression to septic shock and multiorgan failure." (Eugester T *et al.*, 1997, *Swiss Surg*, **3**, 117–120). "... microorganisms rapidly spread along the fascial plane, causing necrosis of the fascia, overlying skin, and vasculature. Septicaemia and systemic toxic effects may lead to death within as short a time as 24 to 96 hours". (Gillen PB, 1995, *J Wound Ostomy Continence Nurs*, **22**, 219–222). The microbes produce hyaluronidase, an enzyme which digests hyaluronic acid and thus liquifies the ground substance of the connective tissue, making it easier to advance and clearing the way for the invasion, spreading out in waves from the point of infection.

Living Systems

In the primordial soup, energy from solar radiation and electromagnetic discharge was trapped by small molecules, exciting them and giving them sufficient energy to interact and eventually create larger and more complex organic molecules. As this process persisted, aggregation and coalescence of large molecules into ordered functional systems facilitated the continuum from chemical to biological evolution.

With increase in molecular complexity came the ability to self-assist chemical reactions. Pockets enfolded within large molecules could impose advantageous electronic surroundings and help guide reactive entities together, acting as catalysts to ensure chemical reactions occurred against unfavourable odds. As these enzymes became more efficient and reliable, self-replicating macromolecules evolved. These were probably only poor copies at first, but as the enzymes refined, more exact replicates would result. However, without a barrier to protect these groups of interactive macromolecules from continual chemical assault, little overall progress could be made. It was necessary to maintain constant and favourable conditions, to isolate a small section of the territory.

The development of molecules possessing areas which either attracted or repelled the ionic aqueous milieu permitted their alignment and integration, so that their hydrophilic areas were always presented to the ubiquitous watery habitat whereas their hydrophobic ends turned inwards towards themselves, repelled from the water, forming a sphere. Like primitive fat droplets, the inside was protected from changes occurring on the outside; primordial membranes emerged. Characteristics of these primitive membranes allowed the selective flow of molecules, enabling certain components to be concentrated inside and other components to be actively excluded. Thus, for the first time the medium in which chemical reactions could take place could be controlled and optimised. If the interactive macromolecules could become trapped within a primitive membrane, a progenitor cell would be formed.

The Cellular Machine

All living entities are comprised of cells. The cell is the basic living unit of all organisms. Smaller systems exist, for example viruses, but these can only become alive when they enter and usurp the cell's normal function.

Most cells possess a nucleus which is bound within a membrane and contains a complex macromolecule, DNA, which encodes instructions for the construction of the complete cell. Mutations in the DNA, which may arise for many reasons, can lead to disruption of cellular function and the production of abnormal and malfunctioning components. Within the cytoplasm, outside the nucleus but bound by the cell membrane, lie many organelles. The endoplasmic reticulum is a series of phospholipid tubes or cavities which traverse the cytosol assisting in intracellular transport and containing enzymes which synthesize proteins. The Golgi apparatus, a specialised region of the endoplasmic reticulum, is responsible for the collection, modification and export of metabolic products from the cell, usually packaged into membraneous vesicles. Mitochondria are elongated peanut-shaped inclusions which have a double membrane. Whilst the outer membrane is smooth, the inner membrane is folded into numerous elongated projections housing a series of enzymes and linked protein complexes which transport electrons. This is the powerhouse of the cell and is concerned with the major catabolic processes of cellular respiration, whereby energy stored in metabolic fuels is made available. In addition, many cells, depending upon their particular specialist function, contain other inclusions such as lysosomes, centrioles and microtubular networks as well as droplets of fat, protein granules, and various crystalline substances.

Both synthesis of new cells and the maintenance of existing cellular function requires the expenditure of energy. It is a constant battle between cellular degradation, sometimes accelerated by the ingestion of toxins, and necessary repair. This energy comes from foods. In effect, the carbon atoms contained within the proteins, carbohydrates and fats eaten in our everyday diet are converted, in a controlled stepwise degradation, into carbon dioxide which is exhaled. Hydrogen atoms are converted to water and nitrogen is lost from the body as urea. The increments of energy released from these complex molecules as they are systematically disassembled is, via the assistance of the mitochondria, stored in high energy molecules called adenosine triphosphate (ATP). This ATP acts like a rechargeable battery, enabling the energy released by all these catabolic pathways to be stored and moved around the body to where it is required. Once the ATP has discharged its energy, it flows back to the mitochondria for recharging.

Routes of Entry

Toxins, or poisons, can enter the body by a number of routes. The more usual, and perhaps accidental, routes are called the "portals of entry". These comprise those areas of the body that are in constant contact with the external environment and over which we have little obvious control, unless special precautions are taken.

The skin is the largest organ of the body, composed of approximately 20 square feet of tough resilient tissue, and it forms the interface between ourselves and the outside world. Usually, although not always, the majority of our skin is covered with clothing but the face and hands, perhaps the arms and lower legs, may normally be exposed. The skin is waterproof and forms an effective barrier to many potentially noxious chemicals. However, certain lipid-soluble compounds may penetrate the skin readily, crossing the outer epidermis and entering the lower dermis with its direct access to the circulatory system. Entry may also be gained through abrasions in the skin surface or via hair follicles, nail beds and sweat glands which traverse the epidermis.

Many substances enter the body via the mouth, taken in deliberately or accidentally with food and drink. A small number of compounds, such as nicotine, may be absorbed directly across the thin membranes lining the nose and mouth and rapidly enter the circulatory system. Others have to travel down into the stomach and small intestine where, dependent upon their physicochemical properties, they are absorbed across membranes and enter blood or lymphatic vessels. The blood draining from the stomach to the top of the rectum enters a special series of blood vessels called the hepatic portal system which delivers the blood directly to the liver. The liver is initially important in protection against toxicity and, unlike other tissues and organs in the body, can regenerate after damage or surgical resection. The liver contains a range of enzymes which are able to metabolise chemicals usually, but not always, decreasing their potential activity thereby limiting their ability to interfere with cellular function, and also increasing their water solubility making it easier to actively remove them from the body via the kidneys into the urine. Thus, because of its unique anatomical position between the incoming blood from the gastroinestinal tract and the outgoing blood into the body's systemic circulation, the liver is able to act as a filter, removing or deactivating potentially toxic compounds. Fat soluble compounds may enter the lymphatic fluid instead of the portal blood. The lymphatic system provides a short-circuit bypassing the liver and draining its fluid directly into the systemic circulation via the thoracic duct.

Although restraint can be shown over what is ingested and it is possible to avoid eating until the surrounding area is safe and clean, there is little choice over what is inhaled. It usually has to be the air around us, unless a portable supply is to hand, and breathing cannot be suspended for more than two to three minutes without deleterious consequences. Assuming at rest, 12 breaths per minute and a tidal volume of 500 ml, 8640 litres of air are taken into the lungs every day. The lungs have a large surface area, equivalent to that of a tennis court, and very delicate membranes which have evolved to act as gaseous exchange surfaces. Compounds, if volatile, can easily and rapidly pass into the blood stream from the lungs and this "portal of entry" initially circumvents the liver with its detoxication capacity.

The Toxicity Process

Once a compound has entered the systemic circulation it can be distributed around the body in a matter of minutes. Whether or not it actually enters a particular tissue depends upon a variety of factors, but some tissues may be particularly susceptible whereas others may have extra protection. For instance, the central nervous system is surrounded by layers of lipid and protein collectively called the "blood-brain barrier" which protects it from water-soluble ionic compounds but it is readily pemeable to many fat-soluble substances. In different circumstances, depending upon the toxicity of the compounds involved, this may be either beneficial or damaging.

Damage to a cell can occur in many different ways. These events may lead to the eventual death of the cell, and if sufficient cells are destroyed this will lead to the death of a tissue or an entire organ. Alternatively, they may cause a proliferative response where the cells may be damaged but nevertheless grow and divide causing organ enlargement or neoplasia. The sequence of events displayed by a cell showing a toxic response can be complicated. It is difficult to tease apart cause and effect, especially if an initial effect leads on to further responses. In this respect, observations of changes within a cell are usually classified under primary, secondary and tertiary stages, although there may be significant overlap.

Those processes included within the primary stage can be thought of as "event initiators" and involve the generation of highly reactive chemical entities such as free radicals or other electrophilic species and the decrease of free thiol levels within the cell, especially glutathione which acts as a vital protective agent. The loss or reduction of blood flow to a tissue depletes oxygen supplies which also aids cellular damage. Once produced, reactive chemical species, especially free radicals, can cause a cascade of peroxidative damage which leads to alteration in the structure and configuration of lipid components. Like a wave caused by a stone thrown into water, the initial chemical reaction is self-propagating, wreaking damage whilst liberating other free radicals to ripple outwards and continue the assault.

These events lead to a general macromolecular disruption which proceeds to secondary consequences. Both organelle and cell membranes, which are composed of a protein and phospholipid mosaic, can be damaged by these peroxidative changes. Alterations in fluidity and permeability lead to changes in intracellular ion concentrations, particularly calcium, and leakage of enzymes, both from intracellular organelles into the cytoplasm and from the entire cell into the intercellular medium. Interference with the correct functioning of mitochondrial membranes will eventually result in a shortage of high energy ATP, leading to a decrease in the ability to repair ongoing cellular damage. The rising levels of intracellular calcium interfere with cytoskeletal function leading to cellular disorganisation and also, perhaps more importantly, the activation of a series of autodestructive enzymes which degrade proteins and phospholipids and in particular the endonucleases which disassemble DNA in an ordered fashion during the process of apoptosis or "programmed cell death".

Finally, during the tertiary stage, gross changes to the cell's appearance occur. Steatosis, or the accumulation of fat, may take place owing to disruption in lipid handling. The intake of water, owing to membrane disfunction, causes swelling of the cell and is termed hydropic degeneration. The cell membrane may appear pitted or blebbed and vacuoles may form within the cytoplasm.

Eventually, the endoplasmic reticulum and mitochondria may become grossly distorted, leak and rupture. The nucleus loses its structure, with the nuclear material condensing and becoming fragmented or just simply fading and dissolving away. These latter processes of irreversible damage are known as necrosis.

Provision of a Toxin

A pollutant or contaminant is usually regarded as that which makes something else impure by contact or mixture, to make it foul or filthy, to defile, sully, taint or infect. To pollute in common usage is generally taken as being "more dirty" than to contaminate. However, this is not always sharply defined. That which may be a contaminant or pollutant in one particular situation may not be so in another. For example, a red poppy in a garden is a desirable thing, a delicate and pretty flower adding its own beauty to the surroundings. A few perhaps out of place but growing wild in a hedgerow are also pleasant, but when the numbers increase vastly and poppies grow unrestricted in a corn field then they present a problem. Poppy seeds harvested with the adjacent corn will contaminate the grain and, if in sufficient concentration, will render it unusable. A contaminant, therefore, appears to be something "out of place" and usually, although not always, has to be in a high enough concentration to produce a problem. To add confusion, opinions also change as to what is a contaminant; some are obvious to recognise whereas others are not. Within a short period of 15 years, five elements (selenium 1957, chromium 1959, tin 1970, vanadium 1971, fluorine 1972) which had been regarded previously as only environmental contaminants were shown to be beneficial micronutrients assisting in the continuation of life. This is only one example which has prompted concern to be expressed over the current continuing trend to "overpurify" the environment.

The population usually tries to live in areas which present the least problems in terms of immediate toxic hazard. However, certain communities do appear to dwell in locations which may be somewhat risky, perhaps unknowingly or because it is accepted as a part of their everyday life. Indeed, some individuals deliberately expose themselves to dangerous materials for a variety of reasons. Nevertheless, the surroundings in which we live and work are not devoid of toxic substances. Chemicals from industry and agriculture are present in all parts of the environment and may eventually accumulate in food chains. Direct exposure in our everyday work may also be a problem.

In general, a contaminant or potentially hazardous material may be presented to the population in one of several ways. It may be uncovered by natural processes such as water erosion or a volcanic eruption or it may be brought to the surface by man-made activities such as mining or quarrying. Alternatively, it may be a new problem in that it is a completely novel synthetic compound not encountered before in nature, or it may be a natural product which is now, owing to man's intervention, produced in quantities not previously seen.

When a potentially hazardous material is widely dispersed it is usually innocuous, and this is the ideology behind diluting a substance out of significance, although the logical extension of dumping waste materials into the vast oceans may not be too wise. For a "toxic event" or "toxic episode" to occur, areas of relatively high concentration of the hazardous material are usually required, although, of course, there are exceptions.

Such factors which bring about high local concentrations within a confined space can again be natural or man-made, or a combination of both. Rain soaking into the ground dissolves low concentrations of materials and then brings together that which has been leached out of perhaps thousands of square miles of land into a river valley, estuary and eventually a delta where the fast moving water hits the relatively static ocean and deposits the silt it has been carrying onto a fertile flood plain. An ideal location for a community to settle and thrive. If man has interfered and this hinterland is covered with mining spoil tips, composed of rejected material brought up from deep below the surface where it has been adequately hidden for aeons, then leaching of materials, especially heavy metals, will be enhanced and potential endemic problems exacerbated. Man-made structures, such as mines, quarries, factories and even cities can also serve to actively concentrate materials which are either extracted or artificially brought into these areas and then handled within confined spaces thereby delimiting their dispersion.

One interesting aspect is that of biomagnification. This is the process whereby a compound which appears in extremely low and harmless amounts, usually within the oceans, is concentrated as it passes through the food chain. Plankton within the seas may accumulate a toxin as they feed on passing particulate matter. Other sea creatures may then ingest this plankton and they themselves are eaten by a series of larger and more predacious creatures until relatively high concentrations reside within the fish population. Provided that the accumulated toxin is not poisonous, or has not yet reached a level where it is poisonous, to the fish, then this may be passed on to animals or humans who eat seafood. Eventually, when the concentration reaches toxic levels, or if a species is more susceptible than others, then damage, disease and possible death ensue. Concentrations in the order of a million-fold or more are not unusual within this process.

The following chapters give examples of substances derived from the surrounding biosphere. Complex organic molecules and simple inorganic moieties have been included, originating from the earth itself, constructed by the fascinatingly complicated intermediary biochemistry of living organisms or designed for nefarious reasons by the hand of man. Some of these may be familiar and others not, but all can be potentially lethal, justifiably earning them the appellation, "molecules of death".

Further Reading

Mason SF (1990). *Chemical Evolution*. Oxford University Press, Oxford.

Mitchell SC, Carmichael PL and Waring RH (2004). The three cornerstones of toxicology. *Biologist* **51**, 212–215.

Timbrell JA (2000). *Principles of Biochemical Toxicology*, 3rd ed. Taylor & Francis, London.

CONTENTS

LIST OF CONTRIBUTORS

S. Aldred
School of Sport and Exercise Sciences
University of Birmingham
Edgbaston
Birmingham B15 2TT
United Kingdom

R. A. Braithwaite
Regional Laboratory for Toxicology
City Hospital NHS Teaching Trust
Dudley Road
Birmingham B18 7QH
United Kingdom

J. Burdon
Department of Chemistry
University of Birmingham
Edgbaston
Birmingham B15 2TT
United Kingdom

J. W. Gorrod
Toxicology Unit
John Tabor Laboratories
University of Essex
Wivenhoe Park,
Colchester CO4 3SQ
United Kingdom

R. M. Harris
School of Biosciences
University of Birmingham
Edgbaston
Birmingham B15 2TT
United Kingdom

A. Hutt
Department of Pharmacy
King's College London
Franklin-Wilkins Building
Stamford Street
London SE1 9NN
United Kingdom

S. C. Mitchell
Faculty of Medicine
Imperial College London
Sir Alexander Fleming Building
South Kensington
London SW7 2AZ
United Kingdom

T. Pawade
Department of Medicine
University of Birmingham
Edgbaston
Birmingham B15 2TH
United Kingdom

D. B. Ramsden
Department of Medicine
University of Birmingham
Edgbaston
Birmingham B15 2TH
United Kingdom

J. C. Ritchie
GlaxoSmithKline Pharmaceuticals
Transnational Regulatory Affairs and Compliance
New Frontiers Science Park
Third Avenue, Harlow
Essex CM19 5AW
United Kingdom

M. J. Ruse
International Programme on Chemical Safety
World Health Organization
1211 Geneva 27
Switzerland

I. C. Shaw
College of Science
University of Canterbury
Private Bag 4800
Christchurch
New Zealand

M. E. Smith
Division of Medical Sciences
Medical School
University of Birmingham
Edgbaston
Birmingham B15 2TH
United Kingdom

G. B. Steventon
Department of Pharmacy
King's College London
Franklin-Wilkins Building
Stamford Street
London SE1 9NN
United Kingdom

R. A. R. Tasker
Department of Biomedical Sciences
Atlantic Veterinary College
University of Prince Edward Island
Charlottetown
Prince Edward Island, C1A 4P3
Canada

M.-C. Tsai
Toxicology Unit
John Tabor Laboratories
University of Essex
Wivenhoe Park
Colchester CO4 3SQ
United Kingdom

R. H. Waring
School of Biosciences
University of Birmingham
Edgbaston
Birmingham B15 2TT
United Kingdom

M. Wheatley
School of Biosciences
University of Birmingham
Edgbaston
Birmingham B15 2TT
United Kingdom

L. J. Wilkinson
School of Biosciences
University of Birmingham
Edgbaston
Birmingham B15 2TT
United Kingdom

J. Woodhouse
Centre for Toxicology
University of Central Lancashire
Preston PR1 2HE
United Kingdom

AFLATOXIN

J. C. Ritchie

> Beneath those rugged elms, that yew-tree's shade,
> Where heaves the turf in many a mouldering heap,
> Each in his narrow cell for ever laid,
> The rude forefathers of the hamlet sleep.
>
> *Elegy Written in a Country Churchyard*
> *Thomas Gray 1716–1771.*

In this famous poem Thomas Gray describes a hamlet's dead forefathers quietly mouldering under "heaves" of turf in a country churchyard. This represents a traditional view whereby moulds were principally associated with decay and disintegration of living matter following death.

However, in recent years scientific investigations have revealed the wider economic, toxicological and public health importance of certain mould species. For example, it is now known that in some circumstances mould-infected foods can be associated with serious toxicity, and sometimes death.

In this chapter the toxicity of a group of compounds called aflatoxins are described in order to illustrate the importance of the toxins produced by moulds.

Description, Occurrence, Uses

Description

Moulds are organisms belonging to the fungal kingdom. They are either saprophytic, growing on dead organic matter, or more rarely parasitic, existing on other living organisms. They are capable of growing on many substances of importance to man (e.g. foodstuffs, wood, clothing), their growth often being highly dependent on the presence of appropriate conditions of humidity and temperature.

Some moulds are beneficial and economically important. These include the cultivated varieties used in cheese making which provide the distinctive aroma, taste and veining which makes these cheeses so attractive and delicious (e.g. Roquefort, Blue Vinney, Stilton). Penicillium is an example of another well-known mould made famous by Sir Alexander Fleming when he discovered the potential of penicillin, produced by the mould, as an antibacterial medicine.

However, many moulds are far from beneficial to man. They may damage stored food, clothing, leather, wood and other materials of economic importance. They may also cause extensive crop losses in the form of blights and rusts. Finally, they may pose health hazards by producing toxic substances called mycotoxins (from the Greek: *mukes*~mushroom, *toxikon*~toxic). The enormous public health and economic implications of mycotoxin contamination are illustrated by the fact that the Food and Agricultural Organisation of the United Nations estimates that up to 25% of the worlds food crops are affected by mycotoxins.

Examples of mycotoxins include the ergot alkaloids produced when the ergot fungus grows on rye (responsible for outbreaks of a disease called ergotism, or St Anthony's Fire), trichothecanes produced by *Fusarium* species (associated with alimentary toxic aleukia fatalities in the Second World War) and the aflatoxins.

Aflatoxins are a group of chemically related mycotoxins which are produced by particular species of moulds. Their name derives from the fungus *Aspergillus flavus* on which much of the early work with these substances was performed (i.e. the genus **A**spergillus, the species **fla**vus and the suffix **toxin**).

Subsequent research revealed that aflatoxins are produced by strains of *A. flavus* and strains of the related species *A. parasiticus, A. nominus* and *A. niger*. Furthermore, it was discovered that there are a number of distinct, but structurally related aflatoxin compounds — the four most commonly seen being designated B1, B2, G1 and G2. The B designation of aflatoxins B1 and B2 resulted from the exhibition by these compounds of **B**lue fluorescence under the ultraviolet (UV)-light, whereas the G designation refers to yellow-**G**reen fluorescence under UV-light.

Aflatoxins M1 and M2, are hydroxylated derivatives of aflatoxin B1 and B2 that may be found in milk, milk products or meat (hence the designation M). They are formed by metabolism of B1 and B2 in the body of the animals, following absorption of contaminated feed. Aflatoxin B1 is the most frequent of these compounds present in contaminated food samples and aflatoxins B2, G1 and G2 are generally not reported in the absence of aflatoxin B1.

Thus, the aflatoxins form a family of highly oxygenated heterocyclic compounds with closely similar chemical structures, that are formed naturally by certain species of moulds.

Occurrence

Human exposure to aflatoxins occurs mainly through growth of the *Aspergillus* species *A. flavus* and *A. parasiticus*. Whether exposure is predominantly to aflatoxin B1, or to mixtures of various aflatoxins, depends upon the geographical distribution of the strains. *A. flavus*, which produces aflatoxins B1 and B2, occurs worldwide, while A. *parasiticus*, which produces aflatoxins B1, B2, G1 and G2, occurs principally in the Americas and in Africa.

Aflatoxins occur both in food crops in the field prior to harvest, and in improperly stored food where mould species have found an opportunity to grow. Fungal growth and aflatoxin contamination are a consequence of an interaction between the mould, the host organic material (i.e. crop, foodstuff) and the environment. The appropriate combination of these factors determines the degree of the colonisation of the substrate, and the type and amount of aflatoxin produced. Humidity, temperature and insect damage of the host substrate are major determining environmental factors in mould infestation and toxin production.

In addition, specific crop growth stages, poor fertility, high crop densities and weed competition have all been associated with increased mould growth and toxin production. For example, preharvest aflatoxin contamination of peanuts and corn is favoured by high temperatures, prolonged drought conditions and high insect activity; while postharvest production of aflatoxins on corn and peanuts is favoured by warm temperatures and high humidity.

Aflatoxins have been detected in milk, cheese, corn and other cereals, peanuts, cottonseed, nuts, figs and other foodstuffs. Milk and milk products, eggs and meat products are sometimes contaminated (generally with aflatoxins M1 and M2) because of the animals consumption of aflatoxin-contaminated feed.

Worldwide, corn contamination is probably of the greatest concern because of its widespread cultivation and its frequent use as the staple diet in many countries. However, due to local practices, customs or conditions, other foodstuffs may represent the greatest problem in certain localities.

One such area is West Africa where contamination of ground nuts (peanuts) is a significant problem. Ground nuts represent an important cash crop and foodstuff for rural farmers in West African countries such as The Gambia and Senegal. However, inappropriate storage conditions in the hot, humid climate can lead to contamination with aflatoxin. Indeed, black powdery moulds can often be seen growing on mounds of ground nuts stored in rural village huts.

Uses

Aflatoxins have no beneficial uses for man — their importance lies in their economic and medical significance in terms of spoilage of foodstuffs and toxicity to animals and man.

However, following the Gulf War in 1991, and the subsequent emergence of "Gulf War syndrome", there has been increased concern regarding the use of biological agents as weapons of mass destruction and/or terrorism. Subsequent investigations have revealed that the Iraqi Government experimented with a variety of biological agents including bacteria, viruses and mycotoxins. Thus, the sinister prospect has been raised of the possible future use of aflatoxins as a biological weapon.

Although there is no firm evidence that aflatoxin was used in the Gulf War, it has been reported that the Iraqis had produced 2200 litres of aflatoxin-containing material, and made seven aflatoxin-containing bombs. Although the properties of aflatoxin are not necessarily ideal as a direct acting agent against military personnel, it has been suggested that their use on foodstores and crops would result in contamination and subsequent economic and logistic disruption in the food supply.

In response to this information the US government added aflatoxins and certain other biological materials to a list of "select agents" covered under "The Antiterrorism and Effective Death Penalty Act of 1996". This law requires the registration of facilities that work with these select agents, and imposes harsh penalties for noncompliance.

Although not a conventional "use", it should be noted that aflatoxins have been incorporated into the medium of popular fiction. Graham Greene in his novel "The Human Factor" (1978) describes a character disposing of a suspected double agent by poisoning him with aflatoxin surreptitiously mixed in his whisky! The agents subsequent death from liver failure is then conveniently ascribed to his propensity for heavy drinking.

Properties — Mechanisms of Biological Interaction

For aflatoxins the liver is the primary target organ for toxicity in all species studied. The precise manifestations of toxicity depend upon a number of factors, including dose and duration of exposure. However it is the potent ability of aflatoxins to induce liver cancer, and the significant economic and public health consequences that follow, that has stimulated much of the work on these compounds over the last 30 years.

Research work has followed a number of different lines of enquiry. Firstly, long term studies have been performed with aflatoxins and mixtures of aflatoxins to characterise the ability of these compounds to induce cancer in a variety of animal species. Secondly, studies to investigate the mechanisms underlying the carcinogenic activity have been performed including genotoxicity, binding and metabolism studies. Thirdly, epidemiological studies have been performed in man to investigate the associations between diet, aflatoxin exposure, occurrence of hepatocellular carcinoma and other factors. These approaches are described in the following sections.

Carcinogenicity studies in experimental animals

Globally, primary hepatocellular carcinoma is among the most common forms of cancer in man. Incidence of the disease varies greatly in different areas of the world, suggesting involvement of environmental etiological factors, and much research has been devoted to the identification of such factors. Because many organic chemicals have been shown to have the capability of inducing primary hepatocellular carcinoma in animals, they have been extensively studied with respect to their possible significance as etiologic agents for primary hepatocellular carcinoma in man. Particular emphasis has been placed on aflatoxins because of their known widespread occurrence as food contaminants.

Mixtures of aflatoxins and aflatoxin B1 have been tested for carcinogenicity in several strains of mice and rats, in hamsters, fish, ducks, tree shrews and monkeys. Following oral administration, these compounds caused hepatocellular and/or cholangiocellular liver tumours, including carcinomas, in all species tested except mice. However, intraperitoneal administration of aflatoxin B1 to infant mice did induce high incidences of liver tumours. Additionally, in some species, the compounds produced tumours at other sites in the body. For example, tumours in the kidney and colon were also found in rats.

Aflatoxins B2, G1 and M1 have been tested separately in rats and induced liver tumours after oral or intraperitoneal administration. However, these compounds appeared to be of lower hepatocarcinogenic potency than aflatoxin B1.

In conclusion, aflatoxin B1, mixtures of aflatoxins and other specific aflatoxins have all shown evidence of carcinogenic potential in animal species.

Mechanistic studies

In order to understand how and why the aflatoxins mediate their toxicity a number of experimental approaches have been taken. One approach has been to investigate the toxicity of these compounds to the genetic material within cells (e.g. mammalian DNA). Most data is available on aflatoxin B1, and this has consistently been shown to possess genotoxic potential in a variety of test systems. For example, in human and animal cells in culture it produces DNA damage, gene mutation and chromosomal anomalies; in insects and lower eukaryotes it induces gene mutations; and in bacteria it produced DNA damage and gene mutation. Other aflatoxins have not been so extensively investigated, but in a variety of studies B2, G1, G2, and M1 have all shown evidence of genotoxicity.

Another approach has been to examine how the aflatoxins are metabolised in the body. Studies using human liver material have shown that aflatoxin B1 is metabolised to a highly reactive chemical compound, called the 8,9-epoxide. Following its formation this compound binds very rapidly to protein, DNA and other important constituents of living cells, forming "adducts". Formation of these adducts disrupts the normal working processes of the cell, and in the case of DNA adducts, can ultimately lead to a loss of control over cellular growth and division. Humans metabolise aflatoxin B1 to the major aflatoxin B1-N7-guanine adduct at levels comparable to those in species which are susceptible to aflatoxin-induced hepatocarcinogenicity, such as the rat.

However, both humans and animals possess enzyme systems which are capable of reducing the damage to DNA and other cellular constituents caused by the 8,9-epoxide. For example, glutathione

S-transferase mediates the reaction (termed conjugation) of the 8,9-epoxide to the endogenous compound glutathione. This essentially neutralises its toxic potential. Animal species, such as the mouse, that are resistant to aflatoxin carcinogenesis have three to five times more glutathione S-transferase activity than susceptible species, such as the rat. Humans have less glutathione S-transferase activity for 8,9-epoxide conjugation than rats or mice, suggesting that humans are less capable of detoxifying this important metabolite.

There is considerable in *vitro* and in *vivo* evidence to support the view that humans possess the biochemical processes necessary for aflatoxin-induced carcinogenesis. Thus, presence of DNA and protein aflatoxin adducts, urinary excretion of aflatoxin B1-N7-guanine adducts and the ability of tissues to activate aflatoxin B1 have all been demonstrated for humans. In addition, studies have suggested that oncogenes are critical molecular targets for aflatoxin B1. A high frequency of mutations at a mutational "hotspot" has been found in p53 tumour suppressor genes in hepatocellular carcinomas from patients residing in areas considered to offer a high risk of exposure to aflatoxins, and where there is a high incidence of hepatocellular carcinoma.

In contrast, this mutational pattern is not found in hepatocellular carcinoma samples from moderate or low aflatoxin exposure countries or regions. Therefore, this hot-spot mutation is believed to be a molecular fingerprint linking the initial event of aflatoxin B1-DNA adduct formation with the ultimate development and progress of human hepatocellular carcinomas.

Human carcinogenicity data

Despite the strong supportive evidence for animal and mechanistic studies, there have been major difficulties in assessing the precise role of aflatoxin in the causation of liver cancer in humans.

Unlike laboratory conditions where exposure of laboratory animals can be accurately defined, exposure of humans to aflatoxins cannot generally be estimated with any great certainty. Exposure to aflatoxin in tropical areas of Africa and parts of Asia and Latin

America can begin very early in life, and episodically thereafter, thus making accurate assessments of exposure extremely problematic. Furthermore, the number of episodes, and the degree of exposure to aflatoxin, varies greatly by country and region, by agricultural and crop storage practices, by season and by other factors difficult to control in any scientific study.

Secondly, there is a high geographical correlation between exposure to aflatoxin, the hepatitis B virus and increased incidence of hepatocellular carcinoma. Prospective epidemiological studies have shown a high incidence of primary hepatocellular carcinoma among hepatitis B virus carriers in endemic areas. Clinical studies have also shown that most primary hepatocellular carcinoma patients are carriers of the hepatitis B surface antigen, and have chronic active hepatitis. Recently, hepatitis B virus sequences have been found to be integrated into the liver cell genome in some, but not all, patients with chronic hepatitis or primary hepatocellular carcinoma. This evidence has identified hepatitis B virus as a major etiological factor for primary hepatocellular carcinoma in certain populations, particularly in Taiwan and the People's Republic of China.

Some epidemiological studies have suggested that aflatoxin poses no detectable independent carcinogenic risk for man, and that it poses risks only in the presence of other risk factors such as hepatitis B infection. Such studies have indicated that the potency of aflatoxins in hepatitis B surface antigen-positive individuals is substantially higher than the potency in surface antigen negative individuals. Clearly, reduction in prevalence of hepatitis B infected individuals through vaccination of those at risk may therefore have an important impact on the risk of liver cancer in these populations. Further studies attempting to define the relationships between the aflatoxin exposure and hepatitis B infection factors are ongoing in Africa and the far East. Studies are also examining the role of hepatitis C virus infection in this complex set of potentially interdependent risk factors for the occurrence of primary hepatocellular carcinoma.

Despite these difficulties, aflatoxin B1 has been classified as a Group I carcinogen (i.e. it is considered that sufficient evidence exists to define aflatoxin B1 as carcinogenic to humans) in humans by IARC (International Agency for Research on Cancer) parameters. Furthermore, the Food and Agriculture Organisation of the United Nations and World Health Organisation Joint Expert Committee on Food Additives concluded in 1997 that they are considered to be human liver carcinogens. However, these expert bodies agree that exact mechanisms of aflatoxin hepatocarcinogenesis have not yet been fully elucidated, and some important points remain to be clarified.

It is to be hoped that better information will be generated as a result of on-going intervention projects, and agricultural development programmes, and by monitoring exposure to aflatoxin and the incidence of liver cancer in areas where hepatitis B virus vaccination is effectively reducing the prevalence of carriers of the viral surface antigen. In addition, initiatives must continue which reduce exposure through measures such as improved farming and storage practices, improved monitoring of foodstuffs and through enforcing food standards both within countries and across borders.

Toxicity Produced — Toxicity Profile

The adverse biological properties of aflatoxin seen in poisoning episodes in animals can be categorised in two general forms:

- **acute aflatoxicosis** which occurs following the ingestion of high doses of aflatoxins over a relatively short period of time. Specific acute episodes of disease may include haemorrhage, acute liver damage, oedema, alteration in digestion, absorption and/or metabolism of food, and possibly death
- **chronic aflatoxicosis** which occurs following the ingestion of low to moderate doses of aflatoxins over a prolonged period. The effects may be subclinical or difficult to recognise. Some of the more frequently described symptoms include impaired food conversion and slower rates of growth, with or without the

occurrence of an overt aflatoxin syndrome as seen with acute poisoning. Underlying these symptoms is a chronic poisoning of the liver leading ultimately to cirrhosis and/or liver cancer (see description of genotoxicity and carcinogenicity data above).

Laboratory investigations in a number of animal species have confirmed that aflatoxins can produce acute necrosis, cirrhosis and carcinoma of the liver. No animal species has been shown to be refractory to aflatoxin toxicity, however, a wide range of acute lethal doses have been observed, indicating different degrees of acute susceptibility. For most species the doses that killed 50% of the animals treated ranged from 0.5 to 10 mg/kg body weight. Species differ in their susceptibility to the acute and chronic effects, and toxicity can be influenced by dose, duration of exposure, age, health, nutritional status and environmental factors.

Further information relating to toxicity profiles are given below in relation to examples of toxic episodes published in the scientific literature.

Examples of Endemic Problems — Toxic Episodes

Examples of toxic episodes in animals

In 1960 more than 100,000 young turkeys on poultry farms in England died in the course of a few months from a mysterious new disease. In view of the lack of an explanation for the disease, it was named "Turkey X disease". Soon, however, it was found that the problem was not limited to turkeys; ducklings and young pheasants were affected, and also showed heavy mortality.

Intensive investigation of the early outbreaks of the disease indicated that they were all associated with particular meals given to the birds. On feeding the meal to poults and ducklings, the symptoms of Turkey X disease were rapidly produced. The suspect feed was imported Brazilian peanut meal and initial speculation was that a fungal toxin might be involved.

Further investigations did in fact demonstrate that the meal was heavily contaminated with *Aspergillus flavus*, that this organism was responsible for producing a toxin (aflatoxins were isolated and the chemical structures identified for the first time), and that the disease was the result of aflatoxin ingestion.

Examples of toxic episodes in man

Northwest India 1974

In the fall of 1974 an epidemic occurred in more than 150 villages in adjacent districts of two neighbouring states in a rural area of Northwest India. The disease was characterised by onset with high fever, rapidly progressive jaundice and ascites. According to one report of the outbreak, 397 persons were affected and 108 people died. One notable feature of the epidemic was that it was heralded by the appearance of similar symptoms in the village dogs.

Liver biopsy specimens from eight cases, and autopsy material from one human case and two dogs were studied. Characteristic features were centrizonal scarring, hepatic venous occlusion, ductular proliferation and cholestasis, focal syncytial giant-cell transformation of hepatocytes, and pericellular fibrosis.

Analysis of food samples revealed that the disease outbreak was probably due to the consumption of maize (corn) heavily infested with the fungus *Aspergillus flavus*. Unseasonable rains prior to harvest, chronic drought conditions, poor storage facilities and ignorance of dangers of consuming fungal contaminated food all seem to have contributed to the outbreak.

The levels of aflatoxin in food samples consumed during the outbreak ranged between 2.5 and 15.6 microgram/g. Anywhere between 2 and 6 mg of aflatoxin seems to have been consumed daily by the affected people for many weeks. In contrast, analysis of corn samples from the same areas the following year (1975) revealed very low levels of aflatoxin (i.e. less than 0.1 microgram/g), and this may have explained the absence of any reoccurrence of the outbreak in 1975. A ten-year follow-up of the epidemic found the survivors fully recovered with no ill effects from the experience.

Kenya 1981

Between March and June 1981, 20 patients (eight women and 12 men aged 2.5 to 45 years old) were admitted to three hospitals in the Machakos district of Kenya with severe jaundice. The patients reported that they had first exhibited symptoms of abdominal discomfort, anorexia, general malaise and low grade fever. After about seven days, jaundice and dark urine had appeared, and the patients had sought admission to hospital.

The patients came from rural areas of mixed woodland and bushed grassland about 150 km Southeast of Nairobi. The rainiest season is from March to May each year, when about 70% of the annual rainfall occurs. 1980 had been an extremely dry year with a poor harvest, but in 1981 the rains had come early, were heavy and prolonged. Maize is the major crop in the area, but some millet, sorghum, beans, cowpeas, pidgeon peas and vegetables are also grown for home consumption.

Interestingly, the relatives and friends of one family told that many of the local doves had died, then the local dogs, and finally the people had become sick. The dogs were known to be consuming essentially the same diet as the local people.

On admission to hospital all patients were jaundiced, some with low grade fever, and extremely weak. Tachycardia and oedema (of the legs and to a lesser extent face and trunk) were seen. The liver was tender in all patients. Eight of the 20 patients improved with a return of appetite, disappearance of jaundice and discharge from hospital in six to 20 days. However, hepatic failure developed in the remaining 12 patients and they died between one and 12 days following admission.

An extensive investigation of the outbreak was performed. Aflatoxin levels in foods were measured and showed high levels of aflatoxin B1 and B2. For example, maize grains from the two homes where severe and fatal illness had occurred contained 12 mg/kg and 3.2 mg/kg of aflatoxin B1, while maize from unaffected homes had a maximum of 0.5 mg/kg aflatoxin B1. Liver samples were obtained from two patients at necropsy and these indicated aflatoxin B1 levels of 39 and 89 µg/kg. Histologically the livers showed evidence of toxic hepatitis — marked centrilobular necrosis with minimal inflammatory reaction. Blood samples from the patient were also tested for possible viral infections and three were found to be positive for hepatitis B surface antigen.

The cumulative evidence suggests that aflatoxin poisoning was the cause of the acute liver disease in this incident. Contributing factors may have included the exceptionally prolonged and heavy rainy season that year which would have provided favourably moist conditions for the growth of aflatoxin producing moulds. Other factors could have been that the previous year's poor harvest had forced some individuals onto a protein deficient diet (this is known to potentiate aflatoxin poisoning in monkeys), and that the severity of the aflatoxin toxicity could have been worsened by the pre-existing liver damage due to hepatitis B viral infection in three of the subjects.

Preventative Measures

There are a variety of strategies which are aimed at minimising the animal and human exposure to aflatoxins. Firstly, reductions in exposure can be achieved through avoidance measures such as improved farming and proper storage practices and/or enforcing standards for food or feed within countries and across borders. Secondly, numerous strategies for the detoxification of aflatoxin contaminated foodstuffs have been proposed. However, it must be recognised that strategies aimed at reducing the risks posed by aflatoxins are dependent upon the resources available, and that this may be a particular constraint in poorer countries and those with a developing infrastructure.

Avoidance strategies

Good farming and storage practices are aimed at eliminating the conditions which encourage the growth of moulds in crops and stored food. For example, ripe crops should not be left in the field too long, and cereal grains, rice and nuts should not be stored under damp, inadequately ventilated conditions.

However, since some degree of aflatoxin contamination is considered unavoidable, even where good manufacturing practices have been followed, many countries have introduced regulatory controls over the levels of these substances allowed in certain high risk foodstuffs.

In the UK, the Ministry of Agriculture, Fisheries and Food (MAFF) have been monitoring the levels of aflatoxins in foods for some years. The "Feeding Stuffs Regulations 1991" set maximum levels for aflatoxin B1 in animal feed, and thus restricts the amount of aflatoxin M1 carried over into milk and milk products. Regulations to limit the levels of aflatoxins in certain human foodstuffs (Aflatoxins in Nuts, Nut Products, Dried Figs and Dried Figs Products Regulations 1992) were introduced at the end of 1992. National limits for aflatoxin content of foodstuffs remain under surveillance, and international regulatory activities are co-ordinated at the regional and WHO level.

In the United States, the Food and Drug Administration (FDA) regulates the quality of food, including the levels of environmental contaminants. The FDA has established guidelines for the levels of aflatoxins permitted in human foodstuffs and animal feed. The maximum permitted level for human food is 20 parts per billion of total aflatoxins. Higher levels are permissible for feed destined for animal consumption.

Detoxification strategies

Because it is impossible to completely avoid some degree of aflatoxin contamination, a variety of strategies for their detoxification in foodstuffs have been proposed. These strategies have included physical methods of separation, thermal inactivation, irradiation, solvent extraction, adsorption from solution, microbial inactivation, chemical methods of inactivation and fermentation. Two of these strategies are described in more detail below.

A wide range of chemicals have been tested for the ability to degrade and inactivate aflatoxins. However, although a number of these chemicals can react to destroy aflatoxins effectively, most are impractical, too expensive or potentially unsafe because of the formation of toxic residues, or the perturbation of the nutrient content of the food. Two chemical approaches that have received considerable attention are ammoniation and reaction with sodium bisulphite.

Studies have shown that treatment of aflatoxin-contaminated corn with ammonia is an effective detoxification approach. Ammonia appears to produces hydrolysis of the lactone ring and chemical conversion of the parent compound to numerous products that exhibit greatly reduced toxicity. Similarly sodium bisulphite reacts with aflatoxins to form water soluble degradation products.

An alternative approach is to attempt to reduce the absorption of aflatoxins from contaminated feed in animals. This may been achieved by the addition of inorganic sorbant materials such as sodium calcium aluminosilicate (HSCAS) in the diet of animals. HSCAS tightly binds aflatoxins in the gastrointestinal tracts of animals, preventing their absorption into the body so that they are passed out unabsorbed in the faeces. This results in a major reduction in the body burden (i.e. exposure) of the animals to the mycotoxin.

Case History of Poisoning

Histories of probable poisoning with aflatoxins are described above in relation to the toxic episodes reported in Northwest India and Kenya. The case described below is of interest in that a full medical investigation and follow-up was performed which indicated that the patient remained well up to 14 years post-exposure. In addition, this case is extremely unusual in that a reasonably accurate estimate of exposure to aflatoxin can be made.

In a deliberate suicide attempt, a 25-year-old female American laboratory worker reported ingesting approximately 5.5 mg aflatoxin B1 over a two-day period, and six months later, approximately 35 mg over a 14-day period. These amounts were consistent with those missing from the laboratories stocks, and can therefore be assumed to be accurate.

After the first episode she was reported have transient rash, nausea and headache, but there were no other ill effects — physical, radiological and laboratory examinations being normal except for sulphobromophthalein retentions of 9% and 7% at 45 minutes (sulphobromophthalein clearance is used as a measure of liver function — clearance in healthy individuals being essentially complete at 45 minutes). Following the second episode the only symptom reported was nausea. Liver biopsies after each episode were normal, and in a 14-year follow-up a physical examination and blood chemistry, including tests for liver function, were normal.

The authors of this report commented that additional factors, such as malnutrition and hepatitis virus, which were absent in this patient, may be necessary for aflatoxin carcinogenesis in humans. Alternatively, the latent period for liver tumour formation may be greater than the 14-year follow-up period, or the exposure levels of aflatoxin in this subject were insufficient to provoke more serious toxicity.

Concluding Remarks

Aflatoxins have only been recognised as a significant issue for human health in the past 35 years. During this time an enormous amount of information has been accumulated on the nature, occurrence, exposure and health effects of these mycotoxins. Clearly, further work is required to clarify their role in the occurrence of primary hepatocellular carcinoma, the molecular, biochemical and pathological mechanisms underlying their toxicity and optimal strategies for minimising both their health and economic impacts on the human populations they affect.

Suggested Further Reading

Eaton DL and Groopman JD (1994). *The Toxicology of Aflatoxins*. Academic Press, New York.

Greene G (1978). *The Human Factor*. The Bodley Head.

International Agency for Research into Cancer (IARC) (1993). Some naturally occurring substances: food items and constituents, heterocyclic aromatic amines and mycotoxins. *IARC Monographs on the Evaluation of Carcinogenic Risks to Humans*, Vol. 56, pp. 245–395.

Krishnamachari KA, Bhat RV, Nagaragen V and Tilak TB (1975). Hepatitis due to aflatoxin. An outbreak in West India. *Lancet* i, 1061–1063.

Ngindu A, Johnson BK, Kenya PR, Ngira JA, Ocheng DM, Nandwa H, Omondi TN, Jansen AJ, Ngare W, Kaviti JN, Gatei D and Siongok TA (1982). Outbreak of acute hepatitis caused by aflatoxin poisoning in Kenya. *Lancet* 1, 1346–1348.

Willis RM, Mulvihill JJ and Hoofnagle JH (1980). Attempted suicide with purified aflatoxin. *Lancet* 1, 1198–1199.

Zilinskas R (1997). Iraq's biological weapons: the past as future? *JAMA* 278, 419–424.

2

BOTULINUS TOXIN

M. E. Smith

Description

Botulism, or "sausage poisoning" as it was originally termed, was first studied seriously after an outbreak in Wildbad, Germany in 1793. Of the 13 people involved six died, and the outbreak was linked with consumption of a locally-produced blood sausage. The regional health officer, Justinius Kerner, described 230 cases in a report published in 1829, most of which were due to eating, badly-processed sausage, hence the name "botulinus" after the Latin word "botulus" a sausage. In the late 19th century, Van Ermengen (1897) first related the disease to a toxin produced by the bacterium *Clostridium botulinum*. Seven immunologically distinct types of botulinus toxin, which are proteins, denoted types A to G, have been identified. They are produced by distinct strains of *C. botulinum*. Humans are susceptible to types A, B, E, F and G and resistant to types C and D. Tetanus toxin is also produced by Clostridium bacteria (*C. tetani*) and it resembles botulinus toxin in many of its structural and functional properties. Both block the release of neurotransmitter but tetanus toxin acts on nerve endings in the central nervous system whereas botulinum toxin acts on peripheral nerve endings.

All types of botulinum toxin are sensitive to low pH such as that existing in the stomach as well as to pepsin, the protease produced in the stomach. However, the toxins produced by Clostridia are complexed with other proteins and the complexes are relatively resistant to digestion in the stomach. They dissociate at the more alkaline pH of the small intestine, where the dissociated toxin can be absorbed into the blood circulation. In infant botulism the intestine is colonised by *C. botulinum* following ingestion of the spores.

The botulinum toxins are large polypeptide molecules (approximately 150 kDa) which contain Zn^{++} ions. They are synthesised as single polypeptides which are cleaved by endogenous protease action to yield the active forms, each of which consists of two polypeptide chains linked by a disulphide bond. The larger of the two chains, denoted the heavy (H) chain, has a mass of approximately 95 kDa and the smaller, the light (L) chain, a mass of approximately 50 kDa. The structure is illustrated in Figure 1. The H chain is composed of two domains of similar mass; the C-terminal domain (H_C) is responsible for the specificity of binding of the toxin to peripheral motoneurones and the N-terminal domain (H_N) for penetration of the toxin into the neuronal cell. In the neuronal cytoplasm, the L chain is released by proteolysis and it is this subunit which is responsible for the disruption of the exocytotic apparatus, which causes blockade of transmitter release.

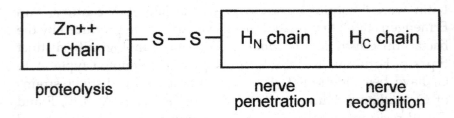

Figure 1. Schematic structure of botulinum neurotoxin.

Occurrence

Several species of Clostridia are able to produce botulinum toxins. These have been divided into groups with distinct physiological properties although a particular botulinum toxin type can be produced by more than one group. However, it has been found that botulinus toxin can be produced by *C. botulinum* strains which are clearly distinct from the hitherto defined species and are related to different species such as *C. butyricum* and *C. baratii* which have been implicated in human botulism. Furthermore, the toxinogenic property is unstable and variants which lack toxicity have been described. Group I strains produce toxins A, B and F, group II, toxins B and F, and group III, toxins C and D. Group IV includes strains which produce botulinum toxin G and strains which are nontoxic. Group IV strains are physiologically distinct from those in the other groups and have now been assigned to a different species termed *C. argentinense*. The botulinum Clostridia produce other toxins besides the botulinum toxins.

Source

Botulinum toxin-producing bacteria are ubiquitously distributed in the environment. They are usually straight or slightly curved rods, and are motile. They reproduce by sporulating and the spores can survive for very long periods of time under extreme conditions of heat, dryness, chemical pollution, radiation and oxygen lack. Spore germination takes place in anaerobic conditions, if the nutritional requirements, such as decomposing animal cadavers and soils rich in organic material, are available. *C. botulinum* is widespread in soils, and sediments in lakes and seas but different strains of *C. botulinum* have a different geographical distribution. Botulinum toxin has been found in fish and other aquatic animals, meat, and meat products, and in fruit and vegetables. In general, the contamination of fish and aquatic invertebrates by *C. botulinum* reflects the bacterial content of the sediments of the areas where they are caught. Thus, for example, *C. botulinum* E is the predominant strain in fish from North America, Europe and Asia. The incidence of *C. botulinum* is lower in meat than in fish. It has been found mainly in meat from cattle, pigs and chickens. The incidence of the bacterium in meat is lower in North America than in Europe. Animals can carry *C. botulinum* in their intestinal tracts, and this can lead to contamination of meat during processing. The predominant types of botulinum toxin found in meat are types A and B. Type C occurs less frequently, and type E rarely. Raw fruits and vegetables, especially those harvested from the soil, can be contaminated with *C. botulinum* (usually A and B) as a result of contamination of the soil from which they are harvested whole. The use of manure as fertiliser can increase the level of contamination. Honey from some regions may contain *C. botulinum* but the incidence of contamination of other foods including prepared foods (vacuum-packed, dehydrated, etc.) is very low.

Uses

In recent years the ability of botulinum toxin to cause muscle paralysis has been harnessed in the treatment of numerous disease conditions in which the abnormality is sustained muscle contraction or spasm. These conditions are collectively termed dystonias. Its usefulness depends on the fact that it can be injected into individual muscles to block neuromuscular transmission, which results in relaxation of the affected muscles. It was first used therapeutically by Scott in 1973, who was able to correct strabismus, or squint, in a patient by injecting it into the muscles of the eye (Figure 2). Since then it has been used to treat a large number of

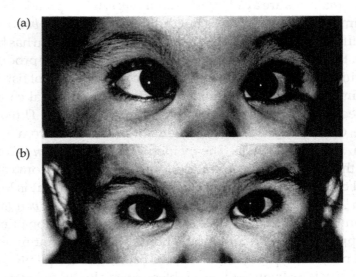

Figure 2. Correction of squint by injection of botulinum toxin into eye muscles. (a) Before treatment and (b) after treatment. (Provided by Mr. H. E. Willshaw, The Birmingham Children's Hospital.)

neurological and non-neurological disorders. Botulinum toxin type A is the form used clinically in most instances, although type F has also been used successfully. Botulinum toxin has also been used to treat other ocular conditions such as thyroid ophthalmopathy (protruding eyeballs), nystagmus (involuntary rhythmic oscillation of the eyeballs) and persistent diplopia (double vision). The main disadvantage of treatment with the toxin, compared to surgery, is the need for repeat injections, which are required for approximately 50% of patients. In some conditions, such as sixth nerve palsy, strabismus may actually be caused by muscle paralysis. In this case, if the muscle paralysis is incomplete, the squint can sometimes be corrected by injection of the toxin into an antagonist muscle.

Dystonias are characterised by symptoms of sustained and patterned contractions of muscles. There are many different types of dystonia and many causes. Bletharospasm, for example, is a form of focal dystonia characterised by intermittent or sustained closure of the eyes due to involuntary contractions of the obicularis oculi muscles. This disability can take the form of an increased rate of blinking, or, in the worst cases, an inability to open the eyes (resulting in functional blindness). In many patients the facial muscles are also affected. Injections of botulinum toxin into the affected muscles have produced an improvement in 70% to 100% of individuals affected with bletharospasm, and it is now considered as a primary form of therapy for the condition. Another type of dystonia, cervical dystonia, is characterised by involuntary contractions of the neck muscles causing repetitive spasmodic head movement, or sustained abnormal postures of the head such as that due to twisting of the neck. Treatment with conventional drugs can be accompanied by disabling adverse side effects, and surgery cannot always be used to correct the problem. Botulinum toxin injected into the affected muscles is an effective form of treatment in these conditions.

Various other focal dystonias can be treated successfully by injections of botulinum toxin in the appropriate muscles. These include writer's cramp, spasmodic laryngeal dystonia which involves the vocal cords and is manifested as a strangled voice and voiceless pauses, oromandibular dystonia which involves the lower facial and tongue muscles and the muscles of mastication (chewing) and leads to problems with chewing and speech, jaw opening or jaw closing dystonia, and anismus or spasm of the external anal sphincter which causes intractable constipation.

Facial dystonias such as hemifacial spasm, characterised by involuntary twitches or sustained contractions of the muscles supplied by the facial nerve, are often difficult to treat successfully by conventional drug therapy. Moreover surgery to section the nerves can lead to permanent facial paralysis. However such individuals have been treated successfully with botulinum toxin, a single injection often abolishing the involuntary movements for as long as five months. Tremor of the hands and head, voice tremor, and some cases of stuttering, and dystonia of the lower limbs, for example that associated with parkinsonism, which can cause pain and gait difficulties, can also be improved by therapy with botulinum toxin. Recently, injections of botulinus toxin "Botox" have been used cosmetically to paralyse small facial muscles and so avoid the appearance of wrinkles. However whether the treatment is effective depends, in all of these conditions, on use of the correct dose of toxin and the accurate targeting of the injection to the affected muscles.

Some patients fail to respond to treatment with botulinum toxin type A because antibodies which neutralise the toxin are present in their blood, presumably because they have been previously exposed to a subclinical dose of the toxin. Moreover some patients develop the antibodies after treatment with the toxin and then become unresponsive to further treatment. Fortunately this is a rare occurrence because focal injection involves delivery of an extremely low dose of the toxin compared to body weight. Moreover, patients who express the antibodies can often be successfully treated with type F toxin. In the future type B toxin may also prove to be useful.

Finally botulinus toxins have become enormously useful tools for scientists attempting to characterise the functions of the presynaptic neuronal proteins which bind the toxins and to unravel the mechanisms involved in transmitter release. Their use is enabling the crucial docking and fusion events in the process of exocytosis remains to be elucidated.

Properties

The toxicity of botulinum toxins is due to their ability to attack nerves specifically and also to catalyse the degradation of selective presynaptic proteins. The toxins bind specifically to the presynaptic membrane of neurones, notably motoneurones (but also autonomic and sensory neurones). After binding to the nerve terminal the toxin gains access to the cytosol of the neurones to block the release of the neurotransmitter. The result is paralysis of the muscles innervated by those nerves.

Neuromuscular transmission is initiated when an action potential travelling down the axon of a motor neurone arrives at the nerve terminal. Vesicles containing the neuromuscular transmitter, acetylcholine, are present in rows which line up either side of an "active zone". Active zones are transmitter release sites. The vesicles arrive at the active zones guided by the cytoskeletal matrix which is made up of microtubules, the contractile protein actin and the smooth endoplasmic reticulum. Attachment to the cytoskeleton is mediated by various proteins including synapsin 1. The active zones are present in the presynaptic nerve terminal localised opposite the tips of the junctional folds of the muscle endplate. Acetylcholine receptor molecules on the muscle membrane recognise the transmitter after its release into the neurotransmitter cleft. These receptors are highly concentrated on the tips of the endplate folds. Depolarisation of the presynaptic nerve membrane at the neurotransmuscular junction by an action potential results in the opening of voltage-gated Ca^{++} channels in the presynaptic membrane. The concentration of Ca^{++} in the extracellular medium is relatively high (in the mM range) compared to that inside the cell (in the range 10^{-7} M). Thus when the voltage-gated channels open, Ca^{++} flows into the nerve terminal down its concentration gradient. This leads to a rapid increase in the Ca^{++} concentration in the cell cytosol to between 0.2 and 0.3 × 10^{-3} M. This triggers, within 200 to 300 microseconds, the synchronous fusion of several hundred small synaptic vesicles with the presynaptic membrane releasing approximately 10,000 molecules of acetylcholine from each vesicle, into the synaptic cleft.

The acetylcholine diffuses across the gap between the nerve terminal and the muscle membrane, (the neuromuscular cleft) and binds to receptors on the muscle surface. This results in the opening of Na^+/K^+ channels and Na^+ flows down its concentration gradient into the muscle. This ion flux causes a localised depolarisation, termed an "endplate potential", in the muscle. This nerve-evoked electrical disturbance can be measured using a microelectrode inserted into the muscle cell at the neuromuscular junction region and compared to a reference electrode. When the amplitude of this depolarisation reaches a threshold level a regenerative electrical depolarisation, known as an "action potential" is triggered in the muscle. This action potential is transmitted into the muscle cell where it triggers contraction of the muscle fibre. At the mammalian neuromuscular junction there is normally a 1:1 relationship between nerve action potentials and muscle action potentials.

Acetylcholine vesicles fuse with the presynaptic membrane at a low rate to release their packets of transmitter even in the absence of nerve action potentials. This spontaneous release is random. It is insufficient to trigger an action potential in the muscle but can cause a small depolarisation of the membrane, termed a miniature endplate potential The smallest depolarisation is caused by release of the contents of a single vesicle, or one quantum of acetylcholine. After release the synaptic vesicle membrane is rapidly taken up into the nerve terminal and reutilised. The acetylcholine is broken down in the cleft to acetate and choline in a reaction catalysed by the enzyme acetylcholine esterase which resides in the neuromuscular cleft and the choline is taken back up into the nerve terminal where it participates in the synthesis of new transmitter.

Botulinum toxin causes a large reduction in the amplitude of nerve-evoked endplate potentials. It does not reduce the amplitude of the smallest spontaneous miniature endplate potentials but it reduces the frequency of these random events (see Figure 3). It therefore reduces the number of packets of transmitter released but not the size of the quantum. Thus under conditions of subtotal blockade the number of fusion events is reduced but those vesicles which do fuse with the membrane contain the normal complement of transmitter. The toxin does not therefore interfere with the process of synthesis of the neurotransmitter. Neither does it interfere with re-uptake and storage of acetylcholine, or with calcium entry at the nerve terminal, or action potential propagation.

The different botulinum toxins all block acetylcholine release, but there are differences in their actions. This can be shown in isolated nerve-muscle preparations in which the evoked release is reduced but the concentration of toxin is not high enough to

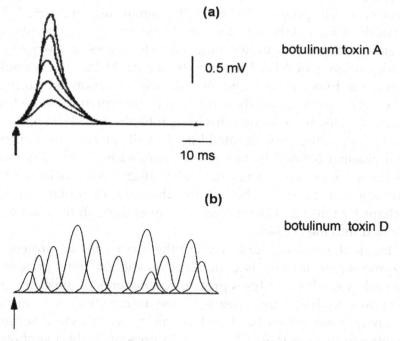

Figure 3. Intracellular recordings of nerve-evoked end plate potentials at neuromuscular functions poisoned with (a) botulinum toxin A and (b) botulinum toxin D. Nerve stimulation at the arrows. Note synchronous recordings in (a) but reduced quanta and asynchronous recordings in (b).

completely abolish the response of the muscle membrane. Thus under these conditions, at neuromuscular junctions poisoned with botulinum A or E, the nerve-evoked release of acetylcholine remains synchronous, so that although the amplitude of the muscle endplate potential is reduced, its time course remains the same. However, intoxication with botulinum B, D or F causes a desynchronisation of the release of quanta of acetylcholine (Figure 3).

Mechanisms of Biological Interaction

There is evidence that the toxin reaches the cell cytosol by the process of endocytosis. The toxin binds to a receptor, as yet unidentified, on the cell membrane, then enters the cell as part of a membrane-bound vesicle which is formed by invagination of the cell membrane. The process is illustrated schematically in Figure 4. The pH of the lumen of the vesicles is acidic because of the operation of an energy-dependent proton pump which derives its energy from the degradation of ATP. The pump transports H^+ into the cytosol of the vesicle. Exposure of the toxin to an acidic environment appears to be necessary to cause the structural transformation of the toxin which enables it to become inserted into the vesicle membrane. The L chain is then translocated into the cell cytosol. This depends on a channel formed in the vesicle membrane by an oligomeric assembly of H chain units (probably from toxin molecules) to form a tetrameric pore. The L chains then in some unknown way, perhaps by folding of the molecules, travel through the pore to be released into the cytosol.

Inside the cytosol, the L chain of the toxin acts as a proteolytic enzyme whose activity is dependent on the Zn^{++} ions present in the molecules. It hydrolyses protein components of the exocytosis apparatus to block the release of the transmitter. It is known that botulinum toxins B, D and G can cleave vesicle associated membrane protein (VAMP), a protein present in the membranes of acetylcholine vesicles. Botulinum toxins A, C and E, on the other hand, act on proteins of the presynaptic plasma membrane. A and E cleave synaptosomal associated protein$_{25}$ (SNAP$_{25}$) and C cleaves

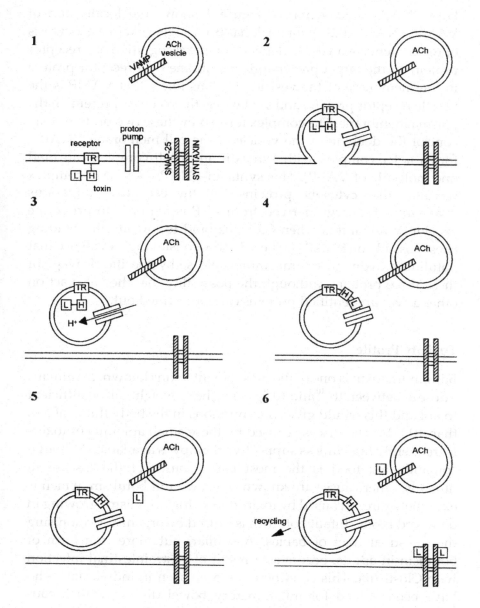

Figure 4. Internalisation of toxin L chain and binding of L chain to presynaptic proteins.

both SNAP$_{-25}$ and syntaxin. Figure 4 shows the localisation of VAMP, SNAP$_{-25}$ and syntaxin. It has been suggested that exocytosis of the contents of a vesicle depends on the recognition of a receptor protein on the target presynaptic membrane by a receptor protein in the membrane of the vesicle. It seems likely that VAMP is the vesicle receptor protein, and syntaxin is the receptor protein on the plasma membrane. The complex formed by these two proteins may control the docking of the vesicle at the cell membrane. SNAP$_{-25}$ forms a 1:1 complex with syntaxin and each complex can bind one molecule of VAMP. This syntaxin–SNAP$_{-25}$ – VAMP complex recruits other cytosolic proteins and the exocytosis apparatus is assembled. Energy derived from ATP is required to prime the exocytotic apparatus when Ca^{++} ions bind to low affinity binding sites, possibly on SNAP$_{-25}$. There is now considerable evidence that botulinum toxins block transmitter release by specific cleavage of these three proteins, although the possibility that they also act on other as yet unidentified proteins cannot be ruled out.

Toxicity Profile

Botulinum toxin is one of the most potent toxins known. In humans a dose of between 10^{-10} and 10^{-9} g per kg body weight can be sufficient to kill and this would give concentrations in the body fluids of less than 10^{-12} M. The disease caused by these small amounts of toxins is frequently fatal unless supportive therapy is available. Although contaminated food is the most usual source of the disease in humans, other causes are known. Thus wound botulism, which is extremely rare, is caused by toxin-producing organisms growing in damaged tissue. Infant botulism is caused by organisms colonising the intestinal tract of babies. A similar, but more rare form of botulism in adults, occurs as a result of gastrointestinal infection with Clostridia. This condition has been seen in individuals who have been treated for inflammatory bowel disease with broad-spectrum antibiotics, or following intestinal surgery.

Ingestion of the toxin causes generalised paralysis of skeletal muscle, and autonomic dysfunction, by binding to receptors on the endings of peripheral motor nerves. The disease affects mammals, birds, and fish but different species vary in their susceptibility to each type of the toxin. Types A, B and E affect humans, and type C birds, dogs, cattle, and mink. In humans wound botulism is usually produced by type A toxin and infant botulism by type A or type B.

The first symptoms in individuals with botulism are typically sudden, onset of weakness in the muscles innervated by the cranial nerves resulting in difficulty in speaking and swallowing and blurred vision. Subsequently weakness or paralysis of all muscles occurs, with a progression of muscle involvement from the upper trunk and limbs to the lower limbs. Death results from respiratory failure. In the foodborne disease, these symptoms are often preceded by vomiting, gastrointestinal upset and constipation, but occasionally by diarrhoea. Patients may also suffer from dry mouth, hypertension, and urinary retention. Many of these symptoms result from dysfunction of the autonomic nerves. Motor and autonomic nerves are affected but sensory involvement is usually absent. There is usually no effect on mental functioning, unless the patient develops respiratory failure. Wound botulism is characterised by the same symptoms except for the absence of the early gastrointestinal features. In infant botulism the child is initially constipated, lethargic, and reluctant to feed, and becomes increasingly weak. The face is expressionless, which is indicative of bulbar palsy, and the cry feeble, the gag, suck and swallowing reflexes are impaired, and muscle weakness is evident.

Botulism is confirmed by laboratory tests which indicate the presence of the C. *botulinus* bacterium or toxin in the patient's blood serum, gastric contents, or stool, or in the case of wound botulism, in wound specimens.

Preventative Measures

As botulism is caused by a protein molecule the disease can be prevented by immunisation with specific antitoxin antibodies which neutralise the toxin. These antibodies can be isolated from an immunised donor, i.e. someone who has been previously exposed to the toxin. Alternatively vaccination with toxoids, toxins that have been modified so that they are no longer toxic but still provoke the formation of specific antibodies, can be performed. Vaccination against botulism is usually only given to individuals working with the toxin, as the disease is rare, especially in the developed countries.

Case Histories

Food generated botulism

In 1998, six people were admitted to the emergency department of a provincial hospital in Thailand. They had difficulty in speaking and swallowing, and blurred vision and dry mouths. Those most severely affected had symmetrical paralysis, diarrhoea and vomiting. Seven more patients from the same villages were found. The symptoms took between six hours to six days to develop and two of the 13 patients died. All were found to have eaten home-canned bamboo shoots which had been processed by boiling in a large container for about one hour. Cultures from these bamboo shoots showed the presence of botulinum toxin type A. The national food safety committee in Thailand recommended that home-canned foods should be processed at high temperatures. The US Department of Agriculture currently recommends that low-acid foods (pH greater than 4.6, including red meat, seafood, poultry, milk and fresh vegetables) should be sterilised at 116°C–121°C in pressure canners at 0.86–0.97 atmospheres. At these temperatures, the time needed to destroy bacteria in low-acid canned food ranges from 20–100 minutes. Spores that survive the cooling process do not usually grow in acidic environments (pH less than 4.6). The toxin is heat-labile and can be destroyed by heating to 80°C for 30 minutes or 100° for ten minutes.

A 40-year-old man decided to "eat up" a home-canned tin of green beans as the ends of the tin were bulging. The beans were unheated, within ten hours he developed blurred vision, slurred speech and dry mouth. On the second day, he had breathing trouble, fixed dilated pupils, facial paralysis, tongue and neck paralysis and severe weakness. The patient was treated with mechanical ventilation and trivalent (A,B,E) equine botulinum antitoxin. Type A toxin was detected in the beans and so was presumably responsible for the symptoms. He made a slow recovery over the next three months.

Botulism in the arctic

For many years, epidemics of illness have been described in which fish or meat products are responsible. This is relatively common in the Arctic, where over 200 outbreaks have been reported since the early 1900s, with an overall fatality rate of about 20%. Annual incidence rates of 30 cases/100,000 have been reported for the Canadian Inuit (Eskimo) population (in contrast, the highest rate in the USA is in Washington State, with 0.43 cases/ 100,000). Foods involved in Arctic botulism are usually putrefied or fermented, either intentionally or unintentionally. Meat is sometimes kept in underground pits for six to nine months; the frozen subsoil is not always cold enough to prevent C. botulinum growth and contamination of food. Traditional dishes of fermented salmon eggs or salmon heads used to be prepared by leaving the protein to decay in pits in the ground over the summer until a pasty mush was formed; this was easily contaminated by the toxin. Modern fermentations are carried out in barrels, plastic jars or bags at ambient temperatures. This increases the risk of toxin production, which is favoured by anaerobic conditions and higher temperatures. Fish is often salted before being allowed to dry and ferment, but although the presence of salt makes C. botulinum growth less likely, this is not always sufficient to prevent toxin formation.

"Wound" botulism

Doctors in Oakland, California, have reported a number of cases of "wound" botulism occurring in drug addicts. This is due to

"tar" heroin, imported from Mexico, which is contaminated with
C. botulinum.

Infant botulism

A 30-day-old infant was brought to the Emergency Department
by her father. She had been in good health until four days before
admission when she had become very constipated and lethargic,
only taking small amounts of milk formula. On the day of admission,
she was unable to swallow and allowed the formula to drip out
of her mouth. On physical examination, the baby appeared grey,
had a very weak cry and was hardly able to move her arms and
legs. She was treated by intubation, intravenous feeds and airway
maintenance; *C. botulinum* toxin type A was isolated from stool
samples. She finally left hospital five weeks later and had totally
recovered six months after discharge. The source of the toxin was
thought to be contaminated honey, which had been used to sweeten
a drink.

Avian botulism

Clostridium botulinum is widespread in soil and only requires warm
temperatures, a protein source and an anaerobic environment to
become active and produce toxin. Decomposing vegetation or
invertebrates in summer can provide ideal conditions. Birds are
most commonly affected by type C toxin, type E toxin being less
important. Usually, fly larvae (maggots) feed on animal carcasses
and store the toxin; birds such as ducks which then eat the maggots
can develop botulism even if only a small number (three or four)
larvae are consumed. Birds can also ingest the toxin directly if
they are carrion-feeders. As *C. botulinum* toxin affects the central
nervous system transmission, affected birds are unable to fly or
walk. Birds with paralysed neck muscles cannot hold up their
heads and may therefore drown. Outbreaks are relatively common
in the US and Canada from July to September and may claim the
lives of thousands of birds.

Suggested Further Reading

Cardoso F and Jankovic J (1995). Clinical use of botulinum neurotoxins. In: Montecucco C (ed.) *Clostridial Neurotoxins*. Springer Verlag, Berlin.

Hatheway CL (1995). Botulism: the present status of the disease. In: Montecucco C (ed.) *Clostridial Neurotoxins*. Springer Verlag, Berlin.

Jancovic J and Hallett M (eds.) (1994). *Therapy with Botulinum Toxin*. Marcel Dekker, New York.

Poulain B, Molgo J and Thesleff S (1995). Quantal neurotransmitter release and the clostridial neurotoxins' targets. In: Montecucco C (ed.) *Clostridial Neurotoxins*. Springer Verlag, Berlin.

Shiavo G, Rossetto O, Tonello F and Montecucco C (1995). Intracellular targets and metalloprotease activity of tetanus and botulism neurotoxins. In: Montecucco C (ed.) *Clostridial Neurotoxins*. Springer Verlag, Berlin.

3

Carbon Monoxide
– The Silent Killer

L. J. Wilkinson

Introduction

It had been a cold day and was obviously going to be an even colder evening so the students decided to leave the gas fire on all night. They'd been lucky to get the flat — it was cheap and the landlord had assured them that the gas fire had been checked by one of his friends. Both students felt they were coming down with the flu; although they'd seemed better earlier that day in the lectures, now they were back home they felt sick, weak and dizzy, too tired to do much and with splitting headaches. Even their cat seemed unwell. Next morning when they didn't arrive for the tutorial, their friends went to the flat and found them both lying unconscious. Fortunately the draughts from the window had been enough to keep them alive; the students were typical victims of carbon monoxide (CO) poisoning.

Approximately 50 deaths per annum in the UK and around 200 sub-lethal poisonings are attributed to carbon monoxide poisoning, but this may be a gross underestimate given that some 250,000 gas appliances are condemned each year. Most experts agree that the true incidence of carbon monoxide-associated morbidity is not known as many sub-lethal poisonings are not recognised or not reported. However, it has been suggested that exposure to CO may be the cause of over 50% of all poisonings in some developed countries.

Properties of Carbon Monoxide

The first preparation of carbon monoxide was achieved by the heating of zinc oxide with charcoal in 1776 by the French chemist, J. M. Francois de Lassone, but the compound was mistakenly identified as hydrogen because it burned with a blue/violet flame. In 1800, the gas was correctly identified as a compound containing carbon and oxygen by an English chemist, William Cruikshank, who named the substance, "*gaseous oxide of carbone*". It is lighter than air, slightly soluble in water and soluble in alcohol and benzene; the molecule has a linear shape with a bond length of 0.1128nm, a melting point of $-207°$ C and a boiling point of $-192°$ C. It can form covalent coordination complexes with metals, particularly nickel, which forms nickel carbonyl, $Ni(CO)_4$, at room temperature so that nickel tubing corrodes when CO is present. Iron carbonyl, $Fe(CO)_5$, is important industrially while chlorination of CO forms phosgene and the reaction of CO with methanol gives acetic acid. Carbon monoxide can also be hydrogenated to liquid hydrocarbon fuels and has many applications in bulk chemical manufacture.

The toxicity of CO may have been first noted during the 1300s by a Spanish scientist, Arnold of Villanova, who observed that the burning of wood without adequate ventilation gave rise to the production of toxic fumes but it was not until the 1850s that the potential of CO as a toxin was really appreciated when a French physiologist, Claude Bernard, revealed that it combined with the oxygen-carrying protein haemoglobin. He poisoned dogs with the gas and noted that the blood in their veins was "cherry pink" (the HbCO complex absorbs at 572 and 535 nm as opposed to oxygenated haemoglobin at 578 and 540 nm). Haemoglobin binds carbon monoxide about 200 times more tightly than it does oxygen so that the energy-generating process of oxidative phosphorylation is blocked when the tissues have insufficient oxygen. The exposure limits are recommended to be less than 35 ppm for an eight-hour exposure, assuming a 40-hour week for a working lifetime.

Sources of Carbon Monoxide

Surprising perhaps, carbon monoxide is produced in the body as a result of the normal breakdown of haemoglobin from red blood cells. The enzyme, haem oxygenase, converts the haem group into bilirubin and releases CO into solution. Current research has even suggested that trace amounts of CO may have a physiological role, possibly acting as a neuromodulator.

The concentration of CO in the atmosphere is usually less that 0.001%. Apart from being an atmospheric pollutant, CO is a product of volcanic activity and is found dissolved in molten rock within the earth's mantle; the CO content of volcanic gases is variable, ranging from 0.01% to 2%. Ambient concentrations of CO are not normally high enough to give rise to CO toxicity and most cases of poisoning occur in a domestic setting, with the incomplete burning of natural gas, propane and butane to power water heaters and central heating systems.

Industrial sources

Carbon monoxide is produced and utilised in significant amounts by a number of industrial processes. The compound is a reducing agent and is commonly used in the reduction of metals, for example iron, from their ores. Other persons at risk include those working in blast furnaces and in the manufacture of chemicals such as alcohols, aldehydes and acids. The use of fuel-powered vehicles in enclosed spaces, such as fork-lift trucks presents a further exposure hazard to workers. Carbon monoxide may also be formed endogenously following exposure to the industrial solvent, methylene chloride, contained in paint strippers and used for degreasing machinery. Inhalation of sufficient quantities of the solvent can give rise to CO toxicity and there is typically a lag time between exposure and the appearance of symptoms due to its accumulation in and slow release from adipose tissue. The toxicity of CO was exploited by the Nazis during the Second World War, when it was used as an extermination gas in concentration camps, although hydrogen cyanide was much more commonly employed.

Domestic sources

The main domestic source of carbon monoxide is from the incomplete combustion of fossil fuels such as petroleum, gas, coal and wood. When oxygen is not limited, the combustion goes to completion, yielding carbon dioxide which is released into the atmosphere but any lack of oxygen results in the production of CO. Typical sources are non-vented kerosene/paraffin stoves and heaters, leaking chimneys and furnaces, blocked or leaking flues, back-draughts from furnaces, water heaters, fireplaces and wood stoves and any equipment which runs on petrol. Using gas stoves or charcoal grills in enclosed spaces such as caravans is also potentially dangerous.

In contrast to the blue flame produced by the burning of fossil fuels in the presence of plenty of oxygen, a yellow flame that leaves a black, sooty residue indicates that the carbon has not been fully oxidised to carbon dioxide. Any fuel burning appliance which has streaks of soot round it should be regarded as suspect, as should the absence of an up-draught in the chimney, water leaking from the vent or flue and excess moisture on cold surface such as windows and walls. Rusting and corrosion on pipe connections and vent pipes, with fallen soot in the fireplace and orange or yellow flames are also all signs of incomplete combustion.

Exhaust from buses, cars and trucks can also be toxic and lethal concentrations of CO have occurred within ten minutes when a car engine has been left to run in a closed garage. Exhaust pipes blocked by snow can also lead to high CO levels. The release of carbon monoxide from car exhausts has been reduced in recent years, with the introduction of catalytic converters that oxidise CO to CO_2. However, faulty exhaust systems and running a car in an enclosed space can still give rise to significant exposure and cars should not be run in the garage but should be backed out to avoid CO build-up. Carbon monoxide poisoning as a means of suicide, typically achieved by running a hose from an exhaust into the interior of the vehicle, is not uncommon, being the cause of over 2000 deaths per annum in the US.

The introduction of energy saving measures, such as better insulation and the installation of double glazing, further increases exposure in domestic settings and this, together with increased usage of central heating in cold weather gives rise to a notable increase in CO poisoning during the winter months. Carbon monoxide detectors are now widely available, either as battery or electrically operated units or as disks that darken when toxic levels of CO are present.

Mechanisms of Toxicity

The majority of inhaled CO is eliminated via the lungs, depending on the diffusion capacity of the lung and alveolar ventilation, but approximately 10%-15% of inhaled CO becomes protein-bound in the body. The compound has an affinity for iron or copper-containing sites, its main target *in vivo* being the iron containing haem group of haemoglobin, the protein responsible for the carriage of oxygen in erythrocytes or red blood cells. Haemoglobin consists of four haem-containing subunits, so each molecule is capable of binding four CO molecules. CO binds haem reversibly, with an affinity over 200 times that of oxygen, to form carboxyhaemoglobin (HbCO). The net result of HbCO formation is a reduction in the supply of oxygen to the body's tissues, but the effect is enhanced as the binding of one molecule of CO to haemoglobin also gives rise to a structural change in the haemoglobin (known as co-operativity), with an increased affinity for oxygen binding at the remaining subunits and a reluctance for Hb to release oxygen at sites in distant tissues.

The competition between CO and oxygen for binding sites on haemoglobin was determined quantitatively by Bernard in 1963 and is represented by the following equation:

$$Hb\,(CO)_4 / Hb\,(O_2)_4 = M\,[P_{CO}] / [P_{CO_2}]$$

where M is a constant equal to 220 at pH 7.4 for human blood. Hence, when $P_{CO} = 1/220 \times P_{O_2}$, the blood will contain 50% COHb and considering that the oxygen content of air is 21%, just 0.1% CO will give rise to this level of saturation of haemoglobin.

The M value shows inter-species variation, for example it is 50% lower for canary blood than human blood. Owing to this susceptibility, canaries were traditionally employed in mines; if the bird became distressed this served as an early warning to workers that the atmosphere was becoming more toxic.

Carbon monoxide also binds to the muscle-bound haem protein, myoglobin. Myoglobin serves to transport oxygen to the mitochondria for oxidative phosphorylation and acts as an oxygen reserve for the muscle in times of high physical activity. The affinity of cardiac myoglobin for CO is around three times that of skeletal muscle myoglobin.

Other cellular targets of CO include cytochrome oxidase (Complex IV) of the mitochondrial electron transport chain, resulting in the failure of the oxidative phosphorylation pathway to reduce oxygen to water and provide ATP, the chemical energy for the cells of the body.

Symptomology

The initial symptoms of CO toxicity are typically vague and usually described as "'flu like". Patients often report general malaise, headache and nausea, which are so common in winter that the diagnosis is often delayed. A worst case scenario would be several members of the same household experiencing similar symptoms, believing that they are suffering from influenza and responding by staying inside and turning up the central heating! It is for this reason that clinicians were recently re-alerted to the symptoms of CO poisoning and advised upon its management. More severe exposure to CO leads to confusion, impaired judgment, memory and co-ordination. Generally, CO poisoning is more likely if patients feel better when out of the house, if symptoms improve when the heating is turned off, if other people in the family are affected and if family pets also seem unwell. Also, if symptoms appear at the same time as faults in operation or ventilation of fuel-burning devices, if the people spending most time in the home are most severely affected and if symptoms do not include the generalised aching, low-grade fever and swollen lymph nodes that would indicate a viral infection, then CO poisoning should be considered.

The brain and heart are particularly susceptible to the effects of CO toxicity, due to their high demand for oxygen for aerobic respiration. More severely affected patients may present with loss of consciousness and cardiac symptoms characterised by chest pain and abnormalities of the heart rhythm. The brain is also sensitive to the effects of CO exposure and can experience neuronal death, especially in the hippocampus, *substantia nigra* and cortex. An added complication is that damage to the blood brain barrier renders the brain even more susceptible to this and other toxins.

Table 1 illustrates the correlation of plasma carboxyhemoglobin levels with severity of symptoms. At 10%–20% COHb, a human may be asymptomatic, or have a slight headache, at 20%–30% COHb, a feeling of malaise and fatigue. At levels of 30%–40% COHb, symptoms will be more severe and once saturation of between 40% and 60% has been reached, victims will experience seizures and may lapse into a coma. At levels above 60%, death is extremely likely. In reality however, this clinical picture is complicated due to inter-individual differences in susceptibility to exposure with age, body size and general health and even mineral deficiencies, such as anaemia.

Table 1. Plasma COHb levels and associated clinical symptoms.

COHb level	Symptoms
>10%	Asymptomatic
10%–20%	Headache
20%–40%	Dizziness, confusion, nausea
40%–60%	Coma and siezures
Above 60%	Death

Another symptom of CO poisoning is the exacerbation of pre-existing medical conditions, such as heart disease, and it is this that may bring the patient to the attention of a physician originally. The much cited "cherry red" colouration of finger tips and lips associated with severe CO poisoning, due to the presence of HbCO in the peripheral circulation is in fact rare, being present in only 2%–3% of symptomatic cases. Often and even when initial symptoms are not particularly severe, neuropsychiatric complications can occur from two to 28 days post-exposure. These include behavioural changes, ataxia and poor memory and arise because the areas of the brain which are most affected are the hippocampus, which controls memory, and the basal ganglia which controls movement. The inhibition of oxidative phosphorylation when CO binds to cytochrome oxidase causes mitochondrial dysfunction and oxidative stress. The release of nitric oxide forms the free radical peroxynitrite that damages mitochondrial enzymes and the brain vascular endothelium leading to lipid peroxidation.

Reperfusion of the brain amplifies the original injury as release of excitotoxic amino acids and proteolytic enzymes gives long-term neuronal damage which leads to delayed defects in memory and learning with the possibility of movement disorders and symptoms of Parkinson's disease in severe cases. Prolonged exposure, especially if it results in coma or altered mental status, may be accompanied by retinal haemorrhages and lactic acidosis, while muscle damage can lead to elevated creatine kinase levels.

Diagnosis

Whenever CO poisoning is suspected, carboxyhaemoglobin levels in arterial or venous blood must be determined with the use of a spectrophotometer (HbCO is cherry red in colour). Background levels of CO, for example from environmental exposure should be less than 2% in non-smokers and less than 5% in smokers, however heavy smokers can have levels up to 13%. The levels of COHb must be interpreted cautiously and based on a predicted lag time between exposure to CO and measurement, during which CO levels may drop considerably. Once a diagnosis of CO poisoning is confirmed, neurological screening may be carried out to ascertain a baseline for measuring changes in mental capacity.

Management

Following removal of a patient from the source of CO poisoning, treatment depends on the severity of symptoms and consideration of other factors, such as pre-existing heart disease. As a general rule, any patient with COHb levels of over 25% is given 100% oxygen, with the net effect of increasing oxygen levels and reducing the half-life of CO in the body. Oxygen supplementation may be required for up to 48 hours in some cases to ensure adequate removal of CO. More severely ill patients may require mechanical ventilation and monitoring of cardiac function.

One bone of contention in the treatment of CO toxicity is the use of hyperbaric oxygen (HBO) at three atmospheric pressures. Haldane initially demonstrated the efficacy of HBO in the treatment of CO poisoning in mice during the late 1800s, but it was not until the 1960s that HBO entered clinical use. HBO is administered in decompression chambers and is reported to reduce the half-life of COHb from four to five hours to around 23 minutes. Treatment with HBO is indicated if an individual is unconscious or is exhibiting severe neurological or cardiac dysfunction or marked acidosis. Aside from the issue of access to hyperbaric chambers prohibiting widespread treatment (most are situated in coastal regions because they are used to treat divers with the "bends"), the evidence for the clinical benefit of HBO use is still uncertain. In one study involving 26 patients, all hospitalised within two hours of discovery, half of whom were treated with 100% oxygen and the other half with hyperbaric oxygen. Three weeks later, the HBO group showed fewer cardiac and neurological effects. A study in 2006 examined the medical history of 230 people exposed to carbon monoxide from 1994–2002, and followed their health to 2005. After seven years, in this previously healthy population, 25% of the original population had died, about three times greater than the average. In those who had suffered damage to the heart muscle, the mortality rate was 38% with half of the mortality being due to cardiovascular problems. Carbon monoxide poisoning is characterised by raised levels of cardiac troponin I and creatine kinase-MB and ECG changes, suggesting that the reduction of ATP due to inhibition of oxidative phosphorylation leads to damage in this very energy-dependent tissue.

Are Some People at Greater Risk?

The elderly are particularly susceptible to the effects of CO toxicity owing to altered physiology with increasing age and a higher incidence of pre-existing medical conditions, particularly cardiovascular problems. Many people within this age group may be partially, or completely, housebound and thus receive greater potential exposure. People with sickle cell anaemia, asthma and emphysema are also more at risk since their oxygenation capacity is already impaired. Foetal haemoglobin has a higher affinity for CO than adult haemoglobin so that the foetus is more susceptible to the effects of CO than the mother. Carbon monoxide not only causes foetal loss but also acts as a teratogen; babies who have been poisoned with CO *in utero* have physical deformities and damage to motor function. The clearance of COHb takes four to five times longer in the foetus than the mother, and because the results of CO poisoning in pregnancy are so disastrous, many centres use HBO in any pregnant patient with a COHb level of >15%.

Prognosis and Conclusion

Prognosis is dependent on a number of factors, not least the severity of symptoms upon presentation. It is estimated that around 30% of "severely poisoned" patients die. Some 11% of survivors report long-term psychiatric problems, including delayed altered personality and or memory. Advancing age and underlying chronic conditions are also associated with a poorer prognosis.

In summary, carbon monoxide is a silent killer that occurs in domestic circumstances and is difficult to detect. As well as causing death, it can have severe long-term side-effects and is particularly damaging to the elderly and to the unborn child.

Suggested Further Reading

Fagin J, Bradley J and Williams D (1980). Carbon monoxide poisoning after accidentally inhaling paint remover. *British Medical Journal* **281**, 1461.

Goldstein DP (1965). Carbon monoxide poisoning in pregnancy. *American Journal of Obstetrics and Gynecology* **9**, 526–528.

Olson KR (1984). Carbon monoxide poisoning: mechanism, presentation and controversies in management. *Journal of Emergency Medicine* **1**, 233–243.

O'Sullivan BP (1983). Carbon monoxide poisoning in an infant exposed to a kerosine heater. *Journal of Pediatrics* **103**, 249–251.

Stewart RD (1975). The effect of carbon monoxide on humans. *Annual Review of Pharmacology* **15**, 409–422.

4

Domoic Acid

R. A. R. Tasker

Toxic episodes associated with the consumption of aquatic species, whether they be bottom-dwelling invertebrates such as clams and mussels, or species higher up the food chain such as crabs or fish, have been present throughout our history. Almost all cultures that have inhabited coastal areas note warnings about the consumption of certain species at certain times of the year or during particular climactic conditions. These stories are found in both the oral and written histories of peoples worldwide, whether they be the ancient Egyptians, the aboriginal peoples of Australasia, or the Beothuk and Mi' Kmaq tribes of eastern Canada.

Study of such toxicities has resulted in the identification of a myriad of causative agents, revealing the rich diversity of these complex ecosystems. Families of toxic compounds identified to date include the Paralytic Shellfish Poisoning (PSP) toxins such as the saxitoxins and the gonyautoxins which produce neurotoxicity primarily by the blockade of neuronal and muscular sodium channels; the brevetoxins and ciguatoxins which cause Neurological Shellfish Poisoning (NSP) and Ciguatera Fish Poisoning (CFP) respectively, not by blocking, but by opening, voltage-sensitive sodium channels; and the compounds associated with Diarrhetic Shellfish Poisoning (DSP)(e.g. okadaic acid) that appear to act primarily by inhibiting intracellular protein phosphatase enzymes within the cells of the intestinal epithelium.

A relative newcomer to this pharmacopoea of marine-derived biotoxins is domoic acid (DOM). Over the span of a scant 19 years, domoic acid has emerged from relative obscurity to become a major economic and health concern on a global scale. This is because DOM was identified in 1987 as the toxin responsible for a new neurologic syndrome called "Amnesic Shellfish Poisoning" that was defined following over 100 cases of severe shellfish poisoning in Montreal, Canada. This chapter will explore the chemistry, toxicity, pharmacology and serious clinical consequences associated with this fascinating compound.

Description, Sources and Uses of DOM

Domoic acid is a naturally occurring compound that is present in a variety of marine species and has historically been isolated from both macro- and micro-algae. Stereochemistry has confirmed the structure of domoic acid to be (2S,3S,4S)-2-carboxy-4-1-methyl-5(R)-carboxyl-1(Z), 3(E)-hexadienyl pyrrolidine-3-acetic acid. In simpler terms this means that domoic acid is a secondary amino acid and a member of the kainoid class of organic compounds which includes another marine-derived toxin, kainic acid, and the toxic acromelic acids A and B which are isolated from mushrooms. Examination of Figure 1 will reveal both the structure of domoic acid and the close structural similarity with kainic acid (KA). It should also be apparent that both DOM and KA have within their rigid structure a domain that is identical to the amino acid glutamate; the significance of this observation will become apparent when we discuss the mechanisms of DOM toxicity. At a molecular weight of only 311.14 daltons, domoic acid is a relatively small molecule, especially when compared to other marine toxins such as saxitoxins and brevetoxins that are complex polycyclic macromolecules with molecular weights in the thousands. The DOM molecule is also highly charged at neutral or physiological pH, causing it to be very water soluble (hydrophilic) and correspondingly to shun lipid environments (lipophobic). The hydrophilic nature of DOM will also be important when we discuss both the contamination of food sources (below) and the actions of this toxin following consumption by humans.

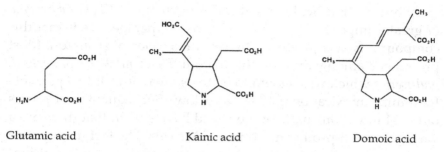

Glutamic acid   Kainic acid   Domoic acid

Figure 1. Chemical structures of glutamic acid, kainic acid and domoic acid.

The original isolation of domoic acid was performed by Japanese researchers who were investigating the insecticidal properties of the red alga *Chondria armata*. Their initial studies derived from the peculiar observation that flies, which had been attracted to and contacted *Chondria armata* drying on the seashore, died shortly thereafter. Subsequent investigations also revealed the presence of minor amounts of several isomers, notably isodomoic acids A, B and C, which are also insecticidal although less potent than the parent compound. Perhaps even more interesting to us, now that we consider domoic acid to be a marine toxin, is the reason this seaweed was drying on the seashore in the first place. It seems that in the islands of southern Japan, *Chondria armata* and another red algae, *Digenia simplex* (the natural source of kainic acid) have been harvested for centuries and used as natural anthelmintics (i.e. deworming medicines) in children. So apparently, traditional Japanese medicine involves feeding two of the most potent neurotoxins on the planet to young children for the removal of intestinal parasites. Based on anecdotal inference, therefore, we must conclude that both DOM and KA are not bioavailable (i.e. are not absorbed into the blood from the GI tract) when administered in this manner. This could be a function of either the seaweed matrix and/or digestive tract development in children, but as we will see, consumption of domoic acid-contaminated shellfish by adults results in serious, and even deadly, consequences.

While domoic acid is a natural constituent of *Chondria armata*, of greater importance from a global health perspective is that this compound is also produced by several species of plankton. Most prominent among these is the diatom *Pseudonitzschia pungens f. multiseries* which was shown to be the source of DOM responsible for human intoxication in 1987 (see below). Subsequent occurrences of DOM in various parts of the world have shown that there are a variety of toxin-producing strains of *Nitzchia*. The fact that various species of plankton appear to possess (or even worse, be acquiring) the ability to produce domoic acid is of obvious concern. With temperatures rising due to global warming and the continuing pollution of our oceans, world-wide emergence and growth of plankton species is on the rise, creating ever increasing risks of toxin production. Moreover, planktonic toxin production represents an avenue whereby these compounds can find their way into the diets of humans and other species, because a wide variety of aquatic organisms use available plankton as a primary food source. Prominent among these are filter-feeders such as clams, mussels and oysters, but as will be described in greater detail later in the chapter, plankton-derived domoic acid has also worked its way into the food chain via fish and crustaceans.

So it appears that domoic acid has been with us for some time and will continue to represent a global health threat in the future. To both understand the nature of DOM toxicity, as well as to develop potentially useful therapeutic strategies for dealing with DOM intoxication, it is necessary to understand how this compound produces its effects on the mammalian nervous system. This is the subject of the next section.

Properties — Mechanisms of Biological Interaction

To understand the actions of domoic acid in the nervous system, we must recognize its structural similarities with glutamic acid (see Figure 1). While most people not familiar with nervous system function think of glutamate as a simple amino acid that participates in biochemical pathways (e.g. feeding carbon atoms into the citric acid cycle), neuroscientists think of glutamate as the major excitatory neurotransmitter in the mammalian CNS. As many as 80% of the excitatory synapses within the brain may use glutamate as a neurotransmitter; either alone or in combination with other molecules. Pathways for the synthesis of glutamate from the precursor glutamine are also a feature of many of the non-neuronal elements of the brain (i.e. glial cells). Based on these findings, it is apparent that glutamic acid is responsible for many of the functions within the brain. These include not only basic cell-to-cell communication, but also many of the complex phenomena such as hippocampal long term potentiation (LTP), a process believed to be involved in learning and memory. Perhaps the most intriguing aspect of the actions of glutamate within the CNS is that while this tiny molecule has a physiological function, just as do other neurotransmitters, glutamic acid is also known to have a role in pathological processes causing cell damage and cell death. Most scientists who work in this field believe that the pathological actions of glutamate are an extension of its physiological function. In other words, cell damage and cell death occur when excessive amounts of glutamate are released, resulting in an "overstimulation" of the target system. This theory is based on a wealth of evidence obtained using both whole animal and cell culture experimental models, in which high concentrations of glutamate or glutamate-like compounds produce neuronal damage. Within the context of human disease there is certainly convincing evidence that excessive release of glutamate during acute trauma such as brain ischemia or hypoglycemia leads to neuronal degeneration. It has also been proposed that glutamate neurotoxicity may be integral to progressive neurodegenerative diseases such as Alzheimer's disease, Huntington's disease or Parkinson's disease. Moreover, it was proposed over 30 years ago that glutamate, in the form of monosodium glutamate (MSG), was responsible for acute neurotoxicity in certain periventricular regions of the brain, thereby resulting in the phenomenon known as "Chinese Restaurant Syndrome".

So what does all this have to do with domoic acid? To answer that question it is necessary to understand that glutamate, like all neurotransmitters, produces its effects by interacting with specific proteins embedded in the cell membrane of the neuron. These proteins assemble in functional units called "receptors". In the case of glutamate receptors, most are tetrameric, which is to say that each receptor consists of four proteins. The nature of the constituent proteins determines the characteristics of the receptor, both in terms of function (which would be the normal response to the appropriate transmitter, in this case glutamate), and in terms of the ability of the receptor to be either stimulated or blocked by specifically defined synthetic or semi-synthetic compounds (ligands) that are used in experimental settings. The use of experimental ligands has allowed us to subdivide glutamate receptors into a variety of subtypes, each of which is presumed to subserve different functions within the nervous system. An overview of this classification scheme is presented in Figure 2. The first major distinction between glutamate receptors is that some are designated "ionotropic" whereas others are "metabotropic". Ionotropic receptors are characterised by being directly linked to a hole (or "channel") in the cell membrane. When in open configuration (e.g. following stimulation of the receptor) these channels allow the movement of ions (e.g. Na^+ or K^+) across the membrane according to their concentration gradients. This movement of ions alters the electrical potential across the membrane thereby either initiating or inhibiting subsequent events. If the ion channel is also permeable to certain divalent cations (e.g. Ca^{2+}), increased intracellular cation concentrations may trigger other events within the neuron such as enzyme activation or gene transcription. In response to normal concentrations of neurotransmitter (glutamate) these ion fluxes produce normal physiological responses. However, if the concentrations of glutamate are abnormally high, or if a ligand with greater potency than glutamate is used to stimulate the receptors, excessively high intracellular concentrations of Ca^{2+} may accumulate and thereby induce cell damage leading to neuronal cell death. In contrast to ionotropic receptors, simple activation of metabotropic receptors is not sufficient for inducing a functional change in the neuron. Metabotropic receptors rely on the enzymatic conversion of constituent molecules within the cell in order to open ion channels in the membrane and produce an effect. These intracellular molecules are referred to as "second messengers", and frequently initiate a cascade of intracellular processes. In simple terms, therefore, ionotropic receptors may be thought of as mediating "fast" events whereas metabotropic receptors are more involved in "slow and complex" events. Over-stimulation of either type, however, may lead to excessive accumulation of Ca^{2+} within the neuron and subsequent neurotoxicity.

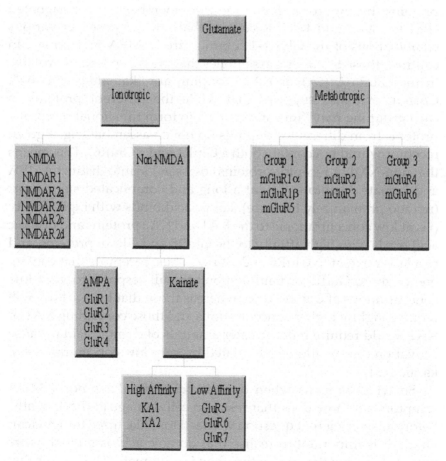

Figure 2. Classification of glutamate receptors and associated subtype-specific receptor proteins.

At present, metabotropic glutamate receptors are not extensively sub-classified. On the other hand, ionotropic glutamate receptors can be further subdivided into Nmethyldaspartate (NMDA) receptors and nonNMDA receptors; the latter grouping containing two subsections designated "AMPApreferring" and "kainate-preferring" (see Figure 2). Both domoic acid and kainic acid are stimulants ("agonists") of non-NMDA receptors. As stated earlier, most glutamate receptors are tetrameric (i.e. are composed of 4 proteins). In the case of non-NMDA receptors, these constituent proteins belong to a family whose members are designated GluR1-7, KA1 and KA2. Receptors that are composed of various combinations of the GluR1-4 subunits are "AMPA-preferring". In contrast, those tetrameric assemblies that are composed of proteins in the GluR5-7, KA1 or KA2 grouping are "kainate-preferring". Current evidence suggests that while the different proteins in each grouping may "mix and match" to form functional receptors, proteins from different groupings do not co-assemble (e.g. a given receptor would not contain both a GluR1 and a GluR6). This means that non-NMDA receptor proteins co-assemble into distinct AMPA and kainate receptors. To cut a long and complicated story short (not too prematurely I'm sure), kainic acid binds with high affinity (i.e. at low concentrations) to the KA1 and KA2 proteins and domoic acid binds with high affinity to the GluR5 and GluR6 proteins, and to a lesser extent to GluR5 and GluR7. Thus, receptors that contain one or more GluR5 or GluR6 subunits will respond to very low concentrations of domoic acid, whereas those that have GluR7 will require slightly higher concentrations and those containing KA1 or KA2 would require much greater amounts of domoic acid to cause activation (but would be stimulated by very low concentrations of kainic acid).

So what happens when domoic acid stimulates non-NMDA receptors and how does that result in cell damage and cell death? Before answering that question we need to make three things clear. Firstly it is important to realise that domoic acid is a much more potent agonist at these receptors than is glutamate. This means that even low concentrations of DOM can produce effects that mimic those produced by large concentrations of glutamate. Secondly you need to understand that domoic acid is a "selective" (rather than a "specific") agonist at non-NMDA receptors in general, and particularly those that contain a GluR5, 6 or 7 protein (i.e. "domoate receptors"). Selectivity means that very low concentrations of DOM will stimulate mainly "domoate" receptors, but that as the concentration of DOM increases there will be a progressive activity at other receptor subtypes in addition to "domoate" receptors. So at the lowest concentrations we get actions via "domoate" receptors, at slightly higher concentrations we get actions via "domoate" receptors plus receptors that contain either KA1 or KA2, at even higher concentrations DOM will stimulate AMPA receptors as well as the previous two types, and so on. This concept of selectivity is important because it not only explains why we get different types of toxicity at different concentrations of DOM, it also explains why a particular receptor blocking drug ("antagonist") may relieve some, but not all of the symptoms of high dose toxicity. Thirdly, remember that not all receptor types are present in all regions of the brain, or even if they are present the relative concentration and/or specific location of each subtype varies. This is why certain brain areas (e.g. the CA3 region of the hippocampus) are more sensitive to the toxic actions of DOM (i.e. lots of GluR6). It also explains why the mechanism of DOM toxicity is not necessarily identical in different regions or different experimental preparations (e.g. cerebellar granule cells in culture versus cortical neurons in culture).

With these thoughts in mind consider the cartoon depicted in Figure 3. Figure 3 represents a fictional neuron that has all of the known sites of domoic acid action; obviously real neurons have only some, or none, of these. Non-NMDA receptors (the solid squares) are depicted at three sites on the glutamatergic neuron: on dendrites (No. 1), on the pre-synaptic terminal (No. 2), and on the post-synaptic terminal (Nos. 3 and 4). Those on the pre-synaptic terminal are exclusively of the "kainatepreferring" subtype, whereas the dendritic and post-synaptic receptors may be either AMPA-type or kainate-type. To complicate things even further, AMPA/kainate receptors come in two models; which I'll call "standard" and "deluxe". The "standard" model kainate receptors are linked to an ion channel that can only flux (allow passage of) Na^+ and K^+. Such receptors produce a simple ion-based depolarisation of the membrane which is analogous to the events preceeding an action potential in a nerve. The non-NMDA receptors depicted on the dendrites only come in this "standard" model, so application of DOM to these receptors results in local depolarisation that may or may not cause the neuron to fire depending on what else is going on (i.e. the overall membrane potential at that time). The post-synaptic receptors (and possibly the pre-synaptic receptors although the evidence is weaker at this time), however, also come in a "deluxe" (edited) model. A "deluxe" non-NMDA receptor contains either a GluR2 (AMPApreferring) or a GluR5 or GluR6 (kainate-preferring) protein that has been modified prior to receptor assembly. This modification determines whether the receptors are able to not only flux Na^+ and K^+, but can also flux divalent cations such as Ca^{2+}. In the case of GluR2, editing prevents Ca^{2+} flux, but for the kainate receptor proteins (GluR5 and GluR6) a "deluxe" channel is a Ca^{2+} permeable channel. So assuming that the excessive influx of Ca^{2+} is the major determinant of neurotoxicity, neurons containing post-synaptic "deluxe" kainate receptors will be more sensitive to domoic acid-induced damage than will neurons that have only the "standard" receptors. Before you begin to believe that "standard" is analogous to "safe", however, consider the post-synaptic NMDA receptor depicted as a polygon in Figure 3. NMDA receptor-linked ion channels are quite capable of fluxing Ca^{2+}, but normally these ion channels are blocked by the presence of a Mg^{2+} ion within the channel. If sufficient NMDA (or glutamate) stimulates the receptor this Mg^{2+} block is relieved and the channel opens. Similarly, and of pertinence to our discussion, the Mg^{2+} block can also be relieved if the membrane is depolarised to a particular potential (about -50 mV). So if enough domoic acid acts at "standard" post-synaptic non-NMDA receptors, the NMDA channels will open and Ca^{2+} will flow into the cell. The route may be a little more complex and the concentrations of DOM required may be higher than if "deluxe" receptors are present, but high concentrations of DOM can lead to

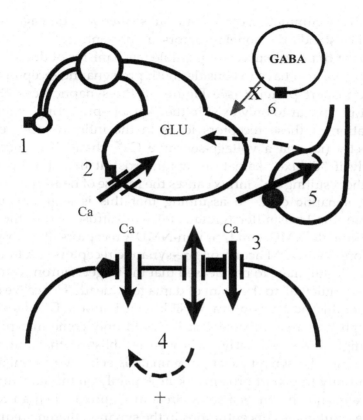

Figure 3. Fictional glutamatergic neuron depicting all known sites of domoic acid action. AMPA/kainate receptors are depicted as squares, NMDA receptor as a polygon and glial glutamate transport protein as a circle. Mechanism 1: AMPA/kainate receptors on dendrites cause depolarisation of the neuron. Mechanism 2: AMPA/kainate receptors on the pre-synaptic terminal open ion channels leading to the influx of calcium and the subsequent release of stored glutamate. Mechanism 3: AMPA/kainate receptors on the post-synaptic terminal open a calcium-permeable ion channel directly. Mechanism 4: AMPA/kainate receptors on the post-synaptic terminal open a sodium/potassium channel that causes membrane depolarization. If the membrane is sufficiently depolarised calcium channels linked to NMDA receptors are opened leading to the influx of calcium. Mechanism 5: Domoic acid inhibits the removal of glutamate from the synapse via an action on high-affinity glutamate transport proteins in surrounding glial cells. Mechanism 6: GluR5 receptors inhibit the release of GABA from interneurons, thereby reducing the activation threshold for glutamatergic neurons. Additional details are provided in the text.

excessive accumulation of Ca^{2+} and subsequent cell damage even if only the "standard" model receptors are present.

To continue our survey of possible mechanisms of domoic acid toxicity, we also have to consider both pre-synaptic receptors and glial transport proteins (see Figure 3). Pre-synaptic non-NMDA receptors appear to only exist in the "kainate-preferring" subtype. Activation of these receptors leads to the influx of Ca^{2+} either indirectly (i.e. via a voltage-sensitive Ca^{2+} channel) or possibly directly if "deluxe" kainate receptors exist at this site. In either case, the resulting Ca^{2+} influx causes the release of neurotransmitter in the synaptic cleft. So assuming that this is a glutamatergic synapse, the transmitter (glutamate) will diffuse across the cleft and stimulate NMDA and/or non-NMDA receptors. It is possible, therefore, that DOM acting on pre-synaptic receptors, could cause sufficient glutamate to be released that the target neuron would be "over-stimulated" to the point of damage or death. Hence we must consider this mechanism as a possible contributor to DOM toxicity, although it seems probable that it would only come into play at very high doses. Similarly, it is well established that following release into the synaptic cleft, glutamate is actively scavenged by high affinity transport proteins located mainly on the surrounding glial cells and shown as a solid circle in Figure 3. Such a process reduces glutamate concentrations in the synapse, thereby reducing the chance of toxicity. At this time, there is a bit of evidence that domoic acid can inhibit these transport proteins. If true, then DOM may increase the active concentration of glutamate in the synapse by preventing removal of transmitter, as well as by increasing release of glutamate and by stimulating the post-synaptic receptors directly. Finally, over the last few years there is increasing evidence that kainate receptors are present on hippocampal interneurons. These interneurons normally release the inhibitory neurotransmitter GABA, thereby raising the threshold for activation of the nearby excitatory glutamatergic neurons. The kainate receptors reduce this tonic inhibition. So in this case, DOM would not only stimulate the excitatory synapse directly, but would also inhibit the on-going inhibition of that excitatory synapse (analogous to pressing on the accelerator in your car while also taking your foot off the brake; a "double hit").

In summary, therefore, there are a variety of ways in which domoic acid appears to produce neuronal toxicity, but all involve either direct or indirect "over-stimulation" of post-synaptic receptors leading to the influx of excessive amounts of Ca^{2+}. The differential sensitivity of different neuronal populations in different regions of the brain, is probably determined both by which of these mechanisms is present, as well as by how many of them are actively involved at the time of toxic insult. Let us now consider the nature of the toxicity produced by DOM, and learn why it's something you want to avoid.

Toxicity Profile

While there are examples in the literature of peripheral toxicities following administration of domoic acid (e.g. gastric and duodenal ulcers in mice and cardiotoxicity in rats) most of what we know concerns toxic events in the central nervous system of mammals. In this respect, domoic acid is a remarkable toxin to study, for several reasons. Firstly, the major brain regions affected and the most prominent behavioural consequences of DOM toxicity are remarkably consistent among different mammalian species. Whether given experimentally to mice, rats, rabbits or monkeys, or whether involuntarily consumed by humans, domoic acid produces a very similar toxicity profile. The anatomical and behavioural sequelae are described below. This represents an enormous opportunity for neurotoxicologists: not only do we know what happens in humans exposed to domoic acid, but we can confirm the validity and relevance of non-human experimental models for studying DOM toxicity or developing therapies for the human population. Secondly, the sequence of events and patterns of damage observed following DOM administration in a given species are both highly reproducible and predictably dose-related. My laboratory has been studying DOM toxicity in mice and rats for over 15 years using an objective seven-category behavioural rating scale which we developed. The dose of domoic acid that produces a given score (e.g. 150 out of a possible 413) is identical today with the dose that produced the same score in 1988; and the neuroanatomical and behavioural response to that dose is also identical. There are species differences with respect to the potency of domoic acid (i.e. the dose that produces a given effect) in that mice are less sensitive than rats that are less sensitive than monkeys and humans, but the remarkable intra-species consistency allows us to use dramatically fewer animals than most standard toxicological testing paradigms. So we have reproducible, dose-related toxicity in experimental models that we know are relevant to the situation in humans; an experimental toxicologist's dream.

Before describing mammalian toxicity it is fitting to mention the natural vehicle for DOM in our food chain; the lowly bivalve (see the next section). There is absolutely no evidence that even massive doses of domoic acid (1000 ppm) produce any toxicity in mussels, clams or oysters. In contrast, the humans admitted to intensive care units in 1987 appear to have consumed less than 4.0 mg/kg of domoic acid (4 ppm). It is not known at this time whether bivalves lack the proper forms of glutamate receptors or whether they simply have no capacity to absorb the toxin from their digestive gland, but in either case they appear to have a distinct advantage over us.

So what happens when we give a mammal domoic acid? Low doses (about 1/20 to 1/10 of the LD50) result in gastrointestinal disturbances (nausea, vomiting, diarrhoea) in humans and monkeys. In laboratory rodents these early effects manifest primarily as sluggishness or reduced mobility (they are almost incapable of vomiting and they can't tell us they feel sick, so we could assume the lack of activity means they don't feel well). As the dosage increases (to about 1/10 to 1/5 of the LD50) muscle fasciculations and intermittent forelimb rigidity characteristically appear, and at slightly higher doses the first truly "neurotoxic" events are seen. These include headaches and confusion in humans, but in mice and rats the "trademark" of domoic acid toxicity at these doses is a repetitive scratching-like motion by a hindpaw and directed at the neck region just behind the ear. "Scratching" behaviour is a definitive sign for domoic acid toxicity when algae or shellfish are screened using conventional bioassay procedures. At slightly higher doses mice and rats will exhibit a variety of stereotypic behaviours such as circling, head bobbing and "wet dog shakes". Interestingly, while these types of behaviours are clearly both centrally-mediated and serious, histopathological examination of animals given these doses of DOM has consistently failed to demonstrate identifiable neuronal damage. In a series of recent experiments, our laboratory and those of others have shown activation of brain regions as measured by either enhanced expression of immediate early genes (e.g. c-fos) or by accelerated glucose utilisation (i.e. 2-DG turnover) at these doses, but whether observed by light or electron microscopy, activation of neurons by domoic acid at these doses appears to be reversible. Signs of cell damage and cell death in the CNS are only observed in animals that receive a slightly higher dose, and cross the threshold to observable motor seizures. This "threshold" effect is also consistent with the clinical profiles of humans who consumed domoic acid; no permanent cognitive damage was noted in any of the patients whose clinical course ended prior to the initiation of seizure behaviour.

Once seizures begin, however, damage is unavoidable. Seizures are characterised by abnormal EEG tracings in all species, but manifest primarily as intermittent myoclonic events that progress to whole body convulsions. Animals and humans can recover after initiation of seizures, but if allowed to continue, death is the next step in this dose-related sequence (obviously for humane reasons this should not be allowed to happen). Histopathological examination of the brains of animals, or humans, who experience DOM-induced seizures show a remarkably consistent pattern of damage. Only certain regions of the CNS are affected. Notable among these are specific subfields of the hippocampus, designated CA3 and CA4 (H3 and H4 in humans), as well as the entorhinal cortex, amygdala, lateral thalamus and hypothalamus. This pattern was first revealed on post-mortem analysis of the human patients who died (see the case report later in this chapter), but has subsequently been confirmed in mice, rats, rabbits and monkeys. If examined using electron microscopy, cells in these regions appear either swollen with conspicuous intracellular vacuoles, or dense and dark. These two descriptions are consistent with changes seen following any glutamatergic excitotoxin, and presumably arise from the receptor-mediated events described in the preceding section.

Finally, of particular interest and note are the effects on the hippocampal subfields and the entorhinal cortex. These regions are known to be interconnected and to participate in memory formation. You will recall that domoic acid intoxication has been designated "amnesic shellfish poisoning" (ASP) because it is characterised by deficits in memory function in affected humans. While some severely affected patients experienced a global retrograde amnesia, all of the patients who presented with abnormal EEG patterns demonstrated a defect in a specific type of memory designated "visuo-spatial". This form of short-term memory allows us to remember a pattern even after it is removed from sight, and also permits us to know where we are in relation to other objects even with our eyes closed. In experimental animals we can measure this type of memory function by using particular types of maze paradigms, notably water mazes where animals have to find a hidden platform using external cues. Studies of domoic acid in mice and rats have consistently shown that doses capable of producing mild seizures also produce deficits in water maze performance. So in both humans and experimental animals both the memory defects and the anatomical pattern of hippocampal cell damage is identical. This is strong evidence that the CA3/CA4 hippocampal subfields and their connection to the entorhinal cortex is crucial to visuo-spatial memory. For this reason domoic acid is becoming an increasingly popular neurochemical for investigating this aspect of cognitive function.

Examples of Endemic Problems — Toxic Episodes

As we have seen domoic acid is produced by a variety of aquatic species worldwide. It can also concentrate in species that have been consumed by humans for millenia. And as just discussed, domoic acid is an extremely potent neurotoxin, requiring ingestion of relatively small quantities to produce severe, and therefore noteworthy, toxicities. Thus it seems curious that domoic acid has only been recognised as having serious implications for human health and as a widespread toxic contaminant in food for the last 19 years. It is obviously possible that toxic events prior to 1987 went undetected or unrecorded. It is equally possible that the *Nitzschia spp.* have only recently acquired the ability to produce domoic acid, meaning that previously the only route of possible exposure was through the consumption of seaweed. Whether either, or both, of these explanations are wholly or partly correct, domoic acid burst on to the marine toxin scene in a highly dramatic fashion in 1987. Within the span of a few days in late November and early December of 1987 over 250 cases of severe illness following consumption of shellfish were reported in Quebec and New Brunswick, provinces in eastern Canada. The overwhelming majority of these were in the Montreal area of southern Quebec. Of these fully 107 were ultimately confirmed as meeting the case description for what was later termed "Amnesic Shellfish Poisoning". Three elderly patients died and nine others, most of whom continued to suffer neurological damage many years later, required confinement in hospital intensive care units. The specifics of the case description are provided in the last section of this chapter.

Following a spectacular sequence of scientific detective work that relied on the cooperation and round-the-clock effort of multiple government and university researchers, the causative agent was identified as domoic acid on 18 December 1987, and the vehicle by which this toxin found its way into the diets of the victims, was the lowly cultured blue mussel (*Mytilus edulis*) harvested from a small region in the eastern part of Canada's smallest province, Prince Edward Island.

To understand how these shipments of live mussels became contaminated with domoic acid, it is necessary to briefly overview the way in which these shellfish are cultivated for commercial distribution. The production of cultured blue mussels for export from Prince Edward Island (PEI) experienced phenomenal growth during the early 1980's. Between 1981 and 1986 the total production tonnage for this industry increased more than 25-fold. The reason for this remarkable growth was a combination of market demand, aquaculture technology, and the geography of PEI which features many wide but shallow estuaries subject to good tidal flushing. Almost all of the mussel industry on PEI uses "long-line" techniques for culturing these shellfish. Some of the major elements of this technique are diagrammed in Figure 4. Blue mussels begin life as a motile form, termed spat, which are subject to the tides and currents flowing in and out of the estuaries. Long lengths of nylon rope are anchored and suspended in the water column, to which are attached polypropylene ropes that "float" in the water and to which the motile mussel spat anchor themselves. Once the spat have been collected they are transferred to a nylon mesh "socks", usually in September of each year. These socks, each loaded with 175 to 200 mussel spat per linear foot (600 per metre) are suspended from the long line, and the mussels simply grow in the socks. During the winter months the long lines are sunk with sandbags to avoid damage from ice, but still remain suspended in the water column. Suspension in the water not only dramatically improves growth rates, as will be discussed below, but also avoids predators such as starfish and crabs that dwell on the ocean bottom. After 12 to 18 months the mussels have grown to market size and are harvested, usually through the ice in a technique that is somewhat unique to PEI. After being cut, washed, and packed the product is shipped live on ice to restaurants throughout North America.

The strength of the PEI mussel industry is dependent upon both the quality of the product and the speed with which it can be produced. Both derive from the culture technique. Wild mussels normally anchor themselves to the ocean floor where predators are present and where the food supply, principally algae, exists at a lower density than it does close to the surface. Mussels are bivalves, and therefore filter-feeders, meaning they extract their food by filtering it from the water and concentrating it in their digestive gland. Having evolved in an environment where food is scant, they not only "feed" continually but they convert all of the available energy into growth, being both osmo-conformers (they

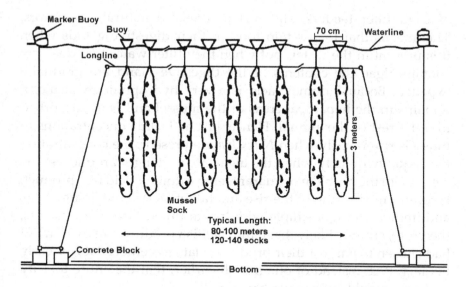

Figure 4. Depiction of the long-line technique used for cultivating blue mussels in Prince Edward Island. Mussels grow in nylon—socks—that are suspended in the water column where plankton are plentiful. In winter the line is lowered to avoid damage from ice. Additional details are provided in the text.

don't maintain a constant osmolarity) and ectotherms (they don't maintain a constant body temperature). They are also not fussy about what they eat, and will filter particles ranging from two to 90 microns in diameter. So what happens when you take an animal adapted to the bottom and suspend it near the surface where algae are plentiful? The answer is that they feed voraciously, siphoning two to five litres of water per hour each, and extracting up to 98% of the available algae; this results in a phenomenal growth rate and a plump, sweet-tasting meat.

Given this background, it is easy to see how mussels suspended in a water column teeming with domoic acid rich *Nitzschia pungens f. multiseries*, would rapidly become highly toxic (up to 900 ppm wet weight). Ironically, and much to the relief of the shellfish producers, the solution to such problems is strikingly simple, and again resides in the physiology of the lowly mussel. In addition to being highly efficient filter feeders, all bivalves are also natural depurators. This means that they retain food in their digestive glands when it is present in the water, but when they filter water that does not contain algae the contents of the digestive gland are gradually expelled. Because domoic acid is a very hydrophilic molecule, it remains in the hepatopancreas of the mussel and is not absorbed into the rest of the animal. Hence, just as DOM concentrations in mussels rise rapidly when *Nitzschia* are present, the concentration falls equally quickly when the algae are no longer around. So the solution to the problem of guaranteeing a domoic acid free product, is careful monitoring of the rise and fall of DOM-containing algae, and implementing a simple withdrawal time before harvest. It is, therefore, entirely likely that if the shellfish producers in eastern PEI had chosen to harvest their product in late December 1987 or early January 1988 instead of one month earlier, that the entire toxicity outbreak would have never happened!

While the 1987 PEI mussel incident is clearly the most famous case of domoic acid poisoning, it is important to note that it was not an isolated event. Another of the more fascinating outbreaks of domoic acid toxicity occurred in the Monterey Bay area of California in September 1991. In this case the victims were not humans but pelicans and cormorants, which literally fell from the sky following consumption of anchovies containing up to 100 ppm of domoic acid. The offending algae in this case was *Pseudonitzschia australis* rather than *Pseudonitzschia pungens*, but the toxin and the outcome were the same. Since these incidents in the late 1980s and early 1990s monitoring agencies worldwide have been on the lookout for domoic acid. Most use a fairly straightforward liquid chromatographic assay for the toxin and suspend harvesting if DOM is detected at greater than 20 ppm (a value that was adopted in Canada early in 1988 but that has never been fully validated as "safe" for all consumers). Since these monitoring programs began, domoic acid has been detected on both the eastern and western coasts of Canada, the west coast of the United States from California to Washington state, in the Gulf of Mexico, and in Denmark, Spain, Italy and New Zealand. The 1987 Canadian outbreak, however, remains the only confirmed incident of serious human toxicity produced by this agent; this is certainly fortunate, given the brevity of the next section of this chapter.

Treatments and Antidotes

At the time of the 1987 poisonings, the attending physicians clearly had no way of knowing what they were dealing with and had no base of information upon which to make rational therapeutic choices. The approaches adopted, therefore, were largely supportive care and, where warranted, standard anticonvulsant therapy (usually with phenytoin [diphenylhydantoin]). Nineteen years later, we now know that the cause of ASP is domoic acid and we have a wealth of literature about the mechanisms underlying different aspects of DOM toxicity. Sadly, however, were another case of ASP to appear tomorrow the advice to the attending physician would be largely unchanged. At this time there are only two pieces of information that are irrefutable and essential to effective patient management in these cases: (1) domoic acid is almost entirely eliminated from the body by urinary excretion so anything that increases renal clearance is useful; and (2) seizure control is essential to the prevention of long-term neurological damage. Both of these findings have been confirmed in numerous laboratory studies involving a variety of species.

While medical advice now may be largely unchanged compared to 1987, there is increasing hope that effective pharmacological tools can be developed. Before briefly discussing some of these findings, however, it is necessary to distinguish between an "antidote" and an "antagonist". Antidotes are agents that reduce toxicity after symptoms have begun. In the case of DOM intoxication, this would mean that a given drug would reduce toxicity in a patient who has presented with the symptoms of ASP. This is clearly what the attending physician requires. In contrast, an antagonist is a drug that prevents or reduces toxicity when administered prior to, or coincident with, the toxin. Such compounds are extremely useful to scientists attempting to objectively define the pharmacological basis of a toxic effect, but unless you are James Bond or Indiana Jones it is unlikely that you would be carrying an antagonist around in anticipation of being poisoned. So for the most part medical practitioners are awaiting the development of antidotes, but most work to date on domoic acid has been concerned with identifying antagonists. On the positive side, the two concepts are not mutually exclusive, especially for the treatment of ASP. As we have seen, domoic acid toxicity manifests as a progressive series of toxic events or behaviours. These occur over the span of a number of hours, days, and even months in human patients. Thus, by concentrating on the development of antagonists that prevent the more severe forms of toxicity, such as seizures, pharmacologists can both understand the mechanisms of DOM toxicity and provide drugs to the clinic that have value in the short term.

The only reported antidote for domoic acid toxicity was described in a series of papers published shortly after the 1987 PEI disaster. These studies by Glavin, Pinsky and colleagues indicated that kynurenic acid, a broad-spectrum excitatory amino acid (EAA) antagonist which is actually present in the brain in small amounts, would reduce both behavioural toxicity and gastrointestinal lesions when administered 45 or 75 minutes after domoic acid in mice. Unfortunately very high doses were required (300 and 600 mg/kg) and the time frame for the effect was fairly narrow. For these and other reasons kynurenic acid has little potential as a human medication. Nonetheless, the potential for developing such a drug still exists.

More encouraging are the results of a number of laboratories, including our own, who have made considerably inroads into identifying receptor-selective EAA antagonists that are effective at partially preventing the consequences of DOM toxicity in animal models. Both competitive AMPA/kainate receptor antagonists (e.g. NBQX) and NMDA receptor antagonists (e.g. CPP) have been shown to partly attenuate DOM toxicity in rodents. Interestingly, data from our laboratory using behavioural testing in mice has independently confirmed findings published by my friend and colleague Antonello Novelli in Spain using isolated cerebellar neurons in culture. We have both found that NMDA potentiates domoic acid toxicity and that antagonists at both NMDA and non-NMDA receptors can reduce such toxicity. These findings support the premise we explored earlier in the chapter, that DOM probably produces toxicity by a combination of mechanisms rather than via a single neurochemical pathway. This may ultimately complicate attempts to develop a single, effective therapy, but it helps to explain why simple solutions have not been forthcoming. Recently, our advances in understanding the pharmacology of domoic acid have resulted in the identification of the most effective agent tested to date. A team of dedicated neuroscientists who formed NeuroSearch A/S (Copenhagen), have developed a currently nameless compound, NS-102, that has high binding affinity for the GluR6 receptor subunit without opening the ion channel. This makes NS-102 a potential antagonist of the "domoate receptor" we discussed earlier. Sure enough, when we administered this compound prior to domoic acid in our mouse models, all indices of toxicity were reduced by up to 50%, and the critically important seizure threshold was dramatically increased. NS-102 will probably never appear in a hospital pharmacy because of its very short biological half-life, but the data are useful in support of the approach. Similarly, there are now several compounds that are selective antagonists of GluR5 receptors, but to date they have not been tested sufficiently to determine their *in vivo* efficacy. So it is apparent that selective agents of this type can be developed, and are promising candidates for the future therapy of ASP, but more work must be done before we have effective clinical tools at our disposal.

Case History of Fatal Poisoning

As stated earlier, only once has domoic acid been definitively connected with fatal poisoning in humans. The original case reports are listed in the "Suggested Further Readings" section at the end of the chapter. Briefly, the clinical events were as follows:

- Of the more than 250 reports of illness associated with the consumption of mussels, 107 met the case definition. No cases were confirmed outside of Canada. Approximately equal numbers of men and women were affected, with 46% of the patients being between 40 and 59 years of age and 36% being 60 or older.

- A standardised questionaire was provided to each patient and 99 responses were received. All but seven of the patients had vomiting, diarrhoea, or abdominal cramps. Headache, often severe, was reported by 43% of patients and loss of memory (mainly short term) was experienced by 25% of the patients. Younger patients were more likely to experience gastrointestinal symptoms whereas older patients and men were more likely to have memory loss. Occurence of other symptoms was not related to age or sex.

- Eighteen percent of patients were hospitalised, usually within two days of the onset of symptoms and the duration of hospitalisation ranged from four to 101 days (median 37.5 days). Hospitalised patients generally presented with confusion, disorientation, and an inability to recall recent past events or experiences. Only four of the 16 hospitalised patients were under 65 years of age and all had pre-existing illnesses (insulin-dependent diabetes, renal disease, hypertension). Twelve of the 16 hospitalised patients were treated in the intensive care unit and had serious neurological dysfunction including coma (nine patients), mutism (11), motor seizures (eight), purposeless chewing or grimacing (six) and reduced or absent response to painful stimuli (eight). Nine patients were intubated because of respiratory secretions and seven had unstable blood pressure or cardiac arrhythmias. Blood and cerebrospinal fluid were negative on bacterial culture and CT scans were normal on nine of the 12 patients (two results unavailable).

- Three patients (71, 82 and 84 years of age) died in hospital 12, 24 and 18 days, respectively, following ingestion of mussels. A fourth patient (84-year-old male) was discharged from the intensive care unit but died of an acute myocardial infarction three months later. Study of the brain tissue obtained from three of these patients revealed neuronal cell loss and astrocytosis mainly in the hippocampus and amygdaloid nucleus.

Acknowledgements

I am deeply indebted to Dr. Tracy Doucette for her input and support during the preparation of this chapter.

Suggested Further Reading

Domoic Acid Toxicity: Proceedings of a Symposium (1990). *Canada Diseases Weekly Report* (Supplement) **16S1E**, Hynie I and Todd ECD (eds.).

Jeffrey B, Barlow T, Moizer K, Paul S and Boyle C (2004). Amnesic shellfish poison. *Food and Chemical Toxicology.* **42**(4), 545–557.

Monaghan DT, Bridges RJ and Cotnam CW (1989). The excitatory amino acid receptors: their classes, pharmacology and distinct properties in the function of the central nervous system. *Annual Review of Pharmacology and Toxicology* **29**, 365–402.

Perl TM, Bedard L, Kosatsky T, Hockin JC, Todd ECD and Remis RS (1990). An outbreak of toxic encephalopathy caused by eating mussels contaminated with domoic acid. *New England Journal of Medicine* **322**, 1775–1780.

Teitelbaum JS, Zatorre RJ, Carpenter S, Gendron D, Evans AC, Gjedde A and Cashman NR (1990). Neurological sequelae of domoic acid intoxication due to the ingestion of contaminated mussels. *New England Journal of Medicine* **322**, 1781–1787.

5

Ecstasy

R. A. Braithwaite

Introduction

"ECSTASY" (also known as "XTC" or "E") is the "street" name of one member of a family of amphetamine related drugs which first became popular in the "rave" or modern dance music culture across Europe in the 1980s (Table 1). Its chemical name is 3,4-methylenedioxymethamphetamine (MDMA); closely related drugs include: methylenedioxyamphetamine (MDA) and methylenedioxyethylamphetamine (MDEA). The chemical structure of this series of drugs and their relationship to older better known stimulants of abuse, amphetamine and methamphetamine, is shown in Figure 1.

Table 1. Major "ecstasy"-like drugs.

MDA	Zen, Love Drug
MDMA	Ecstasy, Adam XTC, E
MDEA	Eve

Although regarded as a modern "designer drug" by the popular press, MDMA (ecstasy) was in fact synthesised in the early part of the 20th century and was patented by the Merck Chemical company in the US in 1914. It was initially developed as a potential appetite suppressant, but never became a "licensed product". The compound was then ignored for almost 60 years, when in the 1970s animal experimental work was carried out comparing it with a number of hallucinogenic mescaline analogues. It was subsequently used experimentally in the US for the treatment of psychiatric disorders. It became particularly fashionable as an adjunct to psychotherapy, much in the same way that LSD was used during the 1960s. It was reported that MDMA produced an "altered state of consciousness with emotional and sensual overtones". One of the major advocates of ecstasy and related drugs around this time was Alexander Shulgin (with help from his wife Ann). Shulgin born in 1925 was a brilliant organic chemist who studied biochemistry at the University of California at Berkeley. He synthesised a large number of "ecstasy like" and other stimulant and hallucinogenic drugs, then personally experimented with each drug and wrote about his experiences in a now famous book — PiHKAL (phenylethylamines I have known and loved). Both ecstasy and amphetamine are phenylethylamine derivatives. He is said to have tried out more experimental drugs than the oldest laboratory rat, but genuinely believed that such drugs were powerful tools that enabled the individual to explore the mind and expand the conscious state. During the 1980s ecstasy became a popular drug of abuse in North America, where it was produced by many illicit laboratories. This illicit use expanded greatly with the synthesis of related compounds, particularly MDA and MDEA (Figure 1). During the 1980s, MDMA became very popular in Europe particularly the UK, Netherlands and Spain. It was made a class "A" drug in the UK in 1971 under the Misuse of Drugs Act.

Deaths related to use of ecstasy were first reported in the US in 1987 and in the UK in 1991. It was only in the early 1990s that publicity surrounding the rise in acute deaths in the UK fuelled its notoriety as a dangerous drug of abuse. Its use also became synonymous with the "rave" or popular dance scene. Although the acute risks (including death) have been repeatedly highlighted by the popular press, the number of reported deaths due to this group of drugs is only a fraction of those caused by other, equally common drugs of abuse, such as heroin or alcohol. There have been less than 200 reported deaths due to ecstasy in the UK over the last decade. What has remained relatively little publicised, is the growing concern regarding the drug's potential for causing chronic

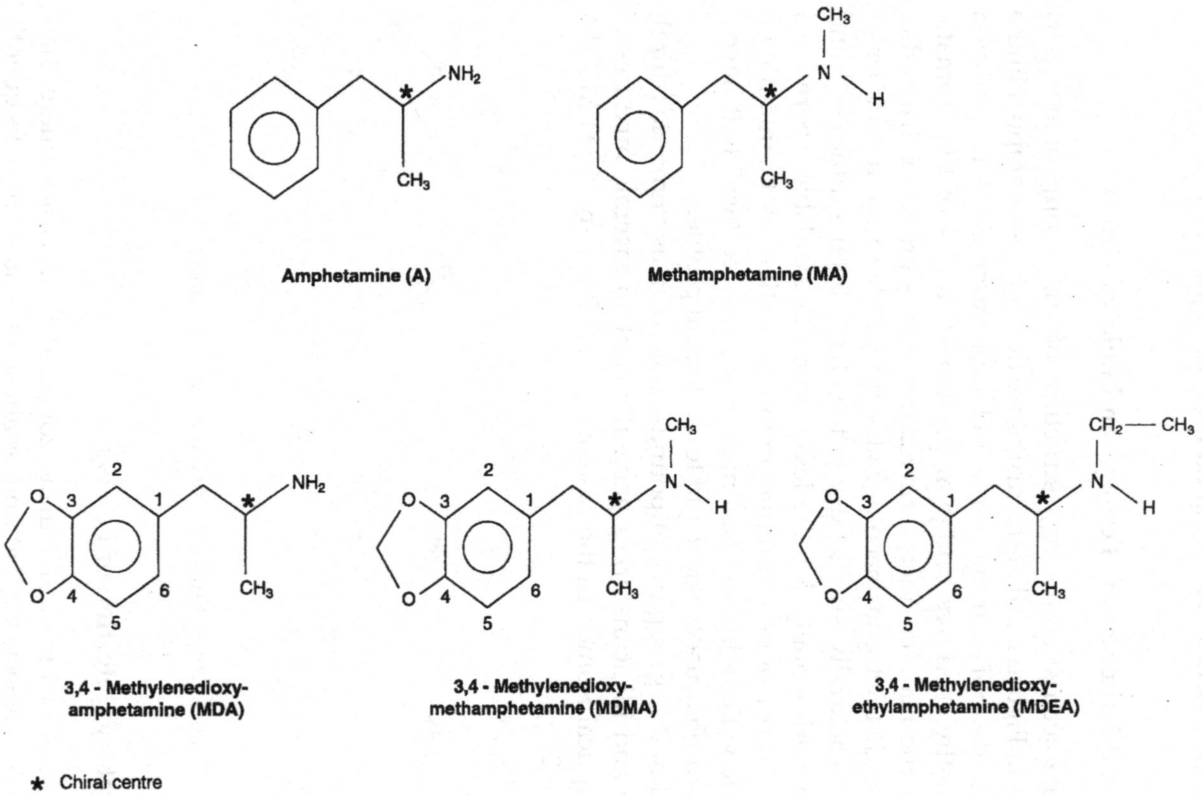

Figure 1. Structure of MDMA and closely related compounds MDA and MDEA in comparison with amphetamine and methamphetamine.

toxicity, particularly that it may be an irreversible neurotoxin, that could cause permanent brain damage in regular or heavy users.

Chemical Structure of "Ecstasy" and Related Drugs

The characteristic chemical structure of this group of drugs is shown in Figure 1. All of the drugs are derivatives of amphetamine and methamphetamine (or ethylamphetamine). The addition of a methylenedioxy bridge at positions 3 and 4 of the aromatic ring is the main characteristic feature of this group of drugs that gives it distinct pharmacological activity. There are a number of other structurally similar drugs that have been synthesised and tried experimentally, but are less commonly available. There are a wide number of other amphetamine derivative, (not containing a methylenedioxy bridge) that may be sold as "ecstasy" that have a different pharmacological and toxicological profile.

All of this group of compounds contain an asymmetric (chiral) centre and illicit materials generally contain racemic mixtures of each optical isomer, in the case of ecstasy, R (–) and S (+) MDMA (Figure 2).

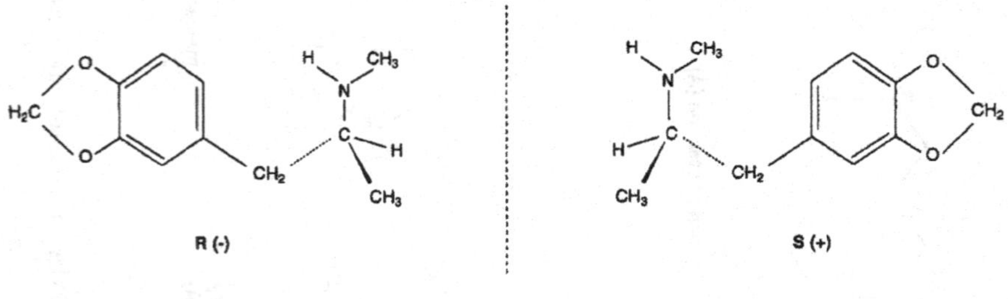

Figure 2. Structure of the two isomers, R (–) and S (+) MDMA.

Laboratory Detection of Ecstasy

Specimens of blood and urine collected from cases where there has been suspected poisoning or abuse of ecstasy may frequently be referred to clinical laboratories for analysis. In suspected drug related fatalities, specimens of post-mortem fluids and tissues may also be taken for analysis. In both clinical and forensic cases, it may be useful to analyse the contents of tablets, capsules or powders that may found at the scene of a death.

A number of different analytical techniques may be used to screen for the presence of ecstasy type drugs, also confirm their precise identity. Immunoassay methods (including some near-patient testing kits) developed for the detection of amphetamine or methamphetamine abuse in urine specimens are frequently employed as first-line screening methods. A "positive" test result would lead to further analysis using more sophisticated methods. However, such general screening methods are unable to differentiate between different amphetamines, and are relatively insensitive to low concentrations of drug that might be found in specimens taken following the ingestion of small doses or MDMA or where there has been a delay in specimen collection, following an episode of illness, or admission to hospital.

More sensitive and specific techniques, using gas chromato-graphy (GC) or GC-mass spectrometry, are able to identify very low concentrations of ecstasy in any type of specimen, and may also differentiate it from closely related drugs (MDA, MDEA) or other commonly abused stimulant drugs. More sophisticated GC-mass spectrometric or liquid chromatographic-mass spectrometric methods are now available to resolve the different isomers (R&S) of ecstasy and their metabolites.

Testing of ecstasy sold at large music events or "raves" has been advocated in the Netherlands and Austria and field laboratories have sometimes been set up. In the UK the analysis of the contents of "amnesty bim" outside clubs or major music festivals has been found to give a useful insight into drug usage , also the strength and purity of "Ecstasy" and related drugs. Testing of seized material is often carried out by police and customs officers, but generally only for evidential purposes. However, this can sometimes provide valuable epidemiological information on the origin of drugs and their distribution across a region, such as the EU.

Problems may sometimes arise when a new synthetic illicit drug appears on the "street" that may be sold as ecstasy. It may take many months before the identity of any new agent is confirmed, which can cause problems when there are cases of "overdosage" or deaths. A recent example of this problem was the appearance of 4-thiomethylamphetamine (4-MTA) which was sold as ecstasy across Europe, particularly in the UK in the late 1990s. 4-MTA was responsible for several deaths across the UK and may have been reponsible for others where the identity of the drug had not been confirmed. Positive identification of this drug in a series of cases across Europe was able to establish that 4-MTA had an acute toxicity profile quite different to ecstasy itself and 4-MTA has now been put on a list of banned substances across the European Community.

Use of Ecstasy

Surveys carried out in the UK and the rest of Europe show that the use of ecstasy (and related drugs) is widespread and probably growing. It has even been suggested that as many as two million doses are taken each week, although there is no accurate way of confirming this figure. The highest use of ecstasy across Europe seems to be in the UK. Population surveys have shown that 4% of young adults (aged 15–34) had used ecstasy within the last year. There are clearly a wide variety of illicit sources of ecstasy. Most drugs appear to be imported into the UK from Holland or Eastern Europe. The quantities stopped by customs represent only a tiny fraction of that coming into the country, perhaps less than 10%. Because of the relative ease of supply, the price of ecstasy is quite low and generally < £5 per tablet. During the 1990s purity of ecstasy tablets was very variable, but recent surveys have shown that non-MDMA containing tablets are quite rare and purity is generally better than 90%.

There is a suggestion from magazine surveys that females start taking the drug younger than males, both usually in their late teens. There is a recent trend towards users taking larger doses also mixing ecstasy with other drugs, particularly cannabis, amphetamine and cocaine. Some users are clearly developing a tolerance to the drug, needing to take bigger doses to get the same effect, which is likely to lead to long-term harm. There is also a trend towards greater experimentation, which involves mixing drugs and trying new combinations without being fully aware of the possible health risks.

Despite earlier concern over earlier ecstasy deaths, there is no evidence to suggest that any of these deaths are due to a "contaminant" or "adulterant". Toxicity and death appears to be directly related to the amount of MDMA contained in the tablets or capsules that were ingested. The interpretation of post-mortem blood concentrations is a complex issue, with differences in concentrations of MDMA and MDA observed between different anatomical sites. MDMA has been shown to undergo post-mortem redistribution and concentrations measured in post-mortem blood may be much higher than concentrations close to the time of death. This can lead to "error" in the estimation of the quantity of drug ingested from post-mortem concentrations in blood.

Pharmacological Action

The pharmacological and clinical effects of this group of compounds are still poorly understood, as they have never been subjected to proper controlled clinical trials. Some of the drug's actions are clearly those of a stimulant, being similar to amphetamine or methamphetamine. However, there appear to be other properties that are more closely related to hallucinogenic compounds such as LSD. A new vocabulary is probably required to describe the properties of this type of drug and understand the reasons for their popularity within modern youth culture. Table 2 lists some of the varied terminology that may be used to describe these effects. Interestingly MDMA and related drugs appear not to produce visual or auditory hallucination (cf LSD or mescaline), but more an alteration in the conscious mind. Users of the drug have described a wide range of pleasurable effects and sensations which are best enjoyed in a social setting, particularly when dancing or listening to loud very rhythmic music (Table 3).

Table 2. Terminology.

Aesthetic enhancer —	Enhances enjoyment of artistic events
Deleriant —	Capable of inducing state of delirium
Empathogen —	Empathy inducing drug
Entactogen —	Improves communicative contact among participants
Entheogen —	Psychodelic drug for ritual purposes
Euphoriant —	Mood enhancer
Psychodelic —	Mind altering agent
Psychotomimetic —	Psychosis mimicking agent
Synaesthesia —	Induces secondary sensation e.g. Sound → Colour

Table 3. "Pleasurable" effects of "ecstasy" reported by users.

- Increased physical and emotional energy
- Heightened sensual and emotional awareness
- Decreased appetite
- Feelings of closeness to others (empathy)
- Increased enjoyment of music and dance
- Increased sexual enjoyment
- Lack of aggressive behaviour

The major pharmacological action of ecstasy involves its effect on the release of neurotransmitters from nerve endings in the brain particularly serotonin and dopamine. More prolonged effects (particularly following repeated doses) involve a loss of serotonin from the brain also a loss of serotonergic neuronal structures. It is unclear if any other neurotransmitter systems are involved.

The drug also has complex cardiovascular and neuroendocrine effects, some of which have been studied in volunteers given small "therapeutic" doses under closely controlled conditions. Single doses of MDMA have been shown to significantly increase blood pressure (systolic and diastolic), heart rate, body temperatures and pupil diameter (mydriasis). Other effects have been described in experimental animals (mice, rats and non-human primates) at different doses and different conditions. (e.g. high ambient temperatures, crowding, etc.).

Some of the general more common adverse effects that have been reported by ecstasy users are shown in Table 4.

Table 4. Undesirable effects commonly reported by ecstasy users.

Desire to and failure to urinate
Dry mouth
Grinding of teeth (bruxism)
Headache
Hot and cold flushes
Insomnia
Jaw clenching (trismus) and spasms
Loss of appetite
Muscle aches
Nausea
Poor concentration
Rapid heart beat
Sweating/sweaty palms

Elimination of Ecstasy by the Body

The breakdown (metabolism) of ecstasy by the body has been studied in experimental animals (rats) and in man (volunteers and overdose cases). This has been shown to be quite complex and is still not fully understood. The proposed metabolic pathway of MDMA is shown in Figure 3. The drug undergoes a series of metabolic reactions and is converted to both active and inactive products. A small proportion of the drug is demethylated (loss of $-CH_3$ group) to produce MDA. The methylene dioxy bridge on the aromatic ring is also opened by a process of demethylenation (loss of CH_2-group) to produce dihydroxymethamphetamine (DHMA). These latter metabolites are further converted to methoxy ($-OCH_3$) compounds (HMMA and HMA). The final stage in the metabolic pathway is conjugation of a number of these metabolites to form sulphate and glucuronide conjugates which are the major products excreted in urine. It is still unclear if further metabolites (possibly quinones) are formed from the dihydroxy metabolites (DHA and DHMA) which might represent an additional pathway towards producing a toxic metabolite(s) responsible for specific organ damage (brain, liver).

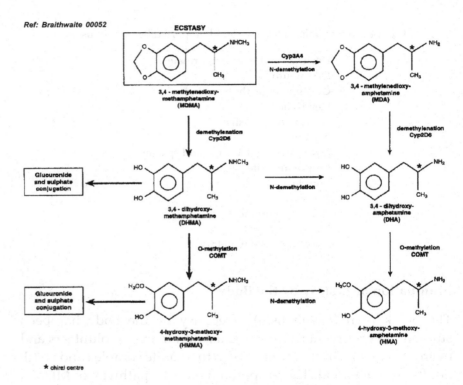

Figure 3. Metabolic pathway of MDMA.

It has been possible to study the rates of drug absorption and elimination (pharmacokinetics) when single doses of MDMA have been given to volunteers (maximum dose generally 125 mg). The drug is well absorbed and reaches its maximum concentration in blood about two to three hours after ingestion. Its elimination half-life (time taken for the drug concentration to fall by one half) has been found to vary according to the dose of drug taken (dose-dependent pharmacokinetics). Following ingestion of smaller doses (40–50 mg) the drug shows a relatively short elimination half-life (three to five hours); however, at bigger doses (100–125 mg) the half-life is prolonged (eight to nine hours). It has also been shown that there is a disproportionate increase in blood concentrations of drug at higher doses. Thus, relatively small increases in dosage, i.e. taking two or three tablets rather than one to two tablets over an evening might cause a dramatic increase in drug concentrations and produce unexpected toxicity.

More detailed pharmacokinetic studies have been carried out where the concentrations of individual optical isomers (R and S enantiomers) have been measured following ingestion of a single dose of racemic MDMA (R and S isomers). Plasma concentrations of the R (–) isomer have been shown to be much greater than those of the S (+) isomer, indicating that the S isomer is eliminated much more rapidly (R to S ratio ~ 2.4). Pharmacological study of the effects of each isomer shows that the S (+) isomer is much more active in terms of acute intoxication and neurotoxicity than the R (–) isomer. Thus, individuals whose elimination of the (+) isomer is impaired, possibly for genetic reasons, may be at greater risk of toxicity. Although the S (+) isomer of MDMA has been found to be more neurotoxic than the R (–) isomer, it is likely that it is in fact a metabolite of this isomer causing the toxicity and *not* the parent drug itself.

Acute Toxicity

Cases of severe poisoning and death have been widely published in the medical literature and popular press. However, considering the vast number of doses of ecstasy said to be ingested every week in the UK, the drug may be somewhat safer than claimed by some "experts". The main causes of death due to ecstasy (or MDA or MDEA) are shown in Table 5.

Table 5. Main causes of death in "ecstasy" users.

- Heat stroke and its complications* (disruption of body's thermoregulation process)
- Cerebral oedema (swelling of brain leading to brain stem death due to fluid overload) and hyponatraemia (lowered sodium concentration)
- Heart rhythm disturbance leading to heart failure due to consequences of rhabdomyolysis and hypernatraemia (raised potassium concentration)
- Stroke (rupture of blood vessels in the brain due to increased blood pressures)
- Fulminant liver failure (destruction of liver — possibly due to unknown toxic metabolite of ecstasy)

* Most common cause of death

Heat stroke

By far the commonest cause of death are the medical complications resulting from "heat stroke". This is often seen in situations where the person has been in a hot environment, such as a club, where there is a high ambient temperature, and the individual has become hot through prolonged exertion, usually by dancing, thus causing the body temperature to rise. This may be exacerbated by a lack of fluid intake. The normal body cooling mechanism is by sweating, but lack of fluid may reduce its effectiveness. There is some evidence that ecstasy and related drugs may actually uncouple the thermoregulatory process in the brain, causing the body to warm up even more than normal following exertion. The consequences of a marked increase in the body's core temperature from 37°C (98.4°F) to over 40°C (104°F) may be disastrous; few patients will survive temperatures greater than 42°C. This may cause collapse and coma, followed by a range of medical complications, including metabolic acidosis, coagulation disorders (disseminated intravascular coagulation) and acute renal failure. Unless the problem is treated as a medical emergency and the body is cooled rapidly the patient may soon die.

One of the unusual complications of such cases, is that a large volume of fluid may be given, or taken by the patient, in the belief that it is an "antidote" to the effects of ecstasy. In one sense, this might be true, particularly if the patient is fluid-depleted, due to excess sweating, or restricted fluid intake. However, if the patient is rapidly given large volumes of water, this may cause a condition known as fluid overload (water intoxication), which may result in hyponatremia (low sodium concentration in plasma) causing the brain to swell and expand, forcing the brain stem into the hole in the base of the skull (foramen magnum) causing it to be compressed, leading to brain stem death. It is now known that ecstasy causes the body to release a specific hormone (antidiuretic hormone; vasopressin) causing a condition known as Syndrome of Inappropriate Antidiuretic Hormone Secretion (SIADH), where excess water cannot be secreted by the kidney and dilute urine cannot be produced. Studies carried out in volunteers have shown that this effect occurs even following the ingestion of relatively small doses (~ 50 mg) of ecstasy.

Heart failure

Other causes of acute toxicity leading to death are much less common. A small number of ecstasy users may die following exertion, unexpectedly causing a cardiac dysrhythmia and heart failure. This rare event is generally only seen in persons with pre-existing heart defects, which may be totally unknown and never diagnosed in life.

Stroke

Ecstasy and many other amphetamines, may cause a rise in blood pressure. This may lead to the risk of stroke, particularly in susceptible individuals, e.g. those with aneuryms (congenital weakspots in arteries in the brain) even following the ingestion of single dose of drug. If there is serious bleeding within the brain, this may be irreversible and lead to brain death or permanent paralysis.

Liver failure

Although relatively rare, ecstasy is known to cause liver damage, leading to rapidly developing (fulminant) liver failure and death in a few individuals. This appears to be associated with a rare idosyncratic reaction to the drug, and not related to dose or frequency of use. The precise mechanism remains unknown, but could be the production of a "toxic" metabolite in certain individuals. In some patients, liver transplantation offers the only chance of recovery.

Chronic Toxicity

The major concern regarding chronic toxicity of ecstasy is related to its possible effects on the brain. A range of neuropsychiatric problems has been reported in ecstasy users, particularly depression and suicidal behaviour (Table 6). Some of these are thought to be linked to the important effect that ecstasy has on neurotransmitters such as serotonin (5-hydroxytryptamine) and possibly dopamine. Serotonin is a particularly important neurotransmitter that is linked to several functions associated with mood control and behaviour (Table 7). In the last decade there has been extensive work carried out using animal models (rats and non-human primates) to look at the effects of ecstasy on brain serotonin and neuronal structures involving this neurotransmitter (Figure 4). The results have been overwhelming in their conclusions, that ecstasy causes major effects on the storage and function of 5-HT in the brain. Following administration of ecstasy, there is a major release of serotonin from nerve endings, with subsequent loss of serotonin from within storage sites within the nerve ending (Figure 4). It also appears that ecstasy is taken up by the transporter process within the nerve ending, leading to a reduction in the synthesis of serotonin due to the rapid inhibition of the enzyme tryptophan hydroxylase to produce 5-hydroxytryptophan, a precursor of 5-hydroxytryptamine, also destruction of nerve fibres through some unknown toxic mechanism. Histological staining of brain slices from dead animals given repeated doses of ecstasy shows a profound and disturbing loss of longer serotonergic nerve fibres (axonal degeneration) with some evidence of re-growth in different areas. In effect, the brain had been "re-wired". These effects in animals appear to be long lasting and possibly irreversible. It is unclear what these effects are on animal behaviour. The threshold for damage in nonhuman primates is thought to be as low as 5 mg/kg.

Table 6. Neuropsychiatric problems associated with "ecstasy" use.

- Anxiety
- Confusional states
- Depression
- Impulsivity and hostility
- Insomnia and restless sleep
- Memory disturbances and higher cognitive dysfunction
- Panic attacks and phobias
- Psychosis (hallucinations and paranoia)
- Suicidal thoughts

Table 7. Role of serotonin in the brain (5-hydroxytryptamine, 5-HT).

- Aggression
- Appetite
- Impulsivity
- Memory
- Mood regulation
- Pain
- Sleep

In the last few years, attention has been focused on the effects of ecstasy on the human brain. A number of studies have been carried out in ecstasy users with application of very sophisticated brain imaging techniques such as PET scanning (positron emission tomography). The results, although controversial are suggestive of damage similar to that seen in animal models. Some other studies have been carried out investigating the effects of MDMA on the metabolism of serotonin in the brain, that have involved taking lumbar punctures to sample cerebrospinal fluid (CSF). Analysis of CSF has shown a reduction in the concentration of 5-hydroxyindoleacetic acid (5HIAA) the main metabolite of 5-hydroxytryptamine (5-HT, serotonin) in the brain, suggesting a reduction in serotonin synthesis. Predicting human toxicity from animal studies is not always simple or reliable. Work carried out in the US suggests that, taking interspecies differences into account, the threshold for neurotoxicity in humans may be as low as 1.5 mg/kg which is equivalent to 100– 120 mg, a single tablet of drug. But this estimate may be too low.

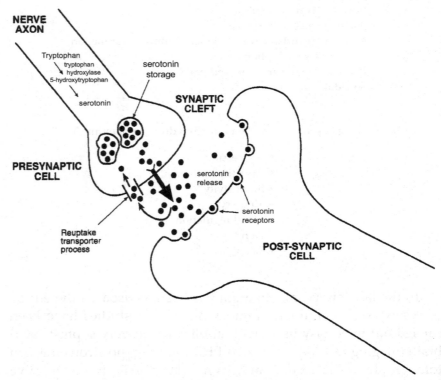

Figure 4. Representation of a serotonergic neurone showing the storage and release of serotonin.

More recently there has been a case report following the death of an ecstasy user, where there was a much greater than expected loss (~ 80%) in the concentration of serotonin in the brain at post-mortem. To some extent the "jury is still out" regarding the full effect of ecstasy on the human brain. However, all of the research findings and clinical case reports do seem to agree that in chronic or heavy users of ecstasy, it is highly likely that irreversible brain damage is being done. This damage may manifest itself in various ways such as faulty information processing or slowly developing neuropsychiatric problems, some of which may be relatively subtle. If these effects are relatively non-reversible, it is possible that serious problems may only manifest themselves later in life on the ageing brain, which may have far less capacity to withstand such damage. This may lead to an increased incidence of suicide, depression or dementia in elderly patients, who were ecstasy users in their youth.

Claims by many users and some agencies that ecstasy is "safer than alcohol" and therefore not a problem may be naïve and probably very premature. If ecstasy was a new drug that had been developed by the pharmaceutical industry for human medical use it would, because of its toxicity profile, never gain a license. Ecstasy users do need to be forewarned of the potential risks to health, particularly of repeated or heavy usage of the drug. Greater publicity over the risks of chronic toxicity may persuade some users to give up using the drug. A much more active clinical research programme, including prospective studies in heavy users, to investigate some of these effects needs to be carried out with some urgency, particularly in view of the very large number of young people regularly using this drug.

Recommendations

For those individuals who continue to use illicit drugs such as ecstasy there are significant risks to health (Table 8). It is unlikely that policies to restrict the production, importation and supply of such illicit drugs will ever be effective within Europe. The only social policy likely to have long term influence is that of education and better warning of the genuine risks of ecstasy use, with the hope of a decline in usage.

Table 8. Health risks of ecstasy abuse.

- Increased risk of acute and chronic toxicity with increase in dose
- Increased risk when mixed with alcohol or other drugs
- Increased risk if overheated (dancing non-stop in a warm environment)
- Increased risk with dehydration (lack of water intake)
- Increased risk with water intake rather than isotonic drinks
- Increased risk with high blood pressure, heart disease or liver disease
- Increased risk for foetus if pregnant
- Increased risk if help is delayed

Acknowledgements

I would like to thank Mrs. S. Samra and Mrs. D. Woolley for secretarial assistance and Dr. Robin Ferner for helpful advice and criticism.

Suggested Further Reading

Boot B, McGregor IS and Hall WA (2000). MDMA ("Ecstasy") neurotoxicity: assessing and communicating the risks. *Lancet* **355**, 1818–1821.

Butler GKL and Montgomerg AMJ (2004). Impulsivity risk taking and recreational 'ecstasy' (MDMA) use. *Drug and Alcohol Dependency* **76**, 55–62.

Curran HV (2000). Is MDMA ("Ecstasy") neurotoxic in humans? An overview of evidence and of methodological problems in research. *Neuropsychobiology* **42**, 34–41.

Elliott SP (2005). MDMA and MDA concentration in antemortem and post-mortem specimens in fatalities following hospital admission. *Journal of Analytical Toxicology* **29**, 296–300.

EMCDDA (1999). Report on the risk assessment of 4-MTA in the framework of the joint actions on new synthetic drugs. European Monitoring Centre for Drugs and Drug Addiction (EMCDDA), Lisbon, Portugal.

Green AR, Mechan AO, Elliott JM, O'Shea E and Colado MI (2003). The pharmacology and clinical pharmacology of 3,4-methylenedioxymethamphetamine (MDMA or "Ecstasy"). *Pharmacological Reviews* **55**, 463–508.

Henry JA, Fallon JK, Kicman AT, Hutt AJ, Cowan DA and Forsling M (1998). Low dose MDMA ("Ecstasy") induces vasopressin secretion. *Lancet* **351**, 1784.

Henry JA, Jeffreys KJ and Dawling S (1992). Toxicity and deaths from 3, 4-methylenedioxyamphetamine ("Ecstasy"). *Lancet* **340**, 384–387.

Lyles J and Cadet JL (2003). Methylenedioxymethamphetamine (MDMA, Ecstasy) neurotoxicity: cellular and molecular mechanisims. *Brain Research Reviews* **42**, 155–168.

McGuire PK, Cope H and Fahy TA (1994). Diversity of psychopathology associated with use of 3,4-methylenedioxymethamphetamine ("Ecstasy"). *British Journal of Psychiatry* **165**, 391–395.

Morgan MJ (2000). Ecstasy (MDMA): a review of its possible persistent psychological effects. *Psychopharmacology* **152**, 230–248.

Ricaurte EA and McCann UD (2005). Recognition and management of complications of new recreational drug use. *Lancet* **365**, 2137–2145.

Shulgin A and Shulgin A (1991). *PiHKAL: A Chemical Love Story*. Transform Press, Berkeley CA (ISBN 0-9630096-0-5).

6

HEROIN

R.A. Braithwaite

Discovery and Early Use

Heroin (diacetyl morphine; diamorphine) in the form of its hydrochloride salt was first introduced into medicine in 1898 by the German Pharmaceutical Company, Bayer. It had been developed by the same research team that had some years earlier discovered Aspirin (acetylsalicylic acid). The drug had been developed as a cough suppressant and "lung stimulant", to be used in patients with severe lung disease. However, the same compound, diacetylmorphine, had also been synthesised by Charles Alder Wright some 24 years earlier whilst working at St. Mary's Hospital Medical School in London. He had continued the work of a colleague, Augustus Mattiessen, who had been appointed as a Lecturer in Chemistry at the Medical School in 1862. Both had been interested in the chemistry of opium alkaloids and in their subsequent researches had made various derivatives of morphine including diacetylmorphine.

In the earlier part of the 20th century heroin was made widely available to the public and was used in many proprietary products and patent medicines. However, this soon became a matter of public concern, both in North America and in Europe. The drug was soon found not to be a "respiratory stimulant" — in fact the exact opposite. It was also found to be very addictive. In the US there were soon reports of people "snorting" heroin, in a similar fashion to cocaine. The American authorities subsequently passed the Harrison Act in 1914 limiting the amount of heroin in proprietary products and an absolute ban on the drug was brought in in 1924.

However, at this time heroin abuse in the UK was relatively rare and was largely confined to the medical profession and others with easy access to the drug. Heroin abuse did not become a major social problem worldwide until some time after the Second World War, in the 1960s.

Opium

Opium is a natural product obtained from the opium poppy, *Papaver somniferum*, also known as the white poppy. This species is not native to Britain. The only species of poppy that is native to Britain is the "red" or corn poppy (Papaver rhoeas). The opium poppy has been grown in many countries and for a short period in the 19th century was grown in East Anglia to provide an alternative source of opium to that imported from India to serve the home market. Historically, most of the world's supply of illicit heroin was derived from poppies grown in the "golden triangle" region of South East Asia, where Burma, Thailand and Laos meet. However, opium poppies are now grown in other countries such as Afghanistan, and parts of the old Soviet Union. More recently there has been a big upsurge in opium production in South America to supply the North American heroin market. Opium is mainly found in the unripe poppy seed capsules and is commonly harvested from the air-dried sap, or latex, which is released by cutting the capsule with a sharp knife. Opium contains 20 or more different alkaloids, the most important of which is morphine, and first identified by Serturner in 1806, which he named after Morpheus, the Greek god of dreams. Other important alkaloids that are found in "raw" opium include codeine, papaverine, thebaine and noscapine and these may also be found in illicit heroin.

The narcotic and pain relieving properties of opium have been known for several millennia and records of its use can be found in ancient Egyptian and Greek writings. Opium was also widely employed by early Greek and Roman physicians and there are extensive written records of its use, also its addictive properties. The products of the poppy were widely known as a useful medicine, and efficient means of painless death, also a means of committing murder.

Homer in the Odyssey, speaking of Helen of Troy wrote: "*She cast into the wine a drug to quiet all pain and strife and bring forgetfulness of every ill*".

Galen, the great Roman physician wrote in the second century AD about opium: "*Opium resists poison and venomous bites, cures headache, vertigo, deafness, epilepsy, dimness of sight, loss of voice, asthma, coughs of all kind, spitting of blood, tightness of breath, colic, jaundice, hardness of spleen, stone, urinary complaints, dropsies, leprosies, the trouble to which woman are subject, melancholy and all pestilences*".

Knowledge of the drug was passed to Arab physicians in the Middle Ages and in later centuries its use spread to Europe, India and China. Paracelsus a renowned German physician working in the early part of the 16th century, was a great believer in the benefits of opium, but was equally aware of the issue of overdosage and toxicity, he wrote:– "*The right dose differentiates a useful drug from a poison*" which is thought to refer to opium.

The use of opium in various medicines became an essential part of medical practice. In the 17th century the great English physician Thomas Sydenham (1624–1689) wrote:

"*Among the remedies which it has pleased almighty god to give to man to relieve his suffering, none is so universal and so efficacious as opium*".

By the 19th century opium was widely used by a large proportion of the population as a universal cure or pick-me-up for any ailment; opium pills being easily bought across the shop counter, even by children, for a few pence. Opium sales in England were completely unrestricted until the first Pharmacy Act was passed by Parliament in 1868. Many eminent Victorians, poets, statesmen, soldiers and writers were dependent on opium in order to function in their normal daily lives (Table 1). Interesting descriptions of the effects of smoking opium and use of patent medicines containing it, such as laudanum, (alcoholic tincture of opium) are well described in literature of the time, such as Thomas de Quincy's "Diary of an

Table 1. Eminent Victorians who were regular users of opium.

- Clive of India
- Elizabeth Barrett Browning
- Florence Nightingale
- John Keats
- Lord Byron
- Thomas De Quincey
- Samual Taylor Coleridge
- Sir Walter Scott
- Wilkie Collins
- William Wilberforce

Figure 1. An opium den in London's East End with a reclining smoker being watched by a group of men. Engraving by A. Doms after G. Dove. Wellcome Institute Library, London.

Opium Eater" and Wilkie Collins's "The Moonstone" published in 1868, one of the first detective stories ever written; it had a complicated plot involving the use of laudanum, and was written when the author was under the influence of opium. Most of the opium was grown and imported from India by the British Government. Disputes over the control of the opium market led to the "Opium Wars" with China, who wished to stop the British importing the drug into mainland China. The outcome resulted in the establishment of Hong Kong as a British Colony and the destruction of the social fabric of Chinese society.

Figure 1 shows a late Victorian engraving of an "opium den" in London's East End. This was a popular stereotypical image of the period and part of the myth created by popular novelists such as Conan Doyle that was also encouraged by the anti-opium movement in England at the time, also anti-immigrant hostility.

Structure of Morphine and Heroin

The chemical structures of morphine and heroin in relation to other closely related opiates such as codeine and dihydrocodeine are shown in Figure 2. Morphine has a "phenolic" –OH group in the 3 position and an aliphatic (alcoholic) –OH group in the 6 position. When both –OH groups (3 and 6 positions) are acetylated, diacetyl morphine, better known as diamorphine or heroin is produced. Codeine is the 3-methyl derivative of morphine by substituting the phenolic –OH group with a methyl group in the 3-position. Dihydrocodeine is closely related to codeine, the only structural difference being that the double bond in the 7–8 position of morphine is "saturated" to produce dihydrocodeine (originally known as DF118). The generally accepted potency of these opiates is diamorphine > morphine > dihydrocodeine > codeine.

Both codeine and dihydrocodeine are widely used analgesic drugs, often formulated in combination with aspirin or paracetamol (e.g. cocodamol, co-dydramol).

Classification and Mode of Action

"Opiates" are drugs that have been directly extracted from opium e.g. morphine and codeine, or are chemical derivatives, such as heroin (diacetylmorphine). All are structurally related to morphine (Figure 2). The term "narcotic" which is derived from the Greek word for Stupor is an archaic term that was used to describe the analgesic action following the administration of a large dose of opiate drug, historically opium.

STRUCTURE OF MORPHINE AND RELATED OPIATES

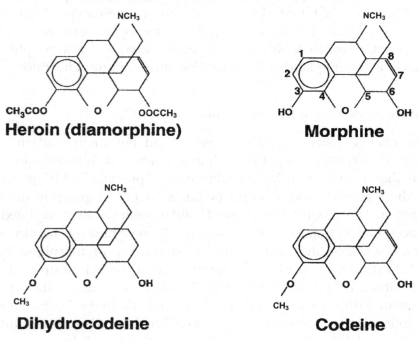

Heroin (diamorphine)

Morphine

Dihydrocodeine

Codeine

Figure 2. Chemical structure of morphine and related opates.

By contrast "opioid" is the more general modern term used to describe all drugs and endogenous compounds that have morphine-like pharmacological activity, which interact with specific receptor sites in the body (Table 2). Three major classes of opioid receptor have been discovered μ (mu), K (kappa) and δ (delta), also many sub-classes of receptor are known. Morphine and closely related drugs produce most of their major effects in the body by their interaction with μ receptors; however, there may be some interaction with other receptors at higher dosages, particularly in overdosage. The main pharmacological effects of opiates are well known and are summarised in Table 3. Interestingly, substitution of the -OH group in the 3 position tends to **reduce** binding to specific morphine (μ) receptors, so codeine acts as a less potent drug than morphine. Moreover, heroin itself is thought to have little or no pharmacological activity, but is a pro-drug, with the majority of its activity being due to the production of active metabolites, particularly morphine.

Table 2. Opioids.

• **Opiates (structurally related to morphine)**
  Morphine
  Codeine
  Dihydrocodeine (DF 118)
  Heroin (Diamorphine)

• **Synthetic opioids (structurally unrelated to morphine)**
  Buprenorphine (Temgesic)
  Dextromoramide (Palfium)
  Dextropropoxyphene (Coproxamol – with paracetamol)
  Dipipanone (Diconal – with cyclizine)
  Methadone (Physeptone)
  Nalbuphine (Nubain)
  Pentazocine
  Pethidine
  Tramadol (Tramake, Zamadol, Zydol)

Medical Use of Heroin

Both morphine and heroin are widely used in medicine for the management of severe, acute and chronic pain. In its pharmaceutical form, heroin is known as diamorphine, and is formulated as the hydrochloride salt, which is a white, crystalline, odourless powder that is freely soluble in water. This is the main advantage over morphine, which is relatively insoluble in water, requiring a greater volume for injection. Because diamorphine is relatively unstable in solution, it is always freshly prepared from ampoules of the solid salt by adding sterile water. The drug is commonly given by the intravenous or intramuscular route, less frequently it is given by the oral, sub-cutaneous or epidural routes.

Table 3. Main pharmacological properties of opiates.

* Anaesthesia
  Analgesia (reduced sensation to pain and its perception)
* Coma
  Constipation
* Convulsions
  Development of tolerance and dependence
  Drowsiness
  Flushing of skin
  Miosis (pin-point pupils)
  Mood changes (very complex, particularly when drugs are abused)
  Nausea and vomiting (less common in recumbent patients)
  Reduced cough reflex
  Reduced gut motility (delayed stomach emptying and digestion of food)
  Reduced urinary voiding reflex
  Respiratory depression (reduces rate of breathing and sensitivity to carbon dioxide at respiratory centre within brain-stem)
* Respiratory failure and apnoea (suppression of breathing reflex within brain-stem)

* Associated with high doses, particularly overdosage

Diamorphine is commonly given to patients in severe pain, such as that following a major accident or following a heart attack. In such situations it is usually given as a single intravenous or intramuscular injection of 5–10 mg, which may be repeated a few hours later if the patient is still in pain. Fatalities have sometimes been seen where bigger doses have accidentally been given, or where the patient is especially sensitive to respiratory depression, as may be seen in those with pre-existing respiratory disease, where the drug has a relatively low margin of safety. However, much larger doses of diamorphine are commonly given to patients with severe pain associated with a terminal illness or following major surgery. In this situation the drug may be given by a constant subcutaneous or intravenous infusion. In patients with continuing severe pain, the dose of diamorphine has to be increased, to obtain the same therapeutic effect, due to the development of tolerance. Interestingly, diamorphine does not appear to cause abuse or addiction in patients requiring the drug for pain relief.

Pharmacokinetics and Metabolism of Heroin

The metabolic pathway of heroin in the body is shown in Figure 3. Within a few minutes of entering the circulation, heroin is rapidly converted to a relatively unstable intermediate, 6-monoacetyl morphine (6-MAM). This intermediate is extremely potent pharmacologically and is particularly important in the forensic investigation of heroin deaths. 6-MAM is relatively rapidly converted to morphine within about 0.5–1.0 hour, completing the process of deacetylation. Morphine is further metabolised in the liver by a process of glucuronide conjugation. In this process, water soluble molecules of glucuronic acid (similar to glucose) are conjugated with morphine to form morphine-3-glucuronide (M-3-G) and morphine-6-glucuronide (M-6-G). These products are more water soluble than parent morphine and are excreted into the urine by the kidney. The main product of this process, M-3-G, is inactive. However, its "sister" metabolite M-6-G is extremely active and adds to the action of morphine in the body. The relative proportions of morphine and its glucuronide metabolites are important in our understanding of the analgesic action of heroin and morphine and interindividual differences in pain relief, they also have an important use in forensic cases, particularly in the investigation of morphine and heroin related deaths.

MAIN ROUTE OF METABOLISM OF HEROIN (Diacetylmorphine)

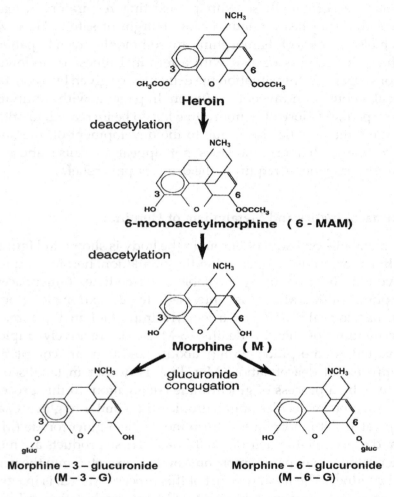

Figure 3. Main route of metabolism of heroin and its metabolites.

Illicit Production of Heroin

Illicit heroin can be made from morphine base or directly from "raw" opium. In addition, there are great variations in the method of manufacture and amounts of chemicals used in the process; the end product may be of variable quality and purity. A common method used to make heroin from opium involves the addition of lime (calcium hydroxide) and water. The suspension is then passed through a coarse filter. Ammonium chloride is then added to the liquid filtrate in order to precipitate morphine base. This is then washed with water and sometimes purified further by dissolving in acid and re-precipitating the morphine base with ammonium chloride. The base material is then dried and refluxed with acetic anhydride for several hours. After cooling the mixture is neutralised with sodium carbonate causing the heroin base to precipitate, which is then washed, filtered and dried. This material can be used directly to make smokable heroin following the addition of cutting agents. If heroin hydrochloride is required, the base material is dissolved in acetone and hydrochloric acid added. As part of the acetylation process, small but variable quantities of acetyl codeine and 6-acetyl morphine are produced. The final product may also contain small but significant quantities of other opium alkaloids such as papaverine and noscopine. Figure 4 shows the results of an analysis of a sample of illicit heroin using capillary gas chromatography. As can be seen, the drug is far from "pure" containing several components. However, heroin with a purity of greater than 90% can be made by illicit laboratories with access to easily obtainable chemicals. The drug has many slang names in common use (Table 4).

Table 4. 'SLANG' names for heroin.

BROWN SUGAR	HARRY
CHI	HORSE
CHINESE H	JUNK
DRAGON	SCAT
GEAR	SKAG
H	SMACK

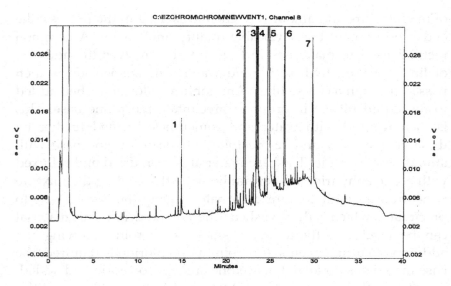

Figure 4. Capillary gas chromatographic analysis of illicit heroin. The major components are shown: (1) caffeine cutting agent, (2) codeine, (3) monoacetylcodeine, (4) monoacetylmorphine, (5) heroin, (6) papaverine, and (7) noscopine.

Purity and Pattern of Heroin Abuse

There are a number of "routes" of heroin administration that abusers may use (Table 5). The chosen route may have important implications for health risks as well as our understanding of the process of drug dependence and addiction. There are also some important differences in the nature and composition of the heroin according to the intended route of administration. The drug may be prepared for illicit use as a free base or hydrochloride salt. The purity of "street heroin" may vary widely, typically being between 40–60%, it is often diluted ("cut") by mixing with other soluble compounds such as glucose or lactose, but may be adulterated further by the addition of different drugs sometimes with the purpose of giving the powder a bitter taste or enhancing its potency (Table 6). The purity of "wholesale" batches of heroin available to dealers is generally much higher than "retail" amounts sold to users. In some cases users may further dilute the bought heroin to sell on to other users. In extreme cases the heroin at the bottom end of the supply chain may be less than 10% purity. Some naïve users may be sold drug that contains little if any heroin (< 1% purity), and composed almost entirely of cutting agent.

Table 5. Routes of heroin administration.

- Intravenous injection ("mainlining")
- Smoking ("chasing the dragon" or "ack-ack")
- Snorting (nasal insufflation)
- Subcutaneous (skin "popping")
- Intramuscular
- Oral
- Rectal

Table 6. Reported "diluents" and "adulterants" in illicit heroin.

Barbitone
Caffeine
Glucose
Lactose
Mannitol
Methaqualone
Paracetamol
Phenazone
Phenobarbitone
Procaine

The frequency of use and dose of heroin taken by established users may vary widely. Some users may take heroin on a once or twice daily basis, others a few times a week, or only at weekends on a recreational basis. The dose taken may range between as little as 50 mg (0.05 g) a day to 5000 mg (5.0 g) a day. Surveys have shown that almost a third of heroin abusers may be taking 250 mg a day or less; just over a third taking between 250–1000 mg daily and almost a third taking more than 1000 mg (1 g) daily. The amount of heroin taken will depend on individual needs, tolerance, drug purity and the route of administration; the intravenous route being the most "efficient" route and snorting being the least "efficient" route used by abusers of the drug. The price of illicit heroin has greatly come down in recent years, but can be variable; a one gram bag" may cost £20–30 and a small "wrap" of 50–100 mg costs only about £5.

Intravenous Administration

Intravenous administration has been one of the most "popular" routes of heroin abuse for many years, both in North America and Europe, including the UK; it causes severe dependence. Intravenous drug administration carries many health risks because of the use of "dirty" injecting techniques and shared needles (Table 7). The greatest risk is that of a transmission of the HIV virus leading to AIDS. However, other viruses, particularly hepatitis B and C, pose additional major risks to health. Heroin used for intravenous administration may be of variable purity, but it is generally in the form of its hydrochloride salt, which is more water soluble. Heroin for injecting is commonly bought as a white, beige or light brown powder (Figure 5) which is dissolved in a small quantity of water (often unsterile) with a dash of lemon juice to aid solubility. This may be mixed on a small spoon over a flame, usually a lighter, candle or gas cooker to sterilise the liquid before drawing the dissolved drug up into a small syringe for injecting. Typical "equipment" used by a heroin addict is shown in Figure 6. Often the liquid is drawn through a filter tip to ensure dissolution of the

Table 7. Some health consequences of injecting heroin.

- Transmission of HIV, hepatitis B and C viruses by sharing dirty needles
- High risk of overdosage and sudden death (particularly combined with alcohol)
- Phlebitis and deep vein thrombosis, leading to loss of use of vein
- Septicaemia (blood poisoning)
- Abscess formation
- Gangrene and loss of limbs due to inadvertent intra-arterial injection of material
- Heart valve damage due to injection of infected material

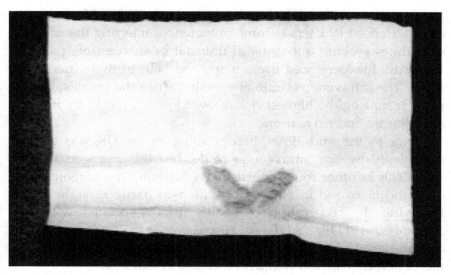

Figure 5. The appearance of illicit heroin – a brown powder in a typical drug wrap.

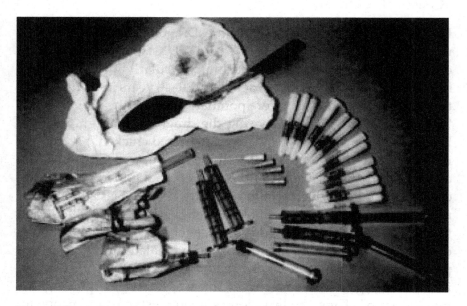

Figure 6. Typical injecting equipment found at the scene of death of a heroin addict showing blackened spoon, also various syringes and needles – some of which have been reused.

heroin powder before injecting into a convenient vein. Most regular or experienced IV users become proficient at injecting themselves, sometimes seeking out veins in unusual or inaccessible parts of the body. Inexperienced users may need help from a "friend" to inject. The intravenous route obviously carries the greatest health risks, including the highest risk of overdosage and death, in both experienced and naïve users.

Prior to the mid 1970s, heroin abuse in the UK was almost exclusively by the intravenous route, but this changed by the mid 1980s as other routes, particularly "chasing the dragon" took over and increased in popularity. This was partly brought about following the recognition of health risks associated with the transmission of HIV and the AIDS epidemic in the 1980s. However, over the last decade there has been a dramatic return to intravenous use of heroin, with a consequent increase in mortality.

"Chasing the Dragon"

Heroin smoking called "Chasing the Dragon", took over as the most popular route of heroin use in the UK from the mid 1980s onwards. The origins of this route are particularly interesting as it started in the Far East. Going back some 30–40 years in Hong Kong, two inhalation techniques were popular; "chasing the dragon" and "ack-ack". In the case of "chasing the dragon" small granules of heroin, which historically were dyed red ("red chicken") would be mixed with a large excess of another drug such as barbitone, an obsolete barbiturate. This would be mixed on a folded piece of tin or aluminium foil which would be gently heated over a flame, the user inhaling the fumes or smoke which were sometimes thought to resemble the shape of a red dragon, hence the term "chasing the dragon". The heroin vapour could also be inhaled through a rolled piece of paper or card (originally a bamboo tube). When fumes were inhaled through a match box cover, this was known as "playing the mouth organ". Thus, the technique spread westward to Europe to become the most common method of heroin abuse. Interestingly, the type of heroin was in the form of "free base". analagous to "free base" cocaine. When in the form of its free base, heroin is much more volatile and can be smoked rather than injected. Free base heroin is generally brown in colour and relatively insoluble in water, therefore, not ideal for injecting, whereas injectable heroin (hydrochloride salt) is not very good for smoking as it easily decomposes when heated. Interestingly, a number of agents initially thought to be simple adulterants, were added to smokable heroin base to improve its volatility and bioavailability. The most commonly seen current "adulterants" are caffeine and paracetamol, but many others have been used (Table 6). Some adulterants are added at source, others in the country of importation. The presence and relative proportions of certain adulterants is used by the Police and Forensic Laboratories around the world to track sources of heroin importation.

A, less common method of smoking heroin is by the so-called "ack-ack" method, where the end of a burning cigarette is dipped in heroin and held upwards at an angle, like an anti-aircraft gun. This method is not as popular (or efficient) as the traditional method of "chasing the dragon".

Snorting (Nasal Insufflation)

Snorting heroin originated in the US early in the 20th century, soon after its abuse potential became noticed. It started because it was similar to the usual method of abusing (snorting) cocaine (hydrochloride salt). Its popularity has returned in the US in recent years, favoured by white middle class males, but it is relatively uncommon in the UK. It is also the least efficient way of taking heroin. Variable quantities of heroin may be swallowed when heroin is "snorted", reducing the amount of active drug reaching the circulation.

Oral Ingestion

Heroin may be taken orally, but it is a very unusual form of abuse. This is because the drug is absorbed slowly without producing any characteristic "hit" or "rush", also the majority of the dose is metabolised by the liver before slowly reaching the circulation. Oral ingestion may sometimes be seen in unusual cases of suicide or murder. The most common situation where the oral ingestion of heroin is involved is that concerning "body packers" or "mules". These are individuals who attempt to smuggle heroin in their body by swallowing the drug in small packages wrapped in plastic or latex bags. Cases have been reported where body packers have been carrying more than 100 individual packages. This can cause intestinal obstruction; or heroin toxicity if the contents start to leak. Less commonly, individuals may try to smuggle heroin by inserting packages in the rectum or vagina.

Heroin Addiction and Its Treatment

Heroin addicts who become physically dependent on regular (daily) amounts of drug may suffer unpleasant symptoms of withdrawal if supplies suddenly become unavailable (Table 8). This can make the addict desperate to obtain heroin or the money to buy it. This is one of the major reasons why regular heroin addicts may resort to crime to be able to feed their habit. In some cases the addict may become so desperate when the first signs of withdrawal start to appear, that they may resort to any means, sometimes violence, to obtain heroin, although this problem is seen more commonly in crack cocaine users. The fear of withdrawal symptoms also makes addicts resort to "hoarding" other drugs, particularly methadone, in case of lack of heroin. It can also make the acute medical management of heroin abusers difficult, particularly in the case of those in police custody or put on remand in prison. A great deal of effort has been put into the clinical management of heroin abusers. In North America and some other countries, particularly South East Asia, many users easily end up in the criminal justice system. The penalties for dealing in heroin can be very severe, including capital punishment in some countries. However, during the 1980s and 1990s the cornerstone of treatment policy in the UK has been "harm reduction", which aims to encourage users to stabilise

Table 8. Signs and symptoms of heroin withdrawal following cessation of heavy use ("cold turkey").

- Craving for drug
- Diarrhoea
- Fever
- Insomnia
- Lacrimation and Rhinorrhea
- Muscle aches
- Nausea and vomiting
- Piloerection – goose pimples ("cold turkey")
- Pupillary dilation
- Sweating
- Yawning

their habits, to stop injecting or to participate in needle exchange schemes. In addition, the most common "therapeutic treatment regime" was the use of methadone substitution. Methadone is a synthetic opioid with a longer duration of action (15–24 hours) that can be taken orally. It is normally given to addicts as a coloured syrup containing 1 mg of methadone per ml of medicine. It can be given on a once daily basis, with a typical maintenance dose of 30–60 ml (30–60 mg), although the recommended starting dose is only 20 mg. Large numbers of heroin abusers have been entered into methadone maintenance programmes, in some cases receiving the drug at a pharmacy or clinic on a daily basis, to reduce the risks of overdosage and to prevent users selling methadone to purchase heroin. Many controlled studies have demonstrated the benefits of methadone maintenance, where it has been shown to be superior to other forms of treatment; one of the major benefits from studies reported in the US being the reduction in i.v. usage and the spread of AIDS. However, in recent years there has been concern about the large rise in deaths caused by methadone overdosage. More recently use of other drugs such as buprenorphine in the form of sublingual tablets (Subutex) has greatly increased in popularity and is considerably safer in overdosage. Many other treatment regimes have been tried including rapid detoxification, abstinence therapy using opioid antagonists, also many complementary therapies such as herbal medicines and acupuncture, but there does not seem to be any simple "cure". However, the outlook for most heroin abusers remains very bleak; most addicts tend to give up their habit when they are ready to do so, if they survive long enough. Several studies have followed up groups of addicts over many years and mortality from various causes remains high at 1%–2% per year. The most significant factor is the psychological dependence on drugs causing a high rate of relapse. Those addicts who are able to change their lifestyles and form stable relationships, also obtain gainful employment, have the best chance of long term survival. Periods of in-patient treatment can be helpful in dealing with acute problems, but therapy and support of the addict in the community offers the only realistic form of effective action that is able to deal with the problems of relapse and long-term survival.

Investigation of Heroin Deaths

In the investigation of heroin related-deaths there are many factors to be taken into account. Most deaths in heroin users are related to intravenous administration and may occur rapidly (Table 9). Survival is unlikely without urgent medical intervention. Artificial respiration is of prime importance also the administration of an opiate antidote (e.g. naloxone) which is able to immediately reverse many of the pharmacological effects of heroin. Urgent medical care is essential when dealing with heroin overdose. When the clinical or forensic laboratory is involved in investigating a death, it is important to gather as much information about the deceased and the circumstances of the death as possible (Table 10). If heroin is found at the scene of death, this can be analysed to look for purity and the presence of adulterants, which may sometimes be matched with those found in body fluids. The post-mortem examination may provide many clues to the death, particularly recent injection marks or scarred veins in older users. At post-mortem, specimens of blood, urine and other fluids and tissues may be taken for toxicological analysis. The presence of heroin and its metabolites

Table 9. Characteristics of deaths in heroin users.

- Most deaths are associated with intravenous administration
- Most deaths in heroin addicts are accidental "overdoses"
- Most deaths are relatively rapid (within two hours)
- Many deaths are associated with a loss of drug tolerance (e.g. time in prison)
- Many deaths are associated with the concurrent use of other drugs (particularly benzodiazepines and alcohol)

Table 10. Factors to consider in the forensic investigation of heroin fatalities.

- Route(s) of drug administration
- Purity of drug and adulterants/cutting agents
- Use of other drugs and alcohol
- Recent drug history, particularly usual drug habits (often very unreliable)
- Medical history of deceased
- Circumstances of the death
- Evidence from death scene, particularly position of body (positional asphyxia)
- Inhalation of food or vomit
- Evidence of disease and other signs at post-mortem examination
- Histological examination of post-mortem tissues
- Collection of post-mortem specimens, particularly blood, for toxicological investigations (site and method of blood collection)
- Findings of toxicological analyses, particularly concentrations of 6-monoacetylmorphine, morphine and its glucuronide metabolites in post-mortem blood. Also, concentration of other depressant-type drugs and alcohol.

will be looked for in the blood and urine. Quantitative analysis of morphine and glucuronide conjugates provides vital information to decide on the likely cause of death. Most rapid heroin deaths are associated with the presence of 6-MAM in blood, urine or brain tissues, also a relatively high concentration of unconjugated (free) morphine to that of conjugated morphine (glucuronides) in blood. In most rapid deaths the proportion of "free" to total morphine is >50%. A wide range in post-mortem blood "free" morphine concentrations may be observed in fatalities; values usually ranging between 50-500 µg/L but values up to 1500 µg/L may be seen in extreme cases of overdosage. Those heroin abusers who have not developed "tolerance" to the drug may die at relatively low drug concentrations; those with well-developed tolerance tend to expire at much higher values. Naïve users, or those addicts recently released from prison have the greatest risk of overdosage due to a reduced or total loss of tolerance. The presence of other drugs, particularly alcohol, benzodiazepines or other opioids, is also associated with a higher risk of sudden death.

Laboratory Detection of Heroin Use

The laboratory detection of heroin abuse is mainly applied to the management of drug users in a clinical environment (general practice, hospital or drug treatment clinic) or in pre-employment or employment screening to exclude drug users from the workforce. Screening can be carried out using specimens of blood, urine, hair, sweat or saliva. However, for most clinical or occupational purposes urine is the most commonly tested fluid. This is because urine specimens can be easily collected in sufficient quantity to be tested. Initial screening is carried out using an immunoassay procedure designed to detect the presence of any opiate-like substance above a pre-determined cut-off value (typically 300 µg/L). If specimens test positive, they are subsequently analysed for the presence of specific opiate drugs and metabolites using techniques such as liquid chromatography- and gas chromatography–mass spectrometry. The presence of 6-MAM in urine is evidence of recent heroin usage, and would be associated with the presence of high concentrations of morphine in urine. The presence of a "high" (>2000 µg/L) concentration of "free" morphine is generally accepted as good evidence of heroin (or morphine) abuse or overdosage. However, the presence of significant concentrations of morphine in urine may result from the ingestion of codeine-containing analgesics or dietary use of poppy seeds. Great care is required in the interpretation of laboratory results when trying to establish proof of heroin abuse, particularly in an occupational setting. There are also established rules and guidelines to ensure good laboratory practice in the performance and interpretation of laboratory investigations.

The Future of Heroin Abuse

The future remains bleak. Earlier efforts to stem the supply of illicit drugs and spread of heroin abuse do not seem to have been successful. For the world's greatest producer (Afghanistan) of opium and heroin, the financial incentives remain high. More could be done to reduce the supply of heroin, with the use of more severe penalties for those convicted of dealing in the drug or involved in its importation. But, the only long-term strategy likely to succeed is to reduce the demand for the drug. This might be brought about by better drug and health education and improved support and medical treatment facilities for existing users particularly those in prison. Increased research on the causes of addiction and improved methods of treatment need to be given a much higher priority.

Acknowledgements

I would like to thank Dr. Stephen George, Dr. Claire George and Dr. Robin Ferner for help in the preparation of this manuscript, also Mrs. S. Samra and Mrs. D. Woolley for secretarial assistance.

Suggested Further Reading

Bernath J (ed.) (1998). *Poppy The Genus Papaver*. Harwood Academic Publishers, Amsterdam.

Berridge V (1999). *Opium and the People – Opiate Use and Drug Control Policy in Nineteenth and Early Twentieth Century England*. Free Association Books, London.

Bird SM and Hutchinson SJ (2003). Male drugs-related deaths in the fortnight after release from prison: Scotland 1996–99. *Addiction* **98**, 185–190.

Darke S and Hall W (2003). *Journal of Urban Health: Bulletin of the New York Academy of Medicine* **80**, 189–200.

de Quincey Thomas (1971). Confessions of an English Opium Eater. Penguin Books Limited, England.

Fudala PJ and Woody EW (2004). Recent advances in the treatment of opiate addiction. *Current Phychiatry Reports* **6**, 339–346.

Fugelstad A, Ahlner J, Brandt L *et al.* (2003). Use of morphine and 6-monoacetylmorphine in blood for the evaluation of possible risk factors for sudden death in 192 heroin users. *Addiction* **98**, 463–470.

Goldsmith M (1939). *The Trial of Opium the Eleventh Plague.* Robert Hale, London.

Hardman JG, Limbird LE, Monlinoff PB, Ruddon RW, Goodman A (eds.) (1996). *Goodman & Gilman's the Pharmacological Basis of Therapeutics,* 9th ed. McGraw Hill, New York.

Kaa E (1994). Impurities, adulterants and diluents of illicit heroin, changes during a 12 year period. *Forensic Science International* **64**, 171–179.

Karch SB Ed. (1998). *Drug Abuse Handbook.* CRC Press, Boca Raton, USA.

Shang J, Griffiths P and Gossop M (1997). Heroin smoking by 'Chasing the Dragon': origin and history. *Addiction* **92**, 673–683.

Sneader W (1998). The discovery of heroin. *Lancet* **352**, 1697–1699.

Webb L, Oyefeso A, Schifano F, Cheeta S *et al.* (2003). Cause and manner of death in drug-related fatality: an analysis of drug-related deaths recorded by coroners in England and Wales in 2000. *Drug and Alcohol Dependence* **72**, 67–74.

White JM and Irvine RJ (1999). New horizons. Mechanisms of fatal opioid overdose. *Addiction* **94**, 961–972.

World Drug Report (2004). *Volume 2: Statistics.* United Nations Office on Drugs and Crime.

7

HYDROFLUORIC ACID

S. C. Mitchell

Introduction

Where to start the story of hydrofluoric acid? Perhaps with an observation reported in 1529 by the German physician and mineralogist, Georg Bauer (also known as Georgius Agricola, 1494–1555). He, amongst his many other contributions, had described how mine workers and metal-smiths used a rock called "fluores" (fluorspar, fluorite, calcium fluoride) to aid in the smelting of ores. This mineral acted as a type of flux, reducing the amount of heat required to liquefy the metal and thus allowing it to be readily separated from the ore. The name "fluores" is derived from Latin *fluor* (or *fluo*) meaning "to flow" ("I flow") as the rock melts quite easily at the relatively low temperature of red heat (1330°C). Another strange property also ranked this material as unusual. It was noticed that after these smelting procedures any remaining heated mineral continued to glow in the dark, an event that is now known as fluorescence (absorbing one form of radiation and emitting it as visible light).

Just after the middle of the 17th century (1670), it is generally accepted that a glass-cutter, Heinrich Schwannhardt of Nuremberg (Nürnberg), had noticed that by treating fluorspar with strong acid he was able to etch glass with the resultant mixture. He developed this process into a useful art to produce decorative glassware, to the envy of his contemporaries. It is also suggested that the Venetians added fluorspar to their glass melt to obtain a semi-transparent (frosted, opaque) glass. However, some texts give the credit of this discovery to an unknown English glass-worker around 1720.

Nonetheless, the mixing of these two components, fluorspar and a strong acid, undoubtedly led to the first synthesis of crude hydrofluoric acid. This property of etching glass was observed and reported by many chemists over the next century as they tried to investigate this substance and identify its components. Notable amongst these investigators was Karl Wilhelm Scheele (1742–1786) who succeeded in obtaining a "peculiar acid" by distilling fluorspar with concentrated sulphuric acid in a glass retort. At the end of his experiments the retort was extensively corroded and the gas formed deposited gelatinous silica on passing into water. He called this substance "fluor acid" (1771). By the beginning of the 19th century, it was almost routine to prepare small quantities of hydrofluoric acid by heating fluorspar with sulphuric acid in lead or platinum vessels thereby avoiding contamination by glass. In 1809, Louis-Joseph Gay Lussac (1778–1850) and Louis-Jacques Thenard (1777–1857) managed to obtain the acid in pure form and published their method of synthesis in the German chemical literature: $(CaF_2 + H_2SO_4 = CaSO_4 + 2HF)$.

It was possible also to obtain nearly anhydrous hydrogen fluoride gas but this material was extremely hygroscopic and was usually prepared by decomposing lead or silver fluorides in platinum tubes with dry hydrogen. Investigations by the English chemist, Humphry Davy (1778–1829), into the composition of hydrofluoric acid led him to conclude "...*that the pure liquid fluoric acid consists of hydrogen united to a substance, which, from its strong powers of combination, has not yet been procured in a separate form, but which is detached from hydrogen by metals...*" ("The Collected Works of Sir Humphry Davy", 1840, Vol. 5, p. 425). This unknown element that existed in fluoric compounds, "...*a peculiar substance, possessed of strong attractions for metallic bodies and hydrogen...*" (Vol. 5, p. 423), was called fluorine, a name suggested by Andre Marie Ampère (1775–1836), and was finally isolated in 1886 by Henri Moissan (1852–1907).

The name hydrofluoric acid had first been suggested by Davy in 1812 (Vol. 4, p. 350) to express the view that the liquid was a hydrated oxy-acid, but was abandoned by him in the following year in favour of the older name "fluoric acid". It was introduced again in 1814 by Gay Lussac to convey the fact that the acid contained, not water as Davy had suggested, but hydrogen. The name has been retained, as we now know that hydrofluoric acid is hydrogen fluoride gas dissolved in water, but the ionic nature of the solution is still uncertain. It was thought that most hydrogen fluoride molecules remained undissociated, with the hydrogen-fluorine bond being so strong that the free energy of solvation of the fluoride ion was unable to compensate for it. However, evidence now suggests that hydrogen fluoride is dissociated, but tight ion pairs ($H_3O^+F^-$) decrease the thermodynamic activity coefficient of H_3O^+. These ion pairs increasingly dissociate on increasing hydrogen fluoride concentrations with the formation of HF_2^-. Clearly this is a complicated situation.

The fluorspar used in the preparation of hydrofluoric acid must be as free of silica as possible to prevent the formation of hydrofluosilicic acid. This is, in part, the reaction by which hydrofluoric acid attacks quartz, glass and other silaceous substances forming silicon fluoride. Industrial amounts may be produced by gently heating powdered silica-free fluorspar at 130°C with concentrated sulphuric acid containing water (10% v/v) in cast-iron retorts with the vapours passing into lead boxes that act as water-cooled condensers. The hydrofluoric acid was transported in lead bottles or those made of ceresine wax, gutta-percha or paraffin coated vessels. The modern day invention of plastics, especially Teflon-like materials, has eased the corrosion situation. Apparatus made of stainless steel, copper or Monel metal (nickel-copper alloys) also appears satisfactory.

Hydrogen fluoride is rarely seen free in nature, although it has been detected in the effluvia from volcanic vents such as the fumaroles of Vesuvius and it may play an unappreciated part in fatalities observed in these areas, along with hydrogen sulphide and carbon monoxide (see appropriate chapters in this volume). The colourless transparent gas may be condensed into a mobile liquid that fumes strongly in air. However, it is an extremely hygroscopic gas and readily dissolves in water in all proportions producing hydrofluoric acid. The more concentrated solutions of hydrofluoric acid also fume strongly in air. All of these fumes have a sharp pungent odour and should be avoided.

Hydrofluoric acid finds many varied uses. The etching and marking of glass, from scientific apparatus to artistic design, is well known, as is its use in polishing crystal glass. A solution of ammonium fluoride in hydrofluoric acid is known in commerce as "white acid" and is used for frosting electric light bulbs. This property of dissolving silica has led also to its use in removing efflorescence (salt encrustation) from bricks or stone, the general cleansing of grime from buildings and monuments, the detachment of sand particles from metal castings and the elimination of surface oxides from silicon in the semiconductor industry.

Other uses include working over (filling) too heavy silks, during paraffin alkylation in oil refining to produce high octane fuels, in the fluorination process in the aluminium industry, in the separation of uranium isotopes to provide fissionable material, for the removal of oxides from stainless steel (pickling), in the dye and ceramics industry and in the production of organic (e.g. Teflon, Freon) and inorganic (e.g. sodium fluoride) fluorine compounds. Some commercial rust removers also apparently contain hydrofluoric acid, it has been extensively used perhaps unwisely to assist in window cleaning and it has been put forward as a fuel source in certain types of engines, particularly for rockets.

Toxicity Profile

The toxicity of hydrofluoric acid did not go unnoticed amongst the early chemists working with this material. The vapours were known to be corrosive and incautious inhalation can produce many damaging effects. Contact with the skin is also gravely dangerous, a problem difficult to avoid as the acid rapidly produced holes in, and escaped from, apparatus designed to produce and contain it.

Humphey Davy was poisoned during his experiments, sufficiently damaging his health to convince him to abandon any further researches into fluorine. The Reverend Thomas Knox and his brother, George J. Knox, both affiliated to the Irish Academy, were severely poisoned whilst trying to repeat and advance Davy's experiments. The Reverend just escaped death and the other spent up to three years recovering from the effects, requiring a lengthy convalescence in Naples. The Belgium chemist, Paulin Louyet, and the French experimenter, Jerome Nickels (Nickles), both died from complications arising from exposure to hydrofluoric acid and inhalation of its fumes. Other workers (notably George Gore and Henri Moissan) were poisoned several times but recovered, although these episodes undoubtedly damaged their health and shortened their lives.

Acute exposure

Dermal contact produces reddening of the skin (erythema), which may be rash-like, accompanied by severe pain. An intolerable prickling and burning sensation may occur with needle-like pains under the nails if the tips of the fingers are contaminated. Slowly healing burn wounds develop that may lead to ulceration and tissue necrosis, the severity of which depends upon the concentration of the acid encountered and the duration of the exposure. Burns may be extensive and severe, involving the underlying bone. Concentrated solutions (over 50%) usually cause immediate pain whereas the onset of symptoms may be delayed for up to 24 hours after contact with more dilute solutions, leading to painful sores that only become apparent the next day.

The liquid or vapour produces a stinging and burning sensation in the eyes and increased lacrimation. Depending upon the concentration, there may be severe irritation of the eyes and eyelids, with conjunctivitis leading to corneal abrasions and ulcers, which may progress to corneal perforation resulting in prolonged or permanent visual defects or total destruction of the eyeball. Inhalation causes irritation of the linings of the mouth and throat with a cough and possible loss of voice. Nosebleeds and sinusitis may develop. Nausea, vomiting and a loss of appetite are also observed. Higher concentrations produce inflammation of the trachea and bronchi (tracheobronchitis) that may progress to pulmonary oedema within 24 hours. Ingestion may lead to necrosis of the oesophagus and stomach, with nausea, vomiting and diarrhoea. Haemorrhagic gastritis and pancreatitis may also occur after significant exposure.

As would be expected, if injury to the skin surface, lungs or gastrointestinal tract is sufficiently severe, then death will ensue from these complications. However, fatal systemic fluorosis has resulted from relatively small cutaneous burns (2.5% total body surface) when exposed to a concentrated solution of hydrofluoric acid. This is a result of the absorption of hydrogen fluoride. Neurological symptoms, including paraesthesia (numbness and tingling), hyperactive reflexes, tetany (positive Chvostek's and Trousseau's signs), clonic-tonic convulsions, muscular pain and weakness may be evident. There is usually a lowering of the serum calcium and magnesium levels (hypocalcaemia; hypomagnesaemia) and cardiovascular hypotension leading to circulatory collapse. Respiration rate is usually initially increased before becoming depressed. Ventricular arrhythmias are common and death usually results from cardiac or respiratory failure.

Chronic exposure

In common with other fluorides, inhalation or ingestion of hydrofluoric acid may cause fluorosis. The symptoms of this condition are weight loss, malaise, anaemia, leucopoenia, discolouration of the teeth and osteosclerosis. It is the latter complication that is the major concern. Fluorine compounds are preferentially retained within the bone. Chronic intake may cause skeletal abnormalities characterised by slow progressive new bone formation usually beginning in the lumbar spine and pelvis. This is the rationale behind health screening via X-ray examination of (male) workers potentially exposed to fluorides. As the period of exposure continues, pain in the joints becomes more intense and movement of the vertebral column and lower limbs becomes more restricted. Calcification (ossification) of ligaments and joints may also occur, with outgrowths or bony spurs in the joints resulting in fusion of the spine ("poker back") and contractures of the hip and knees. This severe stage, called "crippling fluorosis", occurs in cases of heavy industrial fluoride exposure.

The recommended level of exposure of workers to hydrogen fluoride in the atmosphere is 3 ppm (parts per million). A value of 50 ppm exposure for 30 minutes has been quoted as the lowest lethal concentration in man, which would place the material in the "highly toxic" bracket. An ingested dose of 1.5 g has caused death in an adult. An epidemiological survey undertaken two years after an oil refinery accident that released around 20 tonnes of hydrogen fluoride into the atmosphere found both respiratory problems and eye symptoms within the exposed population. It is difficult to be sure of a "safe" concentration — it has been stated that 1 ppq (one part per quadrillion; a quadrillion is 1×10^{15}) of hydrogen fluoride gas in the atmosphere can injure peach trees!

Mechanism of Action

Hydrogen fluoride is one of the most corrosive inorganic acids and a direct irritant of the skin, mucous membranes and linings of the respiratory and alimentary tracts. In common with many acids, a rapid dehydration type of coagulative necrosis occurs from hydrogen ion release; the dilution of concentrated acids produces the local release of large amounts of thermal energy. Later there occurs a slower secondary necrosis that results from the subdermal penetration of the fluoride ions. All fluorides are protoplasmic poisons and they complex with divalent cations, primarily calcium and magnesium, thereby decreasing their concentrations in tissues and blood. This severe electrolyte imbalance (hypocalcaemia and hypomagnesaemia) may lead to subsequent ventricular dysrhythmias and depressed contractility of the myocardium. Neurologically mediated central vasomotor depression also occurs and fluoride may be directly toxic to the central nervous system. The depletion of calcium stores may cause the extracellular release of potassium ions from nerve endings and subsequent severe pain. The fluoride ion also inhibits enzymes involved in glycolysis and oxidative phosphorylation, as well as other metal-containing enzymes, leading to severe disruption of metabolic pathways and interference with energy exchange.

Treatment

Quick treatment is necessary to increase the chance of survival. Termination of exposure is the first priority, together with the removal of affected and contaminated materials. Surface contamination must be washed away with copious amounts of water to reduce local damage and decrease systemic absorption. Eye exposure may be treated by extensive irrigation with a dilute solution of calcium gluconate. If ingestion has recently occurred, a source of calcium such as milk, calcium chloride or limewater, should be given without delay by either drinking or lavage. The calcium in the gastrointestinal tract will react with the fluoride forming the relatively insoluble calcium fluoride that will not be absorbed.

A calcium gluconate or carbonate gel or slurry is either rubbed into the burn or placed around the injury (fill a surgical glove and place the hand inside). Skin lesions resulting from hydrogen fluoride burns usually require the intradermal injection of calcium gluconate. It may also be necessary to split or remove nails to treat the nail beds. Recommendations for the use of magnesium salts in treatment of hydrofluoric acid burns await further clinical trials, with some magnesium salts being effective in some but not all animal models. Where intradermal injection may be hazardous or impractical, an intra-arterial infusion of calcium gluconate in normal saline for several hours has been advocated. This, or an intravenous administration, will also prevent the rapid depletion of plasma calcium or replace it. Calcium chloride is also useful in the treatment of burns to the fingers but can be irritating to the tissues and may itself cause injury. Once in hospital the usual procedures for treatment of burns and poisoning such as maintenance of electrolyte balance, cardiovascular and respiratory support and dialysis, etc. should be available.

Case Histories

"An operative working in an alkylation plant within an oil refinery was splashed in the face with concentrated anhydrous hydrofluoric acid. Despite wearing protective clothing the material still came in contact with his skin. Within 30 minutes he was washed extensively with water and a magnesium oxide preparation was applied to the exposed areas. At the hospital emergency department it was established that there were third-degree burns on his forehead, eyelids, cheeks, nose and upper lip. Some keratoconjunctivitis was reported but no corneal damage. His throat was red and he had trouble swallowing but he appeared to have no respiratory problems. Two hours later he was taken to surgery for the removal of burnt tissue and was progressing satisfactorily. However, after six hours he developed ventricular arrhythmias that responded to cardioversion but he experienced repeated episodes over the next few hours. He subsequently died from cardiac arrest ten hours after exposure. A post-mortem revealed intense congestion of the upper respiratory mucosa and pulmonary oedema".

This case study emphasises that sufficient fluoride may be absorbed from a relatively small area of skin to cause acute systemic fluoride poisoning and death. (*Source*: Temperman PB (1980). Fatality due to acute systemic fluoride poisoning following a hydrofluoric acid skin burn. *Journal of Occupational Medicine* **22**, 691–692.)

"Two male workers were attempting to remove a large bottle of hydrofluoric acid that was seen to be emitting fumes. During the course of this procedure the bottle, perhaps old and fragile and under pressure, exploded spraying them with acid and enveloping them in a dense cloud of acid vapour. They were both immediately showered on site and then taken to the hospital. On admission, one worker was found to be in severe respiratory distress owing to bronchospasm. He was extensively burnt and had a corneal ulcer of his right eye. Numbness and tingling was present in his upper limbs and carpal spasm (flexion of wrist and flexion/extension of fingers — suggests low calcium, tetany?) was observed. Despite treatment, he died of cardiac arrest ten hours after exposure. With the exception of a moist cough and mild wheezing, the second worker showed no signs of respiratory disturbance or pain. He had burns of various sizes on his arms, chest and abdomen covering about 15% of his body surface. He was stable until respiratory distress occurred requiring a tracheotomy. He died four hours after the accident. Similar post-mortem findings were made for both men. Severe burns to the skin. Dilation of the heart with cardiac muscle being pale and flabby and leucocyte infiltration in reddened areas of the ventricle. An acutely inflamed bronchial tree with partially ulcerated mucosa. The lungs showed evidence of pneumonitis and severe haemorrhagic oedema".

These observations attest to the corrosive and infiltrative nature of hydrofluoric acid. (*Source*: Greendyke RM and Hodge HC (1964). Accidental death due to hydrofluoric acid. *Journal of Forensic Medicine* **88**, 383–390.)

"A 46-year-old worker who had been engaged in the preparation of hydrofluoric acid for 16 years was examined for osteosclerotic bone changes. During the course of his work, the man was most likely to have been exposed also to dust from ground and powdered fluorspar and probably aluminium fluoride. Radiological examination of the lower spine and pelvis showed very dense sclerosis. The forearms and lower legs also revealed extensive ligamentous ossification. A 24-hour urine collection demonstrated an urinary fluoride excretion of 15.2 mg/litre, many times the expected normal level (2.0 mg/litre; quoted range 0.4 to 3.4 mg/litre). Perhaps surprisingly, the patient was completely free of symptoms".

This case illustrates the problem of low level chronic hydrofluoric acid exposure and the insidious nature of osteosclerosis (chronic skeletal fluorosis). (*Source*: Wilkie J (1940). Two cases of fluorine osteosclerosis. *British Journal of Radiology* **13**, 213–217.)

"A 33-year-old male developed lower left abdominal tenderness and blood-stained diarrhoea following the self-administration of a concentrated hydrofluoric acid enema whilst under the intoxicating influence of cocaine. Sigmoidoscopy and examination revealed a fulminant acute colitis with severe mucosal ulceration and oedema in the rectum and sigmoid colon. Calcium carbonate enemas were given to bind the fluoride ions and neutralise the rectal acid. Laparotomy revealed an ulcerated, necrotic and purulent sigmoid colon and intraperitoneal pus. The patient recovered following surgery with removal of part of the sigmoid colon, but five months later developed a colonic stricture which was resected".

This case illustrates that the lower end of the gastrointestinal tract is equally susceptible to the caustic effects of hydrofluoric acid. (*Source*: Cappell MS and Simon T (1993). Fulminant acute colitis following a self-administered hydrofluoric acid enema. *American Journal of Gastroenterology* **88**, 122–126.)

Suggested Further Reading

For an example of wide-scale population exposure

Brender JD, Perrotta DM and Beauchamp RA (1991). Acute health effects in a community after a release of hydrofluoric acid. *Archives of Environmental Health* **46**, 155–160.

Dayal HH, Brodwick M, Morris R, Baranowski T, Trieff N, Harrison JA, Lise JR and Ansari GA (1992). A community-based epidemiologic study of health sequelae of exposure to hydrofluoric acid. *Annals of Epidemiology* **2**, 213–230.

For further case histories

Blodgett DW, Suruda AJ and Crouch BI (2001). Fatal unintentional occupational poisonings by hydrofluoric acid in the US. *American Journal of Industrial Medicine* **40**, 215–220.

Shewmake SW and Anderson BG (1979). Hydrofluoric acid burns. A report of a case and review of the literature. *Archives of Dermatology* **115**, 593–596.

8

HYDROGEN SULPHIDE

G. B. Steventon

"I am inclined to think some volatile acid is given off by this carnerine of filth when workers disturb it ...... such effluvia ought, one would think, to impair the lungs".

Ramazzini B (1713) Disease of Workers, De Morbis Artificum Diatriba.

Introduction: Sources and Uses

Natural sources

One of the principal compounds involved in the natural sulphur cycle is hydrogen sulphide. This can be seen in Figure 1. It occurs in volcanic gases and is also produced by the action of bacteria during the decay of proteinaceous (both plant and animal) material. Many bacteria and fungi release hydrogen sulphide into the environment during the decay of material possessing the sulphur-containing amino acids cysteine and methionine as well as the direct reduction of sulphate.

The reduction of sulphate to hydrogen sulphide can be carried out by the members of two genera of anaerobic bacteria (*Desulphovibro and Desulphotomaculum*). The substrates for these organisms are usually short chain organic acids that are provided by the fermentive activities of other anaerobic bacteria on more complex compounds. Thus, hydrogen sulphide production is to be expected in conditions where oxygen is depleted, organic compounds are present and sulphate is available. The production of hydrogen sulphide is balanced by other processes involving a

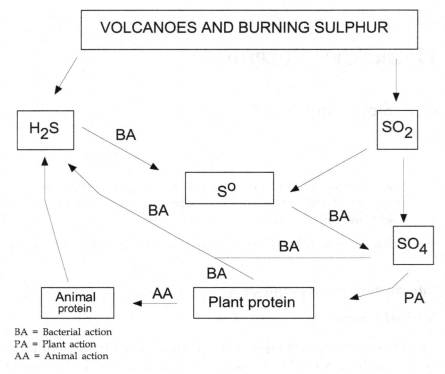

Figure 1. The sulphur cycle.

number of bacteria which are found in both soil and water which can oxidise hydrogen sulphide into elemental sulphur. The filamentous bacteria, *Beggiatoa* and *Thiothrix* are amongst these microorganisms. Photosynthetic bacteria which belong to the families Chromatiaceae and Chlorobiaceae can also oxidise hydrogen sulphide to elemental sulphur and sulphate in the presence of light and the absence of oxygen. Reduced sulphur compounds are also oxidised in nature by members of the genus Thiobacillus. The final result of this activity is the production of sulphate which is extremely stable to further chemical activity. Thus as a result of these various biogeochemical processes, hydrogen sulphide occurs in and around sulphur springs and lakes and is almost continuously present as an air contaminant in some geothermally active areas.

Human Activity

Human activity in various circumstances can release naturally occurring hydrogen sulphide. Natural gas or crude oil deposits can occur in association with hydrogen sulphide in some areas and this will be released during drilling or extraction. The sulphur content of crude oils ranges from 0% to 5% and some natural gas deposits can comprise upto 42% hydrogen sulphide. Coal can also contain up to 80 g/kg of sulphur within some deposits and thus in certain mining and drilling operations (of natural gas and crude oil) hydrogen sulphide can be formed. The development and use of geothermal resources can also be associated with the release of hydrogen sulphide. For example the geothermal generating plant in Baja California (Mexico) produces enough hydrogen sulphide to make the provision of special ventilation essential to protect the electrical systems and alarms are installed for the protection of the personnel.

In industrial processes, hydrogen sulphide can be formed when elemental sulphur or certain sulphur-containing compounds are exposed to organic material at high temperatures. This is usually an undesirable reaction or by-product, though it can be and is an important reagent or by-product in numerous industrial processes, for example, the manufacture of sulphides, sodium hydrosulphide and various organic sulphur compounds. The production of hydrogen sulphide as a by-product occurs in the production of coke from sulphur-containing coal, the production of carbon disulphide, the manufacture of viscose rayon in the Kraft process for the production of wood pulp and the sulphur extraction by the Frasch process.

In the refining of sulphur-containing crude oils, approximately 80%–90% of the divalent sulphur-containing compounds of carbon and hydrogen are converted to hydrogen sulphide. The hydrogen sulphide produced plus that from other industrial, geothermal or natural gas sources can be recovered by a number of processes that are classified as either absorption-desorption processes or as processes that involve the oxidation to oxides or to elemental sulphur. The vast bulk of hydrogen sulphide recovered from industrial processes is used to produce sulphuric acid or elemental sulphur.

Large quantities of hydrogen sulphide have been used in the production of heavy water which is used as a moderator in some nuclear reactors. This is based on the enrichment of the deuterium concentration of water by hydrogen sulphide in a gas-liquid ion exchange system and then separation by fractional distillation. The tanning industry produces hydrogen sulphide in the process of removing hair from the hides. This involves deliming by the addition of ammonium chloride or ammonium sulphate and then pickling of the hides with sulphuric acid in large rotating drums. The hydrogen sulphide and other gases produced are released from these drums on opening to remove the hides or to add additional reagents. The occupations that have a potential for hydrogen sulphide exposure and thus hydrogen sulphide production can be seen in Table 1.

Properties

Hydrogen sulphide (sulphuretted hydrogen, hydrosulphuric acid, hepatic gas, sulphur hydride, rotten egg gas and stink damp) is a colourless, heavier than air gas with an offensive odour of "rotten eggs" and a sweetish taste. It is flammable and burns in air with a pale blue flame and when mixed with air its explosive limits are 4.3% to 46.0% by volume. The relative molecular mass is 34.08 and its density is 1.5392 g/L at 0°C and 760 mm. The melting point is −85.5°C and boiling point is −60.3°C. Hydrogen sulphide is soluble in water, ethanol, ether, glycerol and in solutions of amines, alkali carbonates, bicarbonates and hydrosulphides. The compound in question can also undergo a large number of oxidation reactions with the type and rate of the reaction and the products produced dependent on the concentration and nature of the oxidising species. The main types of these reactions tend to be sulphur dioxide, sulphuric acid or elemental sulphur. Also aqueous solutions of the halogens (chlorine, bromine and iodine) can react with hydrogen sulphide to produce elemental sulphur. In the gaseous state the oxides of nitrogen can react with hydrogen sulphide to form sulphur dioxide but in solution (pH 5–9) the main product is elemental sulphur.

Table 1. Occupations with potential exposure to hydrogen sulphide.

Animal fat and oil processors	Lithographers
Animal manure removers	Lithopone makers
Artificial-flavour makers	Livestock farmers
Asphalt storage workers	Manhole and trench workers
Barium carbonate makers	Metallurgists
Blast furnace workers	Miners
Brewery workers	Natural gas production and processing
Bromide-brine workers	workers
Cable splicers	Painters using polysulphide caulking
Caisson workers	compounds
Carbon disulphide workers	Paper makers
Cellophane producers	Petroleum production and refinery
Chemical laboratory workers	workers
(lecturers, students, technicians)	Phosphate purifiers
Cistern cleaners	Photo-engravers
Citrus root fumigators	Pipeline maintenance workers (lecturers,
Coal gasification workers	students, technicians)
Coke oven workers	Pyrite burners
Copper-ore sulphidisers	Rayon makers
Depilatory makers	Refrigerant makers
Dye-makers	Rubber and plastics processors
Excavators	Septic tank cleaners
Felt makers	Sewage treatment plant workers
Fermentation process workers	Sheep dippers
Fertiliser makers	Silk makers
Fishing and fish-processing workers	Slaughterhouse workers
Fur dressers	Smelting workers
Geothermal power drilling and	Soap makers
production workers	Sugar beet/cane processors
Glue makers	Sulphur spa workers
Gold-ore workers	Sulphur products processors
Heavy-metal precipitators	Synthetic-fibre makers
Heavy-water manufactors	Tank gaugers
Hydrochloric acid purifiers	Tannery workers
Hydrogen sulphide production	Textile printers
and sales workers	Thiophene makers
Landfill workers	Tunnel workers
Lead ore sulphidisers	Well diggers and cleaners
Lead removers	Wool pullers

From NIOSH (1977).

In solution hydrogen sulphide dissociates to form the hydrogen sulphide anion (HS^-) and the sulphide anion (S^{2-}). The pKa for the hydrogen sulphide anion (0.01–0.1 M solutions at 18°C) is 7.04 and for the sulphide anion it is 11.96. At physiological pH (pH 7.4) approximately 33% is present as hydrogen sulphide and 67% is in the form of the hydrogen sulphide anion.

Atmospheric Chemistry

The atmospheric chemistry of hydrogen sulphide involves both photochemical and chemical oxidation with the terminal oxidation products being sulphuric acid (H_2SO_4) and inorganic sulphate (SO_4^{2-}). There have been very few studies on the persistence and interconversions of hydrogen sulphide in the atmosphere or under atmospheric conditions in the laboratory. However, two reports have calculated that the residence of hydrogen sulphide is approximately 1.7 days. Other reports suggest two days in clean air but only two hours in polluted urban air. Studies on the distance traveled by hydrogen sulphide from point of release have shown that it was only 50% of its original concentration within a 2.5 km radius and this decreased to 12%–13% between a 2.5 and 20 km radius. Clearly more work is required before any definitive conclusions about the reactions and persistence of hydrogen sulphide in the atmosphere can be reached.

Toxicity Profile

The offensive odour is the first indication of the presence of hydrogen sulphide gas. However, this should not be used as a warning signal since at concentrations of 150 ppm or greater, a rapid paralysis of the olfactory nerve takes place. Hydrogen sulphide is toxic and irritating when inhaled, in contact with the eyes, nose, throat, skin or if swallowed. If inhaled at concentrations of 1000–2000 ppm then one or two inhalations will cause immediate loss of consciousness and *death*. The results of three post-mortem examinations of individuals who died from hydrogen sulphide asphyxia in sewers showed the following results.

1. Externally, the bodies had a gray-greenish cyanosis (discolouration of the blood and mucus membranes due to excessive concentration of reduced haemoglobin in the blood).
2. Internally, there was haemorrhagic pulmonary oedema (loss of blood into the lungs). The blood and viscera had a greenish colouration while the cerebral and nuclear masses were a greenish-purple colour.

The toxicity of hydrogen sulphide is similar to that of hydrogen cyanide. The mechanism of toxicity is by their reversible binding to cytochrome c oxidase (complex IV) of the mitochondrial respiratory electron transport chain, thus preventing the synthesis of ATP (chemical energy), see Figure 2. The National Institute for Occupational Safety and Health reported (1978) that hydrogen sulphide was a leading cause of death in the work place.

The greenish colouration of the blood and tissues in the hydrogen sulphide poisonings probably results from the formation of a complex between the hydrogen sulphide anion and methaemoglobin to produce sulphmethaemoglobin, which is analogous to cyanmethaemoglobin. The dissociation constant for sulphmethaemoglobin is approximately 6×10^{-6} mole/L, while that for cyanmethaemoglobin is of the order of 2×10^{-8} mole/L. Methaemoglobin is produced when the haem irons of haemoglobin are chemically oxidised by the loss of an electron with a valence change from 2^+ to 3^+. The resulting pigment is greenish-brown to black in colour and cannot combine reversibly with oxygen or carbon monoxide. The process is illustrated in Figure 3.

Despite the lower binding affinity, a nitrite-induced methaemoglobinaemia provides absolute protection and has been used as an antidote against hydrogen sulphide poisoning in laboratory animals as well as a number of human hydrogen sulphide poisonings.

Although avoidance of hydrogen sulphide is the best form of defense, the body does have a limited ability to metabolise this toxic

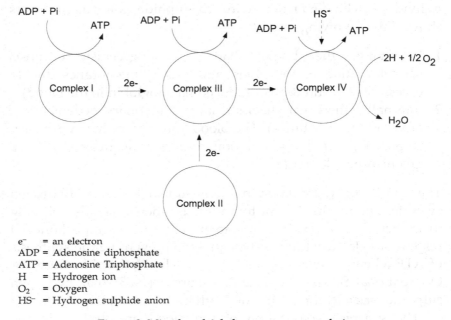

e⁻ = an electron
ADP = Adenosine diphosphate
ATP = Adenosine Triphosphate
H = Hydrogen ion
O$_2$ = Oxygen
HS⁻ = Hydrogen sulphide anion

Figure 2. Mitochondrial electron transport chain.

compound since hydrogen sulphide is produced by the intestinal microflora and during endogenous metabolism, since the sulphide can be released from perisulphides by enzymes like thiosulphate reductase, mercaptopyruvate sulphurtransferase, cystathione-β-synthetase and cystathione-γ-lyase. Indeed recent research has suggested that hydrogen sulphide may in fact play an important role as a secondary messenger in the body in a manner similar to nitric oxide (in smooth muscle relaxation) or as a neuroactive compound that enhances the NMDA receptor in the hippocampus region of the brain. Detoxification is carried out by the enzyme sulphydryl (or thiol) methyltransferase that is present in the intestine, liver, lungs and kidneys. The product of this enzymatic reaction is methanethiol (which is approximately 10x less toxic than hydrogen sulphide). The methanethiol is itself a substrate for the sulphydryl methyltransferase enzyme producing the less toxic

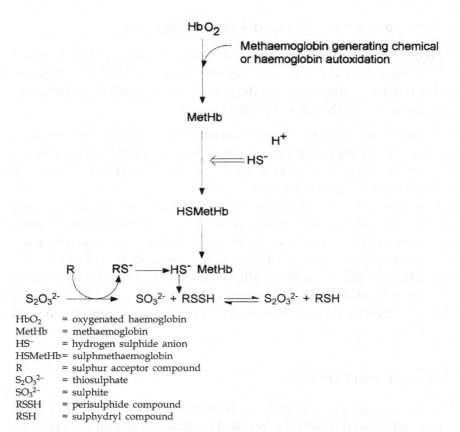

Figure 3. Principles of the therapeutic management of hydrogen sulphide poisoning.

sulphide, dimethyl sulphide, as its product. Hydrogen sulphide is also detoxified by other enzymatic/non-enzymatic processes too, with the liver mitochondria also being very active in its oxidative metabolism. The pathway is believed to proceed via polysulphides, thiosulphate and sulphite to sulphate. Since the mitochondria are the site of the electron transport chain (the target of hydrogen sulphide toxicity) it is not surprising that this organelle is actively involved in its detoxification.

Toxic Episodes of Hydrogen Sulphide Exposure

Hydrogen sulphide intoxication in man has been generally classified into three clinical forms — acute, sub-acute and chronic. These are dependent on the nature of the predominant clinical features and symptoms that the patient presents with.

1. Sub-Acute: Effects of continuous exposure (may be up to several hours) to 100–1000 ppm of hydrogen sulphide. Eye irritation (gas eye) is the most commonly reported effect but pulmonary oedema has also been seen.
2. Acute: Effects of a single exposure (from seconds to minutes) to a massive concentration of hydrogen sulphide (a minimum concentration of 1000 ppm is required to produce acute intoxication). These rapidly produce respiratory distress.
3. Chronic: Effects of intermittent exposure to low to intermediate concentrations (50–100 ppm) of hydrogen sulphide. This is characterised by a "lingering", largely subjective, feeling of illness.

Occupational Exposure

Usually, hydrogen sulphide is encountered in the workplace as an unwanted/undesirable by-product (petroleum refining), or formed as a result of the decomposition of sulphur-containing organic matter in the absence of complete oxidation (sewer workers). Acute hydrogen sulphide intoxication can be a dramatic and often fatal event. Three men were accidentally covered in a cloud of hydrogen sulphide which escaped from a high pressure cylinder. All instantly collapsed and ceased to breath. Only the immediate action of their fellow workers and trained first aiders saved their lives. The two most seriously affected experienced violent convulsions and took around 30 minutes to recover consciousness. None of the three suffered important after effects and not one of them recalled smelling the characteristic "rotten eggs" odour of hydrogen sulphide. The estimated concentration of the gas to which these individuals were exposed was calculated to be 2000 ppm. Twelve workmen in a chemical factory producing benzylpolysulphide were overcome by hydrogen sulphide fumes when a pipeline used to transfer sodium sulphydrate ruptured. The sodium sulphydrate (a liquid) drained away into a sewer (where it reacted with the acidic sewerage) releasing hydrogen sulphide gas from several sewer openings. Two of the 12 died, three stopped breathing but were successfully resuscitated, six lost consciousness but recovered and one workman developed pulmonary oedema. In the mid 1970s, a review of 221 cases of hydrogen sulphide poisonings in the Canadian oil, gas and petrochemical industries showed a 6% death rate, 75% of the victims experienced a period of unconsciousness, 12% were comatose. The vast majority had neurological problems (altered behaviour, confusion, vertigo, agitation); 40% required respiratory assistance and 15% developed pulmonary oedema.

Population Exposure

A devastating exposure of hydrogen sulphide to the community of Poza Rica (Mexico) occurred in 1950. This city of 22,000 people located 210 km northeast of Mexico City was then the centre of Mexico's leading oil-producing district and the site of several oil installations including a sulphur recovery plant. In the early morning a malfunction of the waste gas flare resulted in the release of large quantities of hydrogen sulphide into the atmosphere. The released gas in combination with a low level temperature inversion and light morning winds was carried over a nearby residential area. Most inhabitants were overcome while attempting to leave the area and while helping stricken friends. Within three hours, 320 people were hospitalised and 22 died. Greater than 50% of the patients lost consciousness, many suffered respiratory tract and eye irritation and nine developed pulmonary oedema. Four developed neurological abnormalities, two neuritis of the acoustic nerve and one developed dysarthria.

Treatment

The first and foremost action for a person exposed to hydrogen sulphide is to remove the affected individual to fresh air at once. If the gas has come into contact with the eyes, wash immediately with copious amounts of water. If the gas has come into contact with the skin or is present in the clothing, remove the clothing and thoroughly wash the skin with water. The treatment of acute hydrogen sulphide poisoning is quite varied but normally involves the use of 40% oxygen via a face mask, amyl nitrite inhalations followed by intravenous administration of sodium nitrite and sodium thiosulphate with or without sodium bicarbonate administration. Both amyl nitrite (given by inhalation) and sodium nitrite (i.v. administration) are nitrite compounds that are capable of generating methaemoglobin. Methaemoglobin readily "mops up" hydrogen sulphide anion (in solution, hydrogen sulphide is 70% hydrogen sulphide anion and 30% hydrogen sulphide), thus forming sulphmethaemoglobin and preventing the hydrogen sulphide anion from binding with the complex IV (cytochrome c oxidase) in the respiratory electron transport chain. The sodium thiosulphate is required to produce a sulphydryl anion (RS^-) to react with the hydrogen sulphide anion (HS^-), thus generating a persulphide intermediate (RSSH) and also the sulphite anion (SO_3^{2-}) which under the action of the enzyme rhodanase, converts the persulphide (in the presence of the sulphite) to thiosulphate and a sulphydryl compound.

Hydrogen Sulphide Exposure: An Evaluation of the Health Risks

One of the most important and also recognisable toxic effects of hydrogen sulphide is its ability to induce acute intoxication which results in immediate collapse and is frequently accompanied by respiratory arrest which will cause death if not treated. A second form of toxic insult associated with exposure to hydrogen sulphide is its irritating effect on mucus membranes of the eye and respiratory tract. Gas eye (keratoconjunctivitis) and pulmonary oedema are two of the most common problems caused by localised hydrogen sulphide irritation. The malodorous property of hydrogen sulphide gas, the foul smell of "stink bombs", is well recongised and some individuals believe that this level of hydrogen sulphide is capable of impairing human health.

The vast majority of data available on human health effects associated with hydrogen sulphide exposure have come from incidents involving accidental or industrial exposures. With the exception of Poza Rica, information on general population exposure and health is sparse. There is little if any data on controlled human exposure to hydrogen sulphide and except for the data from odour threshold studies which is 65-70 years old, no new information (<10 years old) is available. There is some information available from animal studies on high concentrations of hydrogen sulphide gas but hardly any information available on long term-low level effects. Epidemiological studies are also lacking concerning the health consequences of long term-low level exposure to hydrogen sulphide in both the general population and industry.

Exposure Levels

Air pollution problems associated with hydrogen sulphide gas and the general population are associated with the foul odour. Sources can be industrial or polluted water. Peak concentrations as high a 0.13 ppm have been reported in the air surrounding industrial sources. At a geothermal site in New Zealand where continuous monitoring was carried out, a concentration of 0.05 ppm was exceeded for 35% of the time over a five-month period.

Concentrations of hydrogen sulphide in the work place vary widely with the shale oil industry and viscose rayon production industry reporting maximum exposure concentrations per day of 15-20 ppm. However, massive accidental exposure to hydrogen sulphide due to equipment failure has been the principal hazard in industry. Since hydrogen sulphide is heavier than air, accumulation to lethal concentrations in low lying or enclosed areas can occur and numerous fatalities have occurred from slow accumulation of this gas in both ambient and industrial environments.

Experimental Animal Studies

The toxic effects of hydrogen sulphide gas have not been studied extensively in experimental animals but tests on a number of species (rat, mouse, cat, dog and goat) have indicated that the central nervous system is the major or primary target. The general pattern is collapse, respiratory arrest and asphyxia caused by the suppression of the higher respiratory centres in the brain. Death usually follows if no intervention is attempted. The experimental data on long term, low level toxicity of hydrogen sulphide in experimental animals is sparse.

Occupational Exposure Effects of Hydrogen Sulphide

Accidental exposure of humans to high concentrations of hydrogen sulphide has occurred in individuals working in petroleum refining, viscose rayon manufacture, sugar beet processing and in tannery works. The exact exposure levels have not been precisely reported in many of these situations, but reported effects range from the lesser conditions of neurasthenic and otoneurological symptoms and keratoconjunctivitis to the serious effects of pulmonary oedema, respiratory failure, collapse and death. The data available enables an estimate of exposure of anything from seconds to minutes to concentrations of approximately 1000 ppm causing acute intoxication, several hours exposure to 100–1000 ppm resulting in keratoconjunctivitis and pulmonary oedema. Intermittent exposure to 50–100 ppm causing "lingering, largely subjective manifestations". This is believed by some workers to be representative of chronic intoxication. A number of studies have associated hydrogen sulphide exposure at concentrations of 10.5–21.0 ppm for several hours with eye irritation in workers. However, the effects of low level, long term exposure to hydrogen sulphide in industry have yet to be systematically evaluated.

General Population Effects from Exposure to Hydrogen Sulphide

Although several incidents of exposure of the general population to hydrogen sulphide have been described, these have fortunately been at the level of malodour, at worst resulting in temporary aliments such as headache, nausea or sleeplessness. However on two occasions, exposure to hydrogen sulphide (Poza Rica, Mexico and Rotorua, New Zealand) has resulted in serious illness and death.

Case Histories of Hydrogen Sulphide Poisoning

Two case histories will be presented here to highlight the life and death aspects of hydrogen sulphide intoxication.

Case 1

A 24-year-old man was found lying unconscious with repetitive and generalised tonic convulsions beside a gas tank in a chemical factory. Approximately four hours before the accident he had poured a hydrochloric acid solution into the tank which contained a mixture of carbon disulphide, hydrazine, sodium hydroxide and methanol.

This mixture was normally allowed to sit in the tank overnight to allow the chemical reaction to proceed slowly (the carbon disulphide will react with the hydrazine and sodium hydroxide to produce sodium sulphide. The concentrated hydrochloric acid is then slowly added to the tank the next day to produce hydrogen sulphide gas which is ventilated through an overhead pipe). The exact details of what happened on the day of the accident are not clear but the victim recalled pouring the acid into the tank very quickly, thus liberating a large amount of hydrogen sulphide very rapidly.

On admission to hospital the patient was markedly cyanotic with repetitive generalised tonic seizures. The seizures were controlled by intravenous administration of phenytoin (700 mg) and diazapam (10 mg) before transfer to the ICU. The patient was comatose with generalised hyperreflexia and was immediately placed in high concentrations of oxygen via a face mask. The rapid and shallow respiration, copious production of frothy sputum and apnea resulted in endotracheal intubation and mechanical respiratory assistance.

Amyl nitrite therapy was initiated with the compound inhaled for 15 seconds every three minutes (this was repeated five times in total) and 300 mg of sodium nitrite was then given intravenously over a period of three minutes followed by 12.5 mg of sodium thiosulphate for another five-minute period. Thirty minutes after the cessation of the nitrite treatment the patient became conscious. The methaemoglobin 12 hours after nitrite treatment was 1.8%. After four days following nitrite treatment the patient was extubated and discharged two weeks after the accident but it took three months for the neurophysiological and EEG tests to return to normal. The patient did recall that he did not smell the odour of rotten eggs before losing consciousness at the accident site.

Case 2

A 16 year-old farm worker began using a high pressure hot water hose to clean manure from the gutters inside a recently emptied calf shed. He was ten metres from a 378,500 litre underground liquid manure storage tank, the contents of which had been mixing for 30 to 60 minutes. After approximately ten minutes the boy began to cough, vomited, collapsed and died. A fellow worker near to an exhaust fan attempted to help the young boy but became light headed, stumbled outside and collapsed. On recovering, he and another worker experienced syncope (temporary loss of consciousness) during a second rescue attempt.

The farm worker had been in good general health with no chronic illness nor was he on any medication or drugs. The results of the post-mortem examination showed tracheobronchial aspiration of food and stomach contents, focal haemorrhages in the hilar portion of each lung, focal pulmonary oedema and small petechial brain haemorrhages. These findings are suggestive of the inhalation of a toxic gas and hydrogen sulphide was implicated. Two days after the fatal accident, air tests carried out under similar conditions produced hydrogen sulphide concentrations of greater than 60 ppm (upper detection limit for the apparatus). The NIOSH recommends a maximum exposure of 10 ppm over ten minutes and immediate evacuation at 50 ppm. No carboxyhaemoglobin was detected at postmortem so carbon monoxide was also ruled out. As no nitric oxide, nitrogen dioxide and sulphur dioxide were detected, this seems to implicate hydrogen sulphide as the toxic agent.

Several factors are believed to have contributed to this tragic accident. These included the overfilling and aggressive mixing of the liquid manure tank contents, inadequate ventilation with only one of the six ventilation fans in use and the placement of the storage tank below and not outside the shed.

Suggested Further Reading

Beauchamp RO, Bus JS, Popp JA, Boreiko CJ and Andjelkovich DA (1984). A critical review of the literature on hydrogen sulfide toxicity. *CRC Critical Reviews in Toxicology* **13**, 25–97.

Klaassen CD (ed.) (1996). *Casarett and Doull's Toxicology: The Basic Science of Poisons*, 5th ed. McGraw-Hill, New York, pp. 357–352.

Haley TJ and Berndt WO (eds.) (1987). *Toxicology.* Hemisphere Publishing Corporation, Washington, pp. 485–487.

9

LEAD: AN OLD AND MODERN POISON

D. B. Ramsden and T. Pawade

Introduction

Throughout history lead has presented a smiling face, offering the proposition that its benefits far outweigh its dangers. With what in hindsight seems perverse determination, man has risen to the challenge of proving this proposition and thus has continually invented new ways of harming himself. This has been aided by the fact that lead's simple salts are often tasteless or even pleasing to the human palate. A once common name for lead acetate was lead sugar. This property has been at the heart of many examples of lead poisoning in the past that went under the name of *Colica Pictonum*, because it was used to convert sour, poor quality wines to apparently sweet, high quality ones.

In historical terms, lead is an ancient metal. Lead artifacts date back to at least 6000 BC. It is one of the original metals of alchemy and as such, associated with the planet Saturn. Metallic lead does not appear on or near the earth's surface but its principal ore does (galena — lead sulphide). This varies in colour and lustre depending on the amount of other constituents, ranging from fresh lead grey to an almost black, dark blue/grey. The use of this and another of its ores — cerussite (basic lead carbonate), which is white (as pigments in eye cosmetics and for other decorative purposes), is thought to be the start of the interaction of humans with lead. Accidental smelting of galena beads in campfires may have yielded the metal itself, which being soft and dull probably would have limited appeal initially. Nevertheless, the ancient Egyptians realised that the metal's high density had some practical advantages and used lead sinker for fishing nets. Others utilised this property for making slingshots. Ancient battlefields, such as Marathon where Greeks and Persians clashed, give plenty of evidence for this use. Later it was discovered that in some instances with lead came silver. Oxidation of the lead yielded rich sources of silver, and the interest in this latter metal fuelled far more interest in lead mining and refining.

The process of producing silver from lead is called cupellation. Silver-rich galena ores yield as much as 600 ounces of silver per ton of lead. The technology is thought to have first been developed along the Black Sea coast of Northern Turkey in a region which was called Pontus, from which it spread rapidly throughout the ancient Mediterranean world. It appears to have been the source of the illustration of good and evil in the Old Testament book of the prophet Jeremiah {Chapter 6, verses 29–30}, which depicted sinners (lead) were consumed in the fire, whereas the righteous emerge as shining silver. Jeremiah lived in Judah (modern day Israel) around 600 BC. Cupellation was particularly important in the rise of ancient Athens. Peisistratus used his wealth from silver (lead) mines at Laurion to convert Athens from a small town to a great city. Later, as Athens prepared for war with the Persians, a new rich vein was discovered, enabling Thermistocles to finance the construction of the large navy which defeated the Persians at the battle of Salamis. The victory helped to ensure the flowering of Greek culture, with all its consequences for modern Western civilisation.

Although lead was used by the Greeks, it was not until the Romans that it was truly appreciated in its own right. They produced the metal on the grand scale for pipe-work to supply water to the cities of the empire. It has been estimated that in constructing one syphon unit for the aqueduct near Lyons in France, Roman engineers used approximately 12,000 tons of the metal. At its height in the late phase of the Roman republic and following the inception of the Empire (from 50 BC to 200 AD), it is estimated that up to half a million people were involved in lead working at any one time, and that the annual amount of lead ore mined may have reached 2,000,000 tons per annum. After the fall of the Roman empire, metal-working on this scale was not reproduced until the Industrial Revolution. It is fitting therefore that the chemical symbol for lead — Pb — is a contraction of its Latin name, plumbum.

Chemistry

Table 1 summarises some of the chemical and physical properties of the metal. As can be seen from this, lead possesses a number of stable isotopes. Because the isotopes are present in different amounts at different points in the Earth's crust, the isotopic ratio can be used to identify the mine of origin of the metal using mass spectroscopy. Lead from older US mines is relatively low in ^{206}Pb, giving a greater $^{207}Pb:^{206}Pb$ ratio that that of lead from younger mines.

Lead forms a wide range of compounds where the metal is present in either of two oxidation states — Pb^{2+} and Pb^{4+}. These compounds range from oxides and simple salts to volatile organometallic compounds such as lead tetraethyl used in petrol. In salts formed by the actions of acids (e.g. acetic and nitric acids), the metal is in the lower of the two oxidation states (Pb^{2+}), and the compounds are ionic in character. In the numerous compounds

Table 1. Chemistry of lead (Pb).

	Atomic no. 82		Atomic weight 207.2
Natural isotopes		204, 206, 207, 208, 210	
Melting point		327.5°C	
Boiling point		1750°C	
Density		11.3 g/cm³	
Abundance in crustal rock 10 ppm	Seawater 0.03 μg/dm³	Solar system 2.9 atoms/10^6 silicon atoms	Universe 0.47 atoms/10^6 silicon atoms
Ionic states	Pb^{2+}		Pb^{4+}
Ionic radius	1.21 Å		0.84 Å
Example of lead compounds	Lead acetate $Pb(CH_3COO)_2$		lead tetraethyl $Pb(C_2H_5)_4$

where the metal is in the higher oxidation state, the metal forms covalent bonds. The Pb^{4+} ion has no real existence.

The element forms stable oxides where the metal is in either lower oxidation state (lead oxide, PbO, commonly known as lithage) or the higher oxidation state (lead dioxide PbO_2) or both (lead sesquioxide, Pb_2O_3, and triplumbic tetroxide, Pb_3O_4, commonly known as red lead). These last two may be considered as mixtures of PbO and PbO_2.

Lead Poisoning (Known in the Past as Plumbism or Saturnism)

Clinical features

The features of lead poisoning depend on a number of factors such as the amount of lead ingested or absorbed, the mode of entry, the speed of intake, the nature of the lead compound and the age of the subject. Thus, each affected individual tends to exhibit a unique spectrum of the various characteristic features of lead poisoning. Acute lead poisoning in the adult as a result of oral ingestion of high levels of divalent lead salts causes gross irritation of the gastric mucosa, resulting in vomiting and diarrhoea, with severe water and electrolyte loss via these routes, going on to collapse, coma and death. In the case of chronic poisoning, the symptoms are often very non-specific and the cause of illness can be attributed to other agents. Features in adults include anorexia, a metallic taste in the mouth, constipation, cramping abdominal pains, peripheral nerve palsies, motor enervation (particularly of the extensor muscles) and a blue line on the gums' margins (Burton's line).

On more detailed examination one can find anaemia and punctate basophilia, heightened bone density (particularly at metaphyseal zones) and raised intracranial pressure. This last will give rise to the coma and convulsions seen more frequently in children. In children the brain appears more severely affected than in adults, where characteristic features of the encephalopathy are headache, lassitude, irritability. Although the immature brain is particularly susceptible to lead, encephalopathy presents in adults also. Chemically, the form of lead that gives rise to these conditions usually is the divalent metal ion — Pb^{2+}. This is true even when the metal itself is initially present, e.g. as shot gun pellets, because bodily fluids leach out the metal and convert it to the divalent ion.

The two most frequent routes of entry to the body for lead are via the gastrointestinal system (soluble lead salts) and via the lungs (particulate lead oxides and salts). Of the two, the lungs are more efficient at taking up lead than the gastrointestinal system. Approximately 50% of lead entering the lungs is thought to be taken up, whereas the figure is only in the region of 25% for the gastrointestinal tract. Lead compounds are also absorbed through the skin. Volatile off-gassing during the lead smelting process gave rise to complex mixtures — lead oxides, chloride, phosphate and sulphate. In a recent study of chemical speciation of lead dust in the lead refining industry, both particle size and chemical composition were shown to affect to extent of absorption by the pulmonary system. Particles greater than 3 μm were deposited in the nasopharyngeal and bronchial regions then transferred to the oesophagus by ciliary transport and swallowed, whereas particles less than 3 μm tended to be deposited in alveolar regions where they were completely absorbed. Particles of lead sulphide, generated primarily in handling the crude ore, were less readily absorbed and less toxic than those of lead oxide or lead carbonate, reflecting their relative solubility in weak acids.

Lead tetraethyl, which until recently was the one readily accessible tetravalent lead compound in the UK, because of its volatility enters almost wholly via the lungs. The efficiency of uptake is higher than that of other lead compounds. Its lipophilicity allows it to cross the blood-brain barrier and gain rapid access to the brain. Thus, the different chemistry, and the route, speed and efficiency of access for this compound are thought to be the reasons why mental confusion is the prominent feature of poisoning with this agent. Because the other components of leaded petrol are themselves noxious, the clinical features seen in petrol sniffers who have overdosed may be attributed to these agents rather than the lead tetraethyl.

From the symptoms described above, it can be seen that lead enters into a number of systems and organs in the body. These include endocrine, haematopoietic, immune, and nervous systems and affects bone, the kidneys, liver, gastrointestinal tract and brain. Although the ion enters chiefly into calcium pathways, hence its deposition in bone, it also takes part in the pathways and interferes with the function of other cations, particularly the ferrous ion, and with compounds containing free thiol groups.

Calcium metabolism and bone

The structural component of bone consists of calcium phosphate and the protein collagen. As mentioned earlier, the Pb^{2+} ion has considerable similarity to the calcium ion, and easily takes the place of calcium in bone's structural matrix. Thus bone may contain up to 95% of the body's total load lead burden. At the menopause, the cortical bone of the average American woman has been estimated to contain 12 µg/g of mineral, and this figure is somewhat higher for trabecular bone. Whilst in bone, lead is thought to be relatively innocuous. Nevertheless, through the actions of osteoblasts and osteoclasts bone is formed and resolved throughout life. This remodelling process means that bone calcium is not static but is continually turning over, consequently releasing any trapped lead into the circulation, whence it may enter into other metabolic pathways, causing adverse effects. Release of previously accumulated lead from bone contributes between 40% to 70% of the blood lead concentration. The half-life for lead in bone is long. For trabecular bone it is greater than one year, and for cortical bone it is between five and ten years. This concomitant release of lead from bone may be particularly important at and after the menopause, when the reduced oestrogen levels cause bone calcium levels to decline. Women on hormone replacement therapy have higher bone lead levels than women not receiving therapy, mirroring the effects on calcium. A second situation where the calcium turnover of adult bone is high is in pregnancy. Lead in the maternal circulation readily crosses the placenta and thus is available to be laid down in rapidly forming foetal bone and to affect other processes such as neurogenesis. Although it is difficult to quantify precisely the effects of lead released in this way on the developing child or the mature adult, all the known effects of lead are deleterious. Thus it can be seen that lead acquired early in life constitutes ongoing risks, particularly for females.

Heme and iron metabolism

Iron absorbed from food is transported in the bloodstream by the protein, transferrin, and heme and haemoglobin from the breakdown of red cells are scavenged and transported by haemopexin and haptoglobin, respectively. Iron is stored within tissues by the protein, ferritin. The primary requirement of iron is the synthesis of heme proteins, such as haemoglobulin, myoglobin (accounting, for approximately 85% of normal iron metabolism) and cytochrome P450 monooxygenases. The red cells of the bloodstream are synthesized in the bone marrow. Under the influence of erythropoietin, erythrocyte progenitor cells are converted into mature erythrocytes and released into the bloodstream. Erythropoietin itself is synthesised as a 39 kDa glycoprotein in the peritubular cells of the kidney. As oxygen levels in the blood decrease, erythropoietin is released from the kidney to stimulate increased heme synthesis and red cell release to increase the blood's oxygen carrying capacity. The lead Pb^{2+} ion enters into haemoglobin pathways. It is a highly potent inhibitor of the second enzyme in the synthesis of heme from from glycine and succinyl CoA, δ-aminolevulinate dehydrase, although it also inhibits δ-aminolevulinate synthase and ferrochetalase. The extent of inhibition of red cell δ-aminolevulinate dehydrase is used as an indicator of the body's lead burden in factory workers suspected of excessive exposure to the metal. However, despite its sensitivity to lead the enzyme is not a good indicator at levels below 400 µg/l.

Lead not only inhibits the synthesis of heme but also that of transferrin. The mechanisms whereby lead is capable of inducing anaemia have not been fully elucidated and appear to be complex. Erythropoietin production increases in response to decreased red cell survival times in lead-exposed individuals even when anaemia is not clinically apparent, but in pregnant women, serum erythropoietin concentration varies as the inverse of the lead concentration. However, ineffective erythropoiesis and a lower red cell half-life do not fully explain lead's effects.

Actions on the kidney

The kidney has two general functions. It filters into the urine waste products in the blood (e.g. creatinine formed by muscle metabolism and urea formed from the breakdown of proteins), and it maintains the hydrodynamic balance of the body by regulating the resorption of water, electrolytes (particularly sodium and chloride ions), glucose and amino acids. The principal functional component of the kidney is the nephron. The glomerulus forms a semi-permeable filter which stops all but very low molecular weight blood proteins entering the urine. Resorption of amino acids and glucose and approximately 85% of sodium ions passing through the glomerulus occurs in the proximal convoluted tubule. The remainder of sodium ion resorption occurs in the distal convoluted tubule. Apart from helping to conserve the body's water, minerals and energy, the function of resorption is to control the osmotic balance between the fluid within cells and that outside them. As the osmotic pressure and sodium ion concentration of blood changes, two hormone systems come into play. Arginine vasopressin regulates water resorption and the renin-angiotensin-aldosterone system acts on the distal convoluted tubule to fine-tune sodium resorption. In a situation where the sodium ion concentration of blood was abnormally high (e.g. because more water than salt is being lost as a result of severe burns), blood pressure would fall and the resorption systems would respond to try and redress this. Thus blood pressure and renal function are intimately related.

Lead is a nephrotoxin, so in lead poisoning renal function is usually severely compromised and hypertension is a cardinal feature. Perinatal exposure to lead can give rise to hypertension in later life. When large groups of patients with relatively low lead burdens are studied, a relationship between serum lead concentration and blood pressure can be seen, although details of the biochemical mechanism by which lead has this effect has not been fully elucidated. In acute lead poisoning proximal tubular function is most affected. The proximal tubule is where most sodium is resorbed. If resorption is impaired, it impacts on blood pressure. The damage to this site also gives rise to glycosuria (loss of glucose into the urine) and amino aciduria (loss of amino acids into the urine). In low dose, chronic poisoning glomerular function is also impaired and tubulointerstitial nephritis occurs. In this condition, low molecular serum proteins and lysosomal enzymes leak into the urine. Another consequence of the poor renal function in chronic lead toxicity is reduced clearance of uric acid (formed from urea), which is deposited as crystals in joints and so gives rise to gout.

The immune system

One of lead's less well understood effects is its involvement with the human immune system. Immune defence occurs via two routes — the humoral immune system involving antibodies synthesised by a type of white cell circulating in the blood (B-lymphocytes), and the cellular immune system involving the action of lymphocytes derived from the thymus gland (T-lymphocytes) and other cells such as macrophages which digest foreign agents. Antibodies are immunoglobulin molecules (five general types termed IgG, IgA, IgM, IgD and IgE), formed by B lymphocytes, which bind to infective agents and clear these from the circulation with the help of a second class of serum proteins called complement (C) components. Lead-exposed workers have been found to have reduced serum concentrations of IgG and IgM and complement components C3 and C4.

The cellular immune system is composed of a number of different forms of T-lymphocytes, two of the principle types being termed T-helper cells and T-killer cells. As the names imply, following viral infection, for example, T-helper cells prime T-killer cells to initiate the process of destruction of infected cells. In this process numerous chemical messengers, termed cytokines, are released to direct and control events. Lead exposure reduces the number of T-helper cells and the serum concentrations of the cytokines interleukin-10 and tumour necrosis factor-alpha (TNF-α). Not only are concentrations of important components of the cellular and humoral immune systems reduced in the unstressed situation, but the functional ability to respond once an insult has occurred is impaired. In a model system where lead oxide particles were introduced into rabbit lungs, which may be related in human terms to production workers exposed to lead fumes and to the general public inhaling combusted leaded petrol fumes, the lead oxide particles were rapidly absorbed. As a consequence, the pulmonary alveolar macrophages contained less TNF-α and were less able to increase TNF-α synthesis in response to a bacterial attack. Thus lead would appear to compromise one of the first lines of defence along the route by which many infectious agents enter the human body. Workers exposed to high levels of lead have been claimed to have a higher incidence of infectious diseases such as influenza, although no large-scale epidemiological studies have been carried out. Because most work has concentrated on the more obviously affected systems of the body, it is not clear precisely what immunological effects lead toxicity has for humans. From animal and *in vitro* cell culture studies it is clear that lead has the potential to interfere with the immune system at numerous points. This interference could not only affect the ability to respond to infectious attack, but also, because cytokines are important signalling molecules in other circumstances, it could reduce the ability to inhibit carcinogenesis and to cope with non-infectious trauma.

Effects on the CNS, "asymptomatic" lead poisoning and lead tetraethyl

The one property that has allowed lead to gain access to most humans is that of lead tetraethyl's ability to improve the running of internal combustion engines, when it is used as an additive in petrol (gasoline). After much debate and legislation, unleaded fuels are the only ones permitted in Western societies, and as a consequence there has been a progressive and continuing decline in the blood levels of children in recent years. The removal of lead from petrol represents one of the most significant events in environmental legislation. This is because the effects of lead re-leased into the atmosphere were not immediately obvious, giving rise to the concept of asymptomatic poisoning.

Causes Of Lead Poisoning

The properties of lead still make it an indispensable material for modern life. An example of this is its resistance to acidic corrosion. Despite much research, the metal still remains the basis of the battery used in all types of automotive vehicles. Its malleability and corrosion resistance are important features in its use as a waterproof roofing agent in the building trade. Its low melting point makes it still the basis of solders used in plumbing. Wherever lead is used the potential exists for lead poisoning. Lead has no natural use in the human body, thus its presence always represents a threat and debate still goes on as to what should be considered a "safe" blood level. Nevertheless, current best practice, involving regular monitoring workers' lead blood levels, in mining and industry should remove the danger of chronic lead poisoning to a very large extent.

Colour is an important property of several lead compounds. As mentioned earlier, lead sulphide is almost black and the naturally occurring ore is lustrous. Lithage is bright yellow; basic lead carbonate is a dense white; the red of red lead is scarlet; lead dioxide is a rich chocolate brown. Several of these compounds, which have limited solubility, have been used in paints both as pigments and for their preservative properties. This use, taking advantage of these properties, has provided one of the most widespread sources whereby lead has gained entrance to the human body. Today, in some States of the US, topsoil replacement for land surrounding domestic building is a major means of reducing human exposure to lead, where the land has become contaminated from paint used in the past to preserve wooden houses. In the past, because most paints were lead-based and modern health and safety consciousness had not been developed, painters often developed chronic lead poisoning, which was known as "painters' palsy".

Lead in some ways can be recognised as a triumph for modern science, because its toxic properties have been recognised and appropriate actions have been taken to either eliminate the danger by eliminating the use where possible, and where not, to keep the danger in check. However, merely recognising the dangers has been itself insufficient to ensure safety. The ancient world knew the dangers of lead, but disregarded them. Roman authors railed against the pollution of wine with sapa, but the cause was ignored and the knowledge lost. In the Middle Ages, outbreaks were regarded as new diseases. Not until Eberhard Gockel, physician to the city of Ulm in the late 17th century was there any test to detect the lead, by precipitating the sulphate following addition of sulphuric acid.

Thus, the question posed is
"Is Lead Really Still a Problem Today?"

The answer appears to be yes, for the very same reasons they appertained in the past. It is incredibly useful. The following extract is a testament to another property, that of density.

Scottish Parliament. Environment and Rural Development Committee. Wednesday 29 September 2004

The Deputy Minister for Environment and Rural Development (Allan Wilson): "I shall be as brief and factual as possible in the circumstances. The simple aim of the regulations is to prevent lead shot from causing pollution of the environment, which is fully consistent with our previous discussion on the water framework directive. I know from some of the views that were expressed in our extensive consultation on banning lead shot that there are people out there who do not believe that lead is a poison. However, there have been decades of research to that effect. Lead is a very harmful substance both to land and to wildlife, which is why we also take steps to reduce its use in other areas of life".

Apart from these wide-ranging examples of the dangers posed by lead, humans, being ingenious creatures have developed a few new ways of introducing it to themselves, or being ignorant or careless have re-instituted some old ways. These are listed in Table 2.

Treatment of Lead Poisoning

As with any form of intoxication, prevention is better than cure and the dangers of lead exposure should receive the public attention it deserves. Particular emphasis should be placed on simple measures which can be employed to reduce exposure to lead dust like wet cleaning and hand washing. It goes without saying that avoiding lead-containing sources should also be encouraged. Such measures are of particular importance in households with children as it is indeed the developing body which is especially vulnerable to the long term effects of lead toxicity.

Chronic lead poisoning tends to be more prevalent in urban children from lower socio-economic backgrounds. Nutritional supplementation in these children is of particular value in combatting lead poisoning as fortification of the diet with calcium and iron has been shown to reduce the absorption of lead. Calcium supplementation may have the additional benefit of preventing the release of lead from bone which accompanies calcium deficiency. Unfortunately however, providing families with such information may have only a short term role in preventing lead poisoning as completely lead-free housing is the only definitive prevention measure. It is not only the public however that need to be educated, effective treatment requires that the physician maintains a high index of suspicion.

The definitive test for establishing lead toxicity is an assay that measures the amount of lead in the blood in micrograms per decilitre (mcg/dl). This is the most universally accepted marker for lead exposure. Further tests include a blood film (looking for punctate basophilia) and X-ray fluorescence which helps to quantify the total body burden. Needless to say, complete

Table 2. Some interesting ways of poisoning oneself and others with lead reported in recent literature.

The Bradford Battery Burners
Bradford is a cold, windy place. Winters in the 1960s were quite severe. Poor families found a source of "free" domestic fuel — old car battery cases. Discarded battery cases dumped invitingly by a gap in the fence of a car wrecker's yard. The cases burned well, but smokily. The outcome was two dead children and around 500 people intensively investigated for lead poisoning.

Painter's Palsy Revisited
(a) Art restorer using ancient pigment — cinnabar (HgS) and other pigments, probably licking the tip of the paint brush.
(b) Hobbyists stripping lead paint with blow torch.

Nipple Shield
An unusual case of severe lead poisoning in a breast-feeding infant girl; the source — a nipple shield made of a lead-containing metal. Despite the severity of the intoxication, by the end of treatment and for a year afterwards the infant has been well and her psychokinetic development has been normal.

Pottery Glaze
Lead eluted from pottery glaze by acids
(a) in fruit punch
(b) in tea

Alternative Medicines (In these cases, STOP taking the pills!)
(a) effects in Omani children
(b) some treatments for multiple sclerosis
(c) some Indian herbal medicines

Bullets Really Are Bad For You
Retained lead bullets dissolve in body fluids.

Chocolate?
Daily Express (31 December 2005)
Headline: "Too much chocolate may harm pregnant mums and children".
Lead content in some cocoa powder 1.9 µg/g; upper safe limit 1.0 µg/g; 20 g chocolate = 12% of safe weekly intake.
Response of Biscuit Cake, Chocolate, Confectionary Association spokesperson — "It's physically impossible to eat enough chocolate for the naturally occurring (lead) levels to be harmful".

Kohl Revisited
A hazardous eyeliner

A Wobbly Millstone
Lead leaking into flour milled with a wobbly millstone, the shaft of which was packed with lead filings.

---

assessment of a patient comprises a full history exploring both the subjects' exposure to lead sources and a thorough enquiry of systemic symptoms (a neuropsychological and behavioural history is important in children). A physical examination may also reveal blue discolouration of the gums (Burton's lines).

Treatment of lead poisoning depends on the blood lead level. Naturally optimal blood levels are zero but less than 10 mcg/dl is considered normal. This is misleading however as even at these seemingly low levels, symptoms of toxicity may still be observed. These include lowered IQ and hyperactivity amongst various other neurophsychological and behavioural problems. Therefore it can be said that there is no true toxic threshold for lead and the full clinical picture needs to be considered.

Blood lead levels of 10–19 mcg/dl indicates mild lead poisoning which may still cause impaired cognitive development in children. BLLs within this range are usually managed by monitoring which entails repeated testing and physical examination on a three- to four-monthly basis. The source of lead needs to be identified and the steps to reduce lead exposure (discussed above) need to be instigated.

BLLs of 20–44 mcg/dl indicate moderate lead poisoning, in addition to preventative measures, this group may warrant treatment with chelation therapy. BLLs above 45 mcg/dl are considered to be severe lead poisoning, and is likely to cause GI symptoms in adults and children. Chelation therapy should be commenced in these patients. Finally, levels above 69 mcg/dl is a medical emergency with high risk of acute CNS symptoms. It warrants chelation therapy and the patient cannot be released from hospital until safe lead free environment is ensured.

The chelation therapy mentioned above works by aiming to reduce BLLs by introducing agents such as DMSA (dimercaptosuc-cinic acid) and EDTA (calcium disodium) which bind to lead and thereby encourage its excretion in the urine. DMSA is particularly useful in younger children as it can be given by mouth, it is not without side effects however which include anorexia, nausea, vomiting and rashes. It is more commonly indicated in patients with BLLs between 45–69 mcg/dl. However at levels above 69 mcg/dl, EDTA is more commonly used, this needs to be given as a continuous infusion. Animal studies have suggested that DMSA is more effective for reducing levels of lead in the brain than EDTA, but that EDTA is more effective at chelating lead from the kidney and has been shown to prevent the progression of chronic renal disease.

It is generally accepted that BLLs above 45 mcg/dl require chelation therapy, but in patients with BLLs between 20–44 mcg/dl, this remains a source of contention. One study investigated long-term neuropsychological changes in children who had been given treatment with DMSA. It showed that chelation therapy in this group produced no significant differences in cognition, behaviour or neuropsychological function when compared to a placebo group. Repeated assessment of the same patients aged seven showed no long-lasting benefit in the same parameters. This may however be a reflection of the duration of DMSA treatment. Therefore, treatment of patients with chelation therapy whose BLLs fall within this range remain at the physicians discretion, particularly as chelation treatment is not without risk. The chemical-lead combination can cause proximal tubular damage of the kidneys, however maintaining adequate hydration should prevent this. Additionally, if chelation therapy is ceased prematurely the blood levels can rebound due to the movement of the metal out of bony stores.

The long-term prognosis of lead toxicity is predictably dependent on the degree of toxicity and the age of the patient. Adults with mild lead poisoning can make a full recovery without long-term problems. However in children, even mild lead poisoning can have a permanent impact particularly on attention and IQ. Naturally higher lead levels pose greater risks of the long-term health problems discussed above. Indeed the effects may take months to years to recover completely.

Suggested Further Reading

Warrell D, Cox TM, Firth JD and Benz EJ (2004). *Oxford Textbook of Medicine.* Oxford University Press.

Alliance for Healthy Homes (2005). *Building Blocks for Primary Prevention: Protecting Children from Lead-Based Paint Hazards.* Centers for Disease Control and Prevention, National Center for Environmental Health.

Some Internet Resources

http://www.keepkidshealthy.com/welcome/lead/leadtreatment.html
http://www.calpoison.org/public/lead.htm1
http://www.nlm.nih.gov/medlineplus/ency/article/002473.htm
http://www.patient.co.uk/showdoc/40000375

10

MERCURY

S. Aldred and R. H. Waring

Alone of all the metals, mercury is liquid at room temperatures and readily volatilises to mercury vapour. Its symbol, Hg, is derived from the Greek word "hydrargyros" meaning "watersilver"; the older English "quicksilver" conveys much the same meaning. The heavy (one teaspoon weighs ~ 70 g) silvery metal easily trickles out when the ores are heated, a process which has been known for some thousands of years as metallic mercury was found in Egyptian tombs dating from 1500 BC. Mercury can be found as the free metal, as inorganic salts and as organic mercury, usually methyl mercury. The free metal occurs naturally in the environment, with about 30,000 tons of mercury being released annually into the atmosphere by degassing from the Earth's oceans and crust. A further eight to 10,000 tons are released into the atmosphere by human activities such as burning household and industrial wastes and fossil fuels such as coal which can contain up to 1 ppm of mercury. Estimates from the Great Lakes basin (US/Canada) suggest that about 5000 tons of mercury are released from these sources in the North American sub-continent. The various forms of mercury are interchangeable. The free metal can be converted to its salts (inorganic) or alkyl or aryl derivatives (organic) by industrial processes but can also be oxidised to inorganic forms in the environment or in the body. Inorganic mercury can be reduced to the free metal in anaerobic or reducing conditions. Anaerobic bacteria can methylate both the free metal and its inorganic salts to methyl and dimethyl mercury (usually combined as MM) and this relatively lipid-soluble form can be taken up and enter the food chain before being converted to inorganic mercury by metabolism in the body. Oxidation/reduction cycles therefore interconvert the various forms, selective toxicity being due to specific uptake into the tissues.

Elemental Mercury

Environmental sources

Metallic (elemental) mercury is widely used in industry and is found in mercury switches and relays. These are components in chest freezers, sump pumps, irons and washing machines (they stop spin cycles or turn on lights and are used in motion-sensitive or position-sensitive safety switches). Mercury-containing relays activate airbags and anti-lock brakes. Mercury is also found in thermostats, thermometers (the small domestic variety contain ~ 0.5 g mercury) and barometers. Mercury-containing manometers, vacuum gauges and semi-conductors are common as are thermostat probes which are part of the safety valve system that prevents gas flow if the pilot light is not lit. Mercury is also a component of mercury vapour lamps, some neon lamps and metal/halide lamps. Industrially, the metal is used in batteries, electroplating, chloralkali production and silver and gold extraction. Many people have mercury amalgams as "fillings" in teeth; as mercury is volatile, the vapour can be detected in minute amounts downwind of crematoria when bodies are incinerated. Mercury and its salts have been used in medicines for centuries, usually as diuretics, antibacterial agents, antiseptics and laxatives. They were the major components in antisyphilitic agents in the 18th century and valuable as the only effective medication available at that time. Metallic mercury is also an ingredient of some folk medicines and practices. Sold under the name of "azogue", metallic mercury is used in "Esperitismo", "Santeria" (spiritual belief systems in Puerto Rico and Cuba, respectively) and in voodoo. The use of "azogue" is typically recommended by spiritualist practitioners and the metal is carried in a sealed pouch, sprinkled in the home or placed on devotional candles.

Clinical aspects

Generally, there is relatively little absorption of the free metallic mercury from the gastrointestinal tract, so that children who break thermometers placed in their mouths and then swallow the mercury are not at any real risk except from fragments of glass. Indeed, an oral dose of 200 g caused no adverse health effects in a child, while in the 18th century drinking 50 ml (~680 g) of liquid mercury was recommended as a laxative. However, digestion of 220 ml (~3.0 kg) has been reported to cause symptoms of immediate tremor, irritability, forgetfulness and fatigue. Absorption from the gut is thought to be ~0.01% and to be related to the presence of mercury vapour in the gastrointestinal tract.

Although absorption across the gut rarely leads to problems, mercury vapour is toxic and easily absorbed when it is inhaled. About 80% of volatile mercury is absorbed across the lungs in humans and cases of poisoning readily arise when mercury enters the body by this route. Acute exposure to inhaled elemental mercury can cause bronchitis and pneumonia as the metal damages the fragile lining of the lungs. Initial signs and symptoms include fever, chills, shortness of breath, a metallic taste in the mouth and can be followed by lethargy, confusion and vomiting, together with further long-term complications of lung damage such as emphysema and lung oedema.

Chronic and intense acute exposure cause skin and neurological symptoms. Chronic toxicity gives a classic picture with three main components. These are tremors, gingivitis (gum inflammation) and erethism, which includes neuropsychiatric findings of insomnia, memory loss, anorexia, perspiration and blushing. Ataxia (a staggering gait) with headache, tunnel vision, numbness and tingling in fingers and toes and salivation can also occur.

Allergic skin reactions sometimes develop following contact with mercury. The elemental liquid metal and the vapour can both be absorbed through the skin (~15%) so contributing to toxicity. Sensitisation to contact leads to outbreaks of dermatitis with redness, itching, rash and swelling which can spread over the body. This can occur in people occupationally exposed to mercury, such as dental assistants and employees in mercury recycling plants. Breathing metallic mercury for four to eight hours at concentrations of 1–44 mg/cubic metre has been reported to result in chest pains, cough, difficulties in breathing and pneumonia. The lowest levels reported as giving toxic effects are 0.15 mg/metre^3 for 46 days; exposure to <0.02 mg Hg/metre^3 for 8 hours/day and 44 hours per week for 23 months did not produce any signs or symptoms of toxicity. Short-term exposure to high levels of mercury appears to be more damaging than long-term exposure to lower levels.

Workers in the chloralkali industry were formerly occupationally exposed to mercury vapour. An increased frequency of intention tremors (time weighted average (TWA) of 0.026 mg/m^3) was found in workers who inhaled mercury vapour for an average of 26 years. Neurobehavioural effects (motor speed, visual scanning, co-ordination and concentration) were affected in individuals with a TWA of 0.014 mg/m^3 while workers exposed to an average of 0.033 mg/m^3 for at least two years also had a higher level of fatigue and tunnel vision.

Low level chronic exposure to mercury vapour can also affect the peripheral nervous system, leading to polyneuropathy (reduced sensory and motor nerve function) and neuropsychological symptoms of stress and behaviour problems. Longer exposures, around 15 years, have been shown in several studies to lead to alterations in pulse rate, blood pressure, memory, sleep disturbance and EEGs, probably as a result of kidney and CNS (central nervous system) dysfunction.

Once inhaled, mercury vapour is relatively lipid soluble and crosses the lung alveoli into the bloodstream and red blood cells. Here, it is largely converted to an inorganic divalent (mercuric, Hg^{2+}) form by the enzyme catalase which is present in the red blood cells. However, small amounts of non-oxidised elemental mercury continue to persist and readily cross the blood/brain barrier into the central nervous system, where toxic effects are caused. These may reflect the subsequent conversion of mercury into ionised forms which are then trapped in the tissues. The characteristic symptoms are of tremor, especially in activities requiring fine control; insomnia and nervousness then appear. As exposure increases, the frequency and magnitude of muscle tremors also increase and this is accompanied by personality and behavioural changes which can include memory loss, excitability, depression and hallucinations, combined with fatigue. These effects correlate with mercury tissue levels of 20–100 µg/gram and probably indicate an occupational exposure to mercury levels of >0.1 mg Hg/metre^3. These neurobehavioural and motor function effects are very long-lasting and have been noted to persist in ex-mercury miners as much as ten years after the exposure had ceased. There may be teratogenic effects; these do not appear to have been reported in humans but exposure of pregnant rats to mercury vapour (0.5 mg/m^3) resulted in an increase in resorptions and also an increase in mortality and congenital defects in the offspring. Studies on women chronically exposed to metallic mercury vapour have been reported as showing a higher frequency of menstrual disturbances, spontaneous abortions and complications of pregnancy. As mercury vapour does not alter endocrine function, it is possible that these effects result from direct damage to reproductive tissues. All forms of mercury cross the placenta to the foetus; at least in rats the foetal uptake of elemental mercury is ten to 40 times higher than that after exposure to inorganic salts, probably as a result of the greater lipid solubility.

Mercury vapour from dental amalgams has been identified as a source of exposure in the population but it is controversial whether this has any clinical consequences. It is possible that a sub-set of the population with many amalgam-type fillings and a high individual susceptibility, may show dysfunction attributable to mercury. Some studies have shown that urinary excretion of mercury has an approximate correlation with numbers of amalgam fillings, but it is uncertain whether this is relevant to human health.

Case History 1
Gold mining in the Amazon basin

Mercury combines with both silver and gold to form amalgams which are liquid at room temperature. Consequently, mercury has been used to extract gold from crushed rock containing the precious metal. The mercury/gold amalgam is collected then heated to remove the mercury which vapourises, leaving pure gold as a residue. As the mercury vapour is toxic and readily inhaled both by the gold refiners and by people in the surrounding locality, classic toxicity symptoms are common in the region.

Case History 2
Mercury exposure in a high school laboratory

On 8 December 1986, 22 students and their teacher in a Connecticut high school chemistry laboratory were carrying out an experiment. This was intended to demonstrate oxidation/ reduction of metals and called for silver oxide to be used. As this was unavailable, mercuric oxide was used instead but the exhaust hoods to remove fumes from the laboratory were not turned on. The students performed the experiment in pairs using 1.75 g mercuric oxide which they placed in a crucible and heated for 15 minutes to drive off the oxygen. When it was realised that the yield of metallic mercury was very low, and that the metal must therefore have vapourised, the experiment was stopped.

The maximum dose to each student was estimated at 9.3 mg, with high levels of mercury in the laboratory air due to vapourisation from the surfaces. When urine samples from the 23 people involved were analysed, on 11 December, eight had levels at or above 30 µg/litre, the maximum acceptable level. Tests on 20 January 1987 showed that six of these eight students still had urine mercury levels above 30 µg/litre and follow-up tests on the remaining 15 people in the class showed that in 14 cases the urine mercury level had increased from the original value, some being above 30 µg/litre, with one sample at 72 µg/litre. On 31 March 1987, although one student had a urinary mercury level of 37 µg/litre, the other 22 people involved remained at or below 30 µg/litre. As the half-life in tissues is up to 90 days, this represents a little over that time; four half-lives (~1 year) would be needed to get the levels back to background values.

Case History 3
Dental amalgam manufacture

Four adults, who were making dental amalgams, were acutely exposed to mercury vapour. Although the levels at the time of the poisoning were not known, the concentrations of mercury in the laboratory were 912 $\mu g/m^3$ when measured 11 days later. All four patients had initial symptoms of nausea, diarrhoea, breathlessness and chest pains. They also complained of a metallic taste in their mouths and produced large amounts of saliva. Later, their gums became sore (gingivitis) and they had tremors with increased excitability. Despite chelation therapy, all four people involved died between 11 and 24 days after their initial exposure with post-mortem blood levels ranging from 58–369 $\mu g/litre$.

Case History 4
Accidental contamination from mercury

Spills and contamination readily occur when children or students remove metallic mercury from laboratories or find mercury "dumped" illegally and take it home to play with. Such incidents usually lead to major clean-ups with large-scale evacuation of local people.

In August 1994, more than 500 people in Belle Glade, Florida were contaminated with metallic mercury after three children found 55 lbs (25 kg) of metallic mercury which had been dumped. The children took the mercury home to play with it and 20 families had to be evacuated while their homes were decontaminated.

Inorganic Mercury

Environmental sources

Inorganic salts of mercury (Hg^+, mercurous or Hg^{2+}, mercuric) have been used widely in industry including antisyphilitic agents, acetaldehyde production, cosmetics, disinfectants, embalming reagents, explosives, mercury vapour lamps, mirror silvering, photography, taxidermy and wood preservation. Possibly the best documented use occurred in the hatting industry. Mercuric nitrate solutions were employed to remove rabbit fur from the skins as part of the process of making felt for the manufacture of hats, essential articles of apparel for both men and women until relatively recently. Chronic exposure to dust or vapours of mercuric nitrate led to the "Mad Hatter" syndrome with its symptoms of tremor, excitability, shyness and nervousness. It is not surprising that the town of Macclesfield in Cheshire, once the centre of the hat-making trade in the UK, also had one of the biggest mental hospitals and that "hatters shakes" was well recognised locally.

Clinical aspects

Inorganic mercury salts are highly toxic and corrosive. If they are swallowed, about 10% is absorbed. They are poorly lipid-soluble and tend to accumulate in the kidney, causing major renal damage. Elimination via the urine is slow, partly due to the induced renal dysfunction. Although the penetration of mercury salts into the CNS across the blood/brain barrier is low, it may become significant on long-term exposure. Faecal excretion of mercury salts is a major route. This may be due to poor absorption if the salts have been ingested or possibly biliary excretion. It has been suggested that the nephrotoxic effects of mercury may be partly due to immunological effects.

Mercuric chloride (corrosive sublimate) was known in the Middle Ages and described as causing severe abdominal cramps, bloody diarrhoea and suppression of urine if taken orally. Mercuric salts have been a favourite agent for committing suicide and when solutions are drunk, they give corrosive ulceration of the intestines, with bleeding necrosis, followed by shock and circulatory collapse. If the victim survives this damage, renal failure occurs within 24 hours. This is due to damage and necrosis in the proximal tubular epithelium of the kidney. As this area controls absorption and excretion of ions in urine, the result is excretion of a high volume of very dilute urine (oliguria) or no urine at all (anuria). If the patient survives all this, some regeneration of the tubular lining of the kidney is possible, but residual scarring will occur. These changes are found in acute toxicity; some tubular damage and proteinuria (protein in urine) may also be seen in workers exposed to chronic low-level mercury salt exposure. This may be an immunological glomerular dysfunction which disappears when the workers are no longer exposed to mercury salts; usually nerve damage (neuropathy) is seen as well.

Inorganic mercury salts have a half-life in the body of about 40 days. Mercury binds to a wide range of enzyme systems and has a particular affinity for proteins containing sulphydryl (-SH) groups. As these are widespread, mercury may bind to microsomes and mitochondria, producing a non-specific cytotoxicity and cell death. In the kidney, mercuric mercury induces synthesis of metallothionein, a metal-binding protein, and is found localised in lyosomes.

Mercurous (Hg^+) salts of mercury are less corrosive and toxic, possibly because they are also less soluble. Calomel (HgCl) has been used in medicine for many years and more recently (in the 1940s and 1950s) as a "teething powder". It is known to be responsible for acrodynia or "pink disease". This was probably a hypersensitivity response to mercury salts on the skin. Affected children developed a pink rash and fever, with swelling of fingers, soles, spleen and lymph nodes and thickened skin. Irritability and insomnia were also, not surprisingly, a feature of the disease.

Case History 5
Poisoning from inorganic mercury — "Crema de Belleza"

Between September 1995 and May 1996, there were cases of mercury poisoning in Arizona, Texas, New Mexico and California. These were associated with a mercury-containing beauty cream called "Crema de Belleza", produced in Mexico, which contained calomel (HgCl) and when analysed had 6%–10% Hg by weight. The cream was advertised for use in skin cleansing and prevention of acne. In response to medical announcements, 238 people reported using the cream; analysis of urine samples from 119 of these individuals showed that 87% had mercury levels above 20 µg/litre, while 27 (26%) of the 104 samples had levels above 200 µg/litre, the highest value being 1170.3 µg/litre. It is of interest that some people in the same household who were close contacts of cream users but who had never used the product themselves, also had elevated mercury levels in urine. The son of one cream user had urine mercury of 50 µg/litre while a woman whose daughter had used the cream for 18 months had a urine mercury level of 31.6 µg/litre showing that contamination of the home environment by the non-biodegradeable mercurial compounds can represent an unexpected source of toxicity. Use of mercury-containing skin-care products may be relatively common in some parts of the world; ~2% of women at a New Mexico post-natal clinic used Crema de Belleza, while another skin-care product made in Mexico "Nutrapiel Cremanina Plus" contains 9.7% by weight of mercury and is widely available in the area.

Organic Mercury Compounds

Environmental sources

Organic mercurial compounds are found in antiseptics, bactericidal agents, embalming agents, fungicides, insecticides, seed and wood preservatives, in histology products and in paper manufacturing. The most important member of this group is methylmercury (MM) although other allyl and aryl mercury compounds are also available. Methyl mercury can be synthesised by methylation carried out by anerobic bacteria in river and lake beds; trace amounts may be formed *in vivo* by methyltransferace enzymes, which use S-adenosylmethionine as a methyl donor.

Clinical aspects

The lethal dose of organomercury compounds is in the range 10– 60 mg/kg body weight although acute toxicity is much less common than chronic damage due to long-term contamination by relatively small amounts of these materials. Organic mercury compounds are much more lipid soluble than the inorganic salts and therefore readily cross lipid membranes including the blood/brain barrier. Methyl mercury concentrates in the red blood cell membranes, so that the blood/plasma ratio is ~20:1. This can be useful in determining whether organic or inorganic mercury poisoning has occurred, as the blood/plasma ratio for inorganic mercury salts ranges from 2:1 on recent exposure to 1:1 after about a week. Aryl mercury compounds accumulate in red blood cells but as they are metabolised to inorganic mercury more rapidly, they show lower blood/plasma ratios than those seen with methyl mercury.

Methyl mercury exposure causes neurological effects when levels in the body are at or above 25–50 mg. High doses of methyl mercury can produce irreversible destruction of neurons in the visual cortex and cerebellum, leading to permanent narrowing of the visual field and to ataxia (staggering gait). The mechanisms of this toxicity are not fully understood; there is a delay between exposure and the appearance of toxicity which suggests that rapid therapy could reduce the symptoms. Mercury is known to inhibit protein synthesis in neuronal cells and cause inhibition of axonal transport of nutrients; this may be enough to cause irreversible damage to susceptible cells. Mercury also affects transmission at the neuromuscular junction by reducing the binding of acetylcholine to its receptor. Methyl mercury itself increases lipid peroxidation in the membranes of the cerebellar neurones and possibly inhibits mitochondrial electron transport. The developing cerebellar cortex appears to be particularly susceptible to methyl mercury which inhibits mitotic spindle microtubules and so alters cell migration. There is evidence that methyl mercury binds to tubulin, a protein component of microtubules; as cell division and cell migration both need intact, functional microtubules for normal development, this may explain why the brains of infants who died after *in utero* exposure to methyl mercury were found to have abnormal organisation of neurons in the cerebral cortex. Other cells are also affected; organic mercury compounds are teratogenic, producing malformations in the developing foetus particularly inducing cleft palate, heart defects and hydrocephalus.

The clearance of methyl mercury from the body is biphasic, with average half-lives of six to seven hours and 52 days. Six volunteers were given a single meal of fish containing 18–22 μg Hg/kg body weight.

The peak blood values of Hg were found between four and 14 hours after ingestion. The blood concentration of total mercury was as high as 60 μg/ml, although the inorganic mercury was not greater than 2 μg/ml. As might be expected, the average red blood cell/plasma concentration ratio was 21, with the hair/blood ratio being 292. Pharmacokinetic studies suggested that the methyl mercury was rapidly absorbed into the blood stream, then cleared and distributed to body tissues before slow excretion. After six months, the mercury levels in the six volunteers had returned to normal (almost zero).

Methyl mercury, like the other alkyl and aryl mercurials, can be converted to inorganic mercury. The aryl-Hg bond is more readily broken than the alkyl-Hg bond. This metabolism appears to take place in the smooth endoplasmic reticulum and may explain the renal damage and inorganic mercury found in the kidney after methyl mercury contamination. It is possible that some of the central nervous system (CNS) effects found with methyl mercury toxicity are due to its conversion to inorganic mercury in brain tissue. As methyl mercury is much more lipid-soluble than inorganic salts, the alkyl group may simply facilitate mercury uptake into the CNS. The characteristic symptoms of tremor, dementia, memory loss, irritability, excitation, sleep disorders and impaired peripheral vision all occur and are possibly due to formation of Hg-SH bonds. This must affect most proteins in all tissues, so the precise targets, if any, are not definable, but probably include enzyme inactivation and alteration of transport processes and structural proteins. Reduction in the activity of antioxidant proteins such as catalase, superoxide dismutase and glutathione peroxidase could lead to tissue damage from the free radicals and peroxides which are continuously formed *in vivo*, particularly in the central nervous system and are detoxified by these enzymes.

Accidental poisonings in Japan in the period 1950–1960 (Minamata and Niigata) and in Iraq in 1972 have given data on the health risks to human populations from methyl mercury. Parasthesia (numbness and tingling) is the first adverse effect to be seen and occurs at intakes of about 300 µg/day for an adult. Prenatal exposure, however, leads to damage at lower concentrations, as the developing foetus is more susceptible to the effects of methyl mercury (in rats, foetal uptake is ~30x higher for organomercurials than for inorganic mercury). Although affected infants may appear normal at birth, there is usually a 12-month delay in learning to walk and talk and an increased incidence of epileptiform seizures. These symptoms occur when the maternal exposure has been calculated to be in the range 800–1700 ng mercury/day/kg body weight, correlating with air mercury levels of 10–20 ppm and red blood cell levels of 40–80 µg/litre. Higher exposure to organic mercurials *in utero* leads to a syndrome similar in some respects to cerebral palsy, with an infant with ataxia, uncordinated movements and mental retardation. As methyl mercury is lipid-soluble, it can be transferred from mother to child via feeding with breast milk. This can therefore lead to further damage to the infant, although maternal milk contains only 5% of the mercury concentration of maternal blood. Adults who are contaminated by methyl mercury initially experience parasthesia, then ataxia accompanied by difficulty in swallowing and articulation. This is then followed by neurasthenic symptoms of weakness, fatigue and inability to concentrate. Loss of vision and hearing then occur, with spasticity and tremor and, finally, coma and death. Autopsy results show necrosis of neurons in the cerebrum and cerebellum, with cerebral oedema and loss of grey matter. When survivors of methyl mercury poisoning are examined, the neurotoxicity effects are generally irreversible although there is a slightly better prognosis for single incidences of poisoning with methyl mercury as compared with long-term exposure.

Organic Mercury in the Environment

Although children occasionally accidentally ingest methyl mercury (MM) preparations such as mercurochrome (a topical antiseptic with 0.2% MM as a aqueous solution) the main source of MM in the general population, at least on the North American continent, occurs from the consumption of fish. Blood levels of 20 µg/litre have been found in communities with a high fish intake, corresponding to 200 µg mercury/day. The FDA have set a limit of 1 ppm of mercury in fish for human consumption (1/10 of the lowest level associated with adverse effects). Non-exposed populations have Hg levels of 8 ppb in blood and 2 ppm in hair; the lowest level thought to give toxic effects in adults is 200 pbb in blood and 50 ppm in hair. Fish at the top of the food chain (shark, swordfish) have the highest mercury levels, and no more than 200 g (one serving) /month is recommended for pregnant women; the first trimester is thought to be the critical period. The levels in tuna, shrimps, salmon, cod, clams and crabs are usually below 0.2 ppm so that problems are unlikely to arise unless over a kilogram of fish is eaten weekly.

Studies in the Canadian Arctic have shown that mercury levels in aquatic mammals are higher in the Western Arctic, where the predominantly sedimentary rocks have relatively high mercury concentrations. On the eastern side of the Arctic Basin, the rocks are igneous or metamorphic, with low mercury levels. However, mercury values are rising, presumably due to man-made contamination (paper pulp manufacture), the liver of Western Arctic ringed seals contained 22.9 ± 28.7 µg/g net weight in 1972, but sampling in 1989–1993 gave average values of 32.9 ± 36.9 µg/g net weight. Levels in seals in the Eastern Arctic were about 1/3 of these figures.

Methyl Mercury Minamata Disease

Case histories − Minamata disease

(a) Minamata bay in Japan does not usually experience high tides. In 1953, a local factory involved in the manufacture of plastics was using mercuric chloride as a catalyst in the conversion of acetylene to acetaldehyde. The mercury-containing wastes were released into the land-locked bay, where they sank to the sea bed. Once there, they were methylated by the anaerobic micro-organisms living in the sludge and methyl mercury was formed. This, being fat-soluble, was incorporated into the food chain and finally contaminated the local fish. The people of Minamata ate the fish, a staple of their diet, and soon showed signs of methyl mercury poisoning. At least 52 people died almost immediately, while many more developed degenerative neurological problems, with symptoms such as paraesthesia, ataxia, and vision and hearing loss. The rate of cerebral palsy in the region is still (in 2000) high, at least 6% of births being so affected.

(b) In 1972, large quantities of wheat and barley treated with methyl mercury as a fungicide before planting were accidentally released as food quality grain in Iraq. Many villagers ground the grain and made it into bread, which they and their families ate. At least 450 people died of mercury poisoning and over 6500 people were treated in hospital. Methyl mercury concentrations in the flour ranged from 4.8 –14.6 µg and the chief clinical symptoms were paraesthesia, visual problems, and deafness and dysarthia. All the deaths occurred in patients with blood mercury levels greater than 3 mg/l.

(c) Minamata disease in the Amazon. Over the last ten years, a number of studies have reported mercury contamination of fish in the rivers of the Amazon basin, together with relatively high levels of mercury in the hair of fish-eating people living by the rivers. Mercury, released from gold-extraction processes (see case histories for elemental mercury) and from "slash and burn" techniques of cultivation of the local soils has contaminated the water supply. Methyl mercury is then formed in the anaerobic sludge on the river beds. Herbivorous fish contain very little mercury, but fish which eat other fish have higher levels. Although clinical manifestations of the full Minamata symptoms have not yet appeared, methyl mercury has still affected the health of the local (fish-eating) populations.

Motor and visual functions have been found to decrease with increasing mercury concentrations in hair, even at levels below 50 ppm, showing that toxic effects can be seen at relatively low levels of contamination.

Analysis of Mercury

The primary method used to analyse for mercury is CVAAS (cold vapour atomic absorption spectroscopy). This has a sensitivity of parts per trillion for urine while determination of mercury in fish, shellfish, pharmaceuticals and foodstuffs has a sensitivity in the low parts per billion range, as has the analysis of mercury in samples of tissues and hair. Mercury levels in newly formed hair reflect those in blood; the concentration however is about 250 times greater in hair. Once the mercury has interacted with the sulphur-containing proteins in hair it is essentially "fixed" and so provides an accurate monitor of exposure and accumulation. Analysis of the hair length can therefore be used to estimate the dates of contamination by mercury and also the peak blood levels achieved.

X-ray fluorescence (XRF) has been used to measure mercury in the wrist bones of dentists exposed to heavy metal contamination in the workplace. XRF allows simultaneous analysis of a number of different metals in the tissues and can measure mercury levels in the low ppm (parts per million) range. There is some evidence that bone levels of mercury are more useful in identifying chronic exposure to relatively small amounts of the metal.

GC/AFS (gas chromatography/atomic fluorescence spectrometry) can be used to separate individual mercury species so that the relative contributions of metallic, inorganic and organic mercury in tissues can be established.

Treatment of Mercury Poisoning

Metallic mercury is radio-opaque so that if it is swallowed, it can be visualized in the gastrointestinal tract.

Therapeutic agents

1. **BAL** (British Anti-Lewisite, 2,3-dimercapto-1-propanol, dimercaprol in an oil solution) should not be used for methyl (or organic) mercury toxicity as it can raise levels in the brain, so increasing neurotoxicity. Intramuscular injections of ~3 mg/kg are given every four to six hours. Its use is contra-indicated for patients with glucose-6-phosphate dehydrogenase deficiency or an allergy to peanuts (the excipient for the dimercaprol is peanut oil). The safety of BAL in pregnancy has not been established. Side effects include nausea, vomiting, abdominal pain, headache, lacrimation, conjunctivitis and burning sensations in lips and throat. BAL is not recommended for detoxication of organomercury compounds.

2. **D-Penicillamine** (ß,ß-dimethyl cysteine, D-pen, PCN) is used for oral medication. It forms a complex with mercury which is excreted in urine so cannot be used safely on patients with renal failure. Doses of 15–40 mg/kg/day can be given; the drug is unsafe in pregnancy and should be avoided in patients with a penicillin allergy. Its side effects include gastrointestinal problems, rashes, proteinurea and blood clotting problems. Like BAL, PCN is not used for treating organic mercury poisoning because it initially leads to higher blood levels of mercury.

3. **DMSA** (2,3-dimercaptosuccinic acid, "Succimer") is used to treat poisoning by both inorganic and organic mercurial compounds. It is currently the agent of choice, with fewer side effects than BAL or PCN, and should be used immediately if mercury poisoning is suspected as the laboratory confirmation may take as long as a week.

    The usual dose is 10 mg/kg orally three times a day for five days then 10 mg/kg orally twice a day for 14 days. The safety for use in pregnancy has not been established; side effects can include mild gastrointestinal disturbances and a transient rise in hepatic enzymes in plasma. Complications such as neutropenia which can arise in therapeutic use are reported to resolve when the therapy is finished.

4. **Resins**

   Polithiol and related compounds are non-absorbable resins used orally to improve the removal of methyl mercury and short-chain alkyl organic mercurials. These are secreted in the bile before excretion in faeces; use of polithiol prevents their re-absorption lower down the gastrointestinal tract.

5. **Other therapy**

   Haemodialysis can be used when renal function has declined after severe mercury toxicity. As mercury redistributes itself between red blood cells and plasma, haemodialysis is more effective if L-cysteine is added as a chelating agent, since the free thiol group combines with mercury and so removes Hg ions from the system.

   Desferrioxamine and glutathione have some protective effects against mercury-induced lipid peroxidation; vitamin E therapy has also been suggested as a possible protectant against membrane oxidation. Use of antioxidants in general may reduce some of the long-term chronic symptoms of mercury poisoning.

Table 1.

	Mercury metal	Mercury vapour	Inorganic mercury (Hg^+ and Hg^{2+})	Organic mercury
Biological half-life	35–90 days	35–90 days	~40 days	~70 days
Absorption	<0.1% from gut, 15% from skin	90–100% from lung, 15% from skin	~7% from gut	80–100% from gut
Major site of toxicity	Kidney, brain	Lungs, kidney, brain	Kidney, brain, GI tract	Brain, kidney, foetus
Major route of excretion in man	Urine	Urine	Urine	90% in faeces via bile

Prognosis

The outcome depends on the form of mercury compound which is the toxicant and on the severity of exposure. Mild exposure to elemental, inorganic and organic mercury can be followed by a complete recovery. Death may result from a severe exposure while neurological complications such as mood and behaviour alteration, memory loss and fine tremors often remain even when patients with moderately high exposure have been removed from the source of contamination.

Suggested Further Reading

Bates B (1998). Heavy metals and inorganic agents. In: *Clinical Management of Poisoning and Drug Overdose*, Vol. 55, pp. 750–756.
Young J (1994). Mercury. In: *Goldfrank's Toxicology Emergencies*, Vol. 74, pp. 1051–1062.

11

MUSHROOM TOXINS

R. H. Waring

Amanita Mushrooms — Toxic Cyclic Peptides

It was a pleasant autumn day when the man saw the mushrooms. They looked so like the ones he used to cook at home in Vietnam. Delighted, he collected all six and took them back for dinner. The mushrooms had a wide smooth cap, between 6–16 cm across and were a pale slightly greenish yellow colour, smelling of raw potatoes. Some of the smaller ones had a "veil" of tissue extending from the edge of the cap to the upper stalk. In the larger mushrooms, this thin white membrane was draped around the upper stalk, revealing the thin white gills, attached to a pale stalk which was up to 15 cm long and had a large rounded bulb at the base with frills of a white sac-like membrane.

The meal went well and the mushrooms made a tasty addition to the food. However, early the following morning, the man developed vomiting, abdominal pain and severe watery diarrhoea. His friends became worried and took him to the hospital where tests showed that his serum transaminase levels were elevated (AST, aspartate aminotransferase 84 U/h , normal 0–48 U/h; ALT, alanine aminotransferase 100 U/h, normal 0–53 U/h) as was the bilirubin level (0.8 mg/dl as opposed to 0–0.3 mg/dl). All this pointed to liver damage, although his prothrombin time (PT) of 12.2 seconds was within the normal range (11.0–12.8 seconds). Twenty-four hours after eating the mushrooms, the diarrhoea stopped and the patient felt sure that he was going to get better. Although he did not realise this, the PT, AST and ALT values were rising and on the third day after eating the mushrooms, the PT was greater than 70 seconds , the AST and ALT values were 4000 U/h and 2300 U/h, respectively while the bilirubin level was 10 mg/dl. Three days later, although the AST value was 350 U/h, the bilirubin level had risen to 17 mg/dl and the patient developed metabolic acidosis and low blood pressure. Next day, his skin was cold and he had hepatic encephalopathy and oliguric renal failure, finally dying from multiple organ failure nine days after the mushroom meal. He was a typical victim of eating *Amanita phalloides*, the Death Cap mushroom.

Ingestion of *A. phalloides* is probably responsible for ~90% of deaths worldwide from mushroom poisoning and this case history is typical, with a delayed onset of symptoms usually ~12 hours after eating the mushrooms, then a symptom-free period which may last up to five days before jaundice and often death. As little as one bite of these mushrooms can kill an adult as the median lethal dose is ~0.1–0.3 mg *A. phalloides*/kg body weight. The mushrooms typically weigh 20–25 g so that potentially ~1/3 of a single mushroom could be fatal to the average 70 kg man. They are more common in the autumn in damp weather and can be found in the UK, Europe and the cool coastal regions of the US, though not in South East Asia. Ecologically, *A. phalloides*, which was brought to the US on imported cork tree seedlings, is a beneficial mycorrhizal fungus, living on tree roots and providing nutrients to the tree. To humans, *A. phalloides* looks harmless and has no unpleasant taste or smell. Nevertheless, the toxins (amanitins, see Figure 1) cause liver necrosis and kidney damage and are not destroyed by cooking or drying. Other *Amanita* species are also toxic; the Destroying Angel (*A. virosa*) and *A. bisporigera* and *A.verna* similarly contain amatoxins as do some other gilled mushrooms, particularly *Conocybe filaris*, *Galerina autumnalis*, *G. venenata*, *Lepiota josserandii*, *L. chlorophyllum* and *L. helveola*. These, however, are less tempting and less likely to be eaten, although in 1986 and again in 1994, amanitin-containing *Lepiota* mushrooms poisoned five experienced amateur mushroom hunters in Long Island (New York), causing two deaths from fulminant hepatic failure. *A. virosa* is very similar to the Death Cap mushroom but usually has no green or yellow pigmentation, all

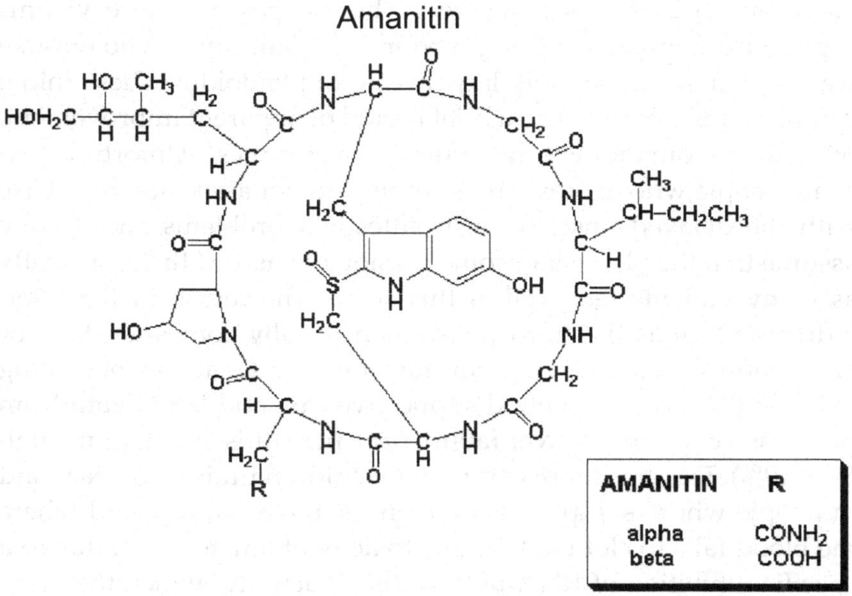

Amanitin

Figure 1. Amanitin, toxin of the Death Cap mushroom.

parts of the mushroom being pure white. The base of the stalk and the giveaway volva are often buried in the soil in both species of *Amanita*, making them more difficult to recognise. *Amanita* species are not found in South East Asia, so that many cases of amanitin poisoning occur in the US when immigrants from this area mistake these toxic species for the "Paddy straw" mushrooms (*Volvariella volvacea*), that they are used to collecting back home.

Amatoxins such as the amanitins are bicyclic octapeptides; several different isoforms are known, although the α- and β-amanitins are most common (Figure 1). They all have a molecular weight of ~900 and are thermostable so that they remain poisonous even after cooking. Phallotoxins and virotoxins (both cyclic heptapeptides) are also present and are toxic to rodents. They appear not to be absorbed and do not exert acute toxicity in man, although the phallotoxin phalloidin may reduce mucosal cell membrane integrity. Cyclopeptide toxicity occurs in three phases. Typically, about six to 24 hours (average 12 hours) passes before victims experience nausea, vomiting, abdominal pain and a cholera-like watery diarrhoea, possibly largely due to phalloidin. These initial symptoms are followed by a brief period of apparent improvement, with asymptomatic rises in serum liver enzymes. Unfortunately, some people who eat toxic mushrooms do not associate this phase with the delayed onset of gastrointestinal problems and merely assume that they have had some version of gastric "flu", especially as many patients feel well at this stage. The return of liver and kidney failure as the third phase then usually comes too late for any useful treatment. The mortality rate for amatoxin poisoning is 10%–60%, even with good supportive care and later death from progressive fulminant liver failure or pancreatitis is still quite high (5%–20%). Death is more common in children under ten years and in people whose symptoms come on relatively rapidly and where the blood fails to clot readily. The toxicity of amatoxins is due to a specific inhibition of RNA polymerase II activity which therefore inhibits DNA transcription and protein synthesis, leading to cell death. Lesions occur particularly in tissues with rapid protein turnover and transport functions, such as gut, liver and kidney proximal tubules.

Obviously, prompt treatment of A. phalloides poisoning is essential but it is not always easy to find out which mushrooms have been eaten, particularly after several days. If there are any specimens left, a mycologist should be able to provide identification. If a drop of liquid is squeezed out of the mushroom onto lignin-containing paper such as newspaper, then dried and a drop of concentrated hydrochloric acid placed on the spot, then the appearance of a blue colour (Meixner test) confirms the presence of amatoxins (or psilocin, see later).

Amatoxins are rapidly absorbed and have been detected in urine as early as two hours post-ingestion. It is rare to find amatoxins in plasma after 36 hours unless the patient has pre-existing kidney damage. Amatoxins do not appear to be metabolised and are bound to renal tissue then excreted over two to three days so that in the early stages the concentration of amatoxins is much higher in urine than in blood. Some studies have suggested an enterohepatic recycling of the toxins as they can be found in faeces several days after the initial dose and are known to bind to liver tissue as well as the kidney. Amanitin reaches the liver by a non-specific transport system and is filtered by the glomerulus and re-absorbed by the renal tubules, resulting in acute tubular necrosis.

Gastric lavage is only useful if performed within six hours of the dose of mushrooms. As most patients only arrive at hospital after the onset of symptoms, this is rarely helpful although multi-dose activated charcoal can prevent absorption of further toxic material from the stomach and bile. Symptomatic and supportive treatment is essential, with rehydration and correction of metabolic disturbances. Some studies recommend the use of silibinin (from silymarin, extracted from the milk thistle), high dose benzylpenicillin and aucubin, an iridoid glycoside of *Aucuba japonica*, all of which appear to reduce hepatocyte uptake of the amatoxins. Several centres have reported successfully using the Molecular Adsorbent Recirculating System (MARS) with albumin dialysis as an intermediate stage treatment. Liver transplantation has been tried in a number of patients and seems to be useful as a last resort for cases with a poor prognosis if they can be identified − criteria used include prothrombin time 20% greater than normal, serum creatinine > 1.4 mg/dL, serum bilirubin > 4.6 mg/dL and progressing hepatic encephalopathy. Amatoxin poisoning following ingestion of *Lepiota* mushrooms can, like *Amanita phalloides* consumption, also lead to liver failure and death. In a report on ten Spanish patients who had eaten *Lepiota* species, five had liver problems and two of these died while one later developed chronic active hepatitis. Of the eight survivors, five subsequently developed chronic polyneuropathies; several reports have suggested that amatoxins in fact produce whole-body toxicity and that chronic tissue damage, especially to cardiac muscle, is common in those individuals who do not succumb to the acute dose.

Cortinarius Species — Orellanine

The *Cortinarius* mushrooms contain orellanine and the Deadly Webcap (*C. rubellus*) and the Fool's Webcap (*C. orellanus*) account for most of the fatalities. Young mushrooms of this species often have a pale web-like tissue between the cap and the stem; this appears as a yellow ring on the stem or at the edge of the cap in older specimens. The cap itself is 3–7 cm across and usually rusty brown to orange, sometimes with a steeper and darker-coloured centre. These mushrooms are easily mistaken for funnel chanterelles which share the same habitat in pine forests.

Poisoning from these mushrooms usually begins with anorexia, headache, gastritis and chills within 36 to 48 hours of eating the meal. There is then an extremely long asymptomatic latent period of three to 14 days (mean 8.5 days in a meta-analysis of 245 patients from across Europe). This is followed by an intense burning thirst with excessive urination then a return of the nausea, headache, muscular pain and chills. In severe cases, patients experience spasms and loss of consciousness. Patients with *Cortinarius* poisoning may be mis-diagnosed as having glomerulonephritis or appendicitis if the significance of the delayed nephrotoxicity is not appreciated. Death occurs in ~ 15% of cases and is due to severe renal tubular necrosis and kidney failure, usually weeks after the initial poisoning; recovery in those who survive may take several months. On autopsy, as well as damaged kidneys, there is fatty degeneration of the liver and severe inflammation in the gastrointestinal tract.

Orellanine (Figure 2) has a bipyridyl structure with positively charged nitrogen atoms and is related to the herbicides paraquat and diquat. These too are renal toxins in man, probably because the quaternary nitrogen atoms initiate redox reactions *in vivo*, removing electrons from oxygen and generating highly toxic and reactive chemical species containing oxygen such as peroxides and superoxides. An orthosemiquinone radical metabolite has also been postulated.

Orellanine is a potent nephrotoxin; the intraperitoneal LD50 in mice is 15–20 mg/kg and the human lethal dose is thought to be 100–200 g fresh mushrooms as *C. orellanus* usually contains

Orellanine

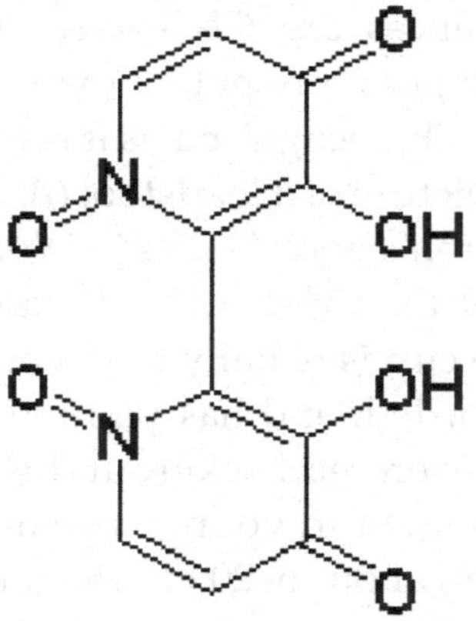

Figure 2. Orellanine, toxin of the Webcap mushrooms.

~14 µg toxin/g tissue. No specific treatment is available; most patients are given supportive care and dialysis for the interstitial nephritis. About 10% of patients require renal transplantation and most studies find that only 50% of patients are fully recovered five years later. Patients with a shorter time between eating the mushrooms and noticing toxic symptoms have a worse prognosis. Some people appear to be resistant to the toxin; in one study of 26 patients only 12 had acute renal failure while in another report of three people who shared a mushroom casserole of *C. speciosissimus*, two developed end-stage renal failure while the third had only mild symptoms. Females may be more resistant than males. Generally, the prognosis can be estimated by measuring the serum creatinine level before treatment and knowing the number of days since ingestion of the mushrooms. Using an empirical formula of serum creatinine (µmol/L) + 316 /number of days elapsed × 100, an index of > 2.1 is linked with a poor outcome and a requirement for kidney transplantation.

Amanita Species — Ibotenic Acid and Muscimol

Ibotenic acid–muscimol poisoning is produced when some *Amanita* species (Fly Agaric, *A. muscaria* and Panthercap, *A. pantherina*) are eaten. Both substances are CNS-active isoxazoles although muscimol is about six times more potent than ibotenic acid. The Fly Agaric mushroom is easily recognised as it is brightly coloured and looks like the popular picture of a toadstool (*A. muscaria* is the magic mushroom in the children's book "Alice in Wonderland" and small children sometimes eat the mushrooms because they are familiar with the pictures). The cup is usually 5–15 cm across, although up to 30 cm has been recorded, and has white removable spots on a pale orange-brown or even red background which fade to yellow with age. The cap is domed in young specimens, but flatter with older samples; the stem, about 6–20 cm long, usually has a ring of tissue at the top and at the base, remnants of the universal "veil" which encloses the juvenile form. The mushroom is commonly found in pine and fir woods and a number of varieties with slightly different colourings occur throughout the world.

Ibotenic acid is a central glutamate receptor agonist while muscimol is a GABA (γ-aminobutyric acid) agonist (Figure 3) so that effects on brain neurotransmission account for the symptoms. In adults these usually occur within 30 minutes to two hours of eating the mushrooms and the muscimol-mediated GABA-ergic effects predominate; there may be some nausea and vomiting but drowsiness and dizziness are always seen, followed by a period of hyperexcitability, hallucinations, euphoria and delirium. Periods of drowsiness and excitability may alternate but the effects usually

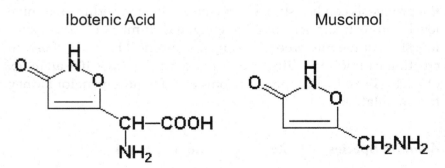

Figure 3. Toxins of the Fly Agaric and Panthercap mushrooms.

fade within six hours. Fatalities from Fly Agaric mushrooms are rare in adults. In children, the excitatory glutaminergic effects are more common and they may experience ataxia, hyperkinetic behaviour and seizures for up to 12 hours; the symptoms such as convulsions and twitching can be controlled with benzodiazepines.

Fly Agaric mushrooms are not normally mistaken for other types; although small children occasionally eat the mushrooms out of curiosity, ingestion in adults is deliberate and done by those individuals seeking an "out of mind experience". The Panthercap has the same toxins as the Fly Agaric but in higher amounts. Symptoms often occur later, after eight to 12 hours, and the mortality rate is 15%–20%.

Fly Agaric mushrooms have been used for centuries, probably millennia and it is possible that *A. muscaria* is the "Soma" of the Hindu scriptures. Traditionally they were eaten in Siberia by shamans engaged in rituals to establish contact with the worlds of the ancestral spirits. Ibotenic acid and muscimol are excreted unchanged so that in some tribes, the shaman ate the mushrooms and the rest of the people drank his urine. The most important people drank first, then produced urine in their turn which was then drunk by the less socially advantaged members of the group. This gives partial detoxication as both *A. muscaria* and *A. pantherina* contain small amounts of muscarine which gives rise to sweating and drooling. Any muscarine present at the start of the séance is removed on passage through a human body so that those at the end of the chain have a less toxic, if less aesthetically pleasing, experience. The sun-dried Fly Agaric has a higher concentration of hallucinogens and is believed to be a better route to reaching the ancestors. Some writers have suggested that Santa Claus with his house at the North Pole, his red and white suit and his flying reindeer may be an offshoot of the ancient Siberian shaman cult of the red and white hallucinogenic Fly Agaric mushrooms.

Gyromitra Species — Gyromitrin

The False Morel mushroom (*Gyromitra esculenta*) is easily confused with the true morel (*Morchella esculenta*). In some parts of the world, these fungi are considered a delicacy and non-toxic when they are prepared by par-boiling which removes most of the volatile toxic gyromitrin. The mushrooms must be cut into small pieces and boiled at least twice in a large amount of water for at least five minutes. The water must be thrown away after each boiling; as the gyromitrin is volatile, the fumes are also toxic, giving rise to headache, dizziness and nausea if the cook carries out the process in a confined space. The mushrooms are sometimes called "beefsteak mushrooms" as they can resemble a small piece of well-cooked steak and are particularly popular in Scandinavia. Generally, they are irregularly shaped with a surface rather similar to a brown brain (true morels look more like a brown sponge); unlike the true morels, false morels have solid stems.

Gyromitrin toxicity does not affect everyone — the reasons why some people are not susceptible are not known. Those affected show problems about four to six hours after eating the mushrooms, experiencing feelings of fullness in the gut, headache, nausea, vomiting and diarrheoa. More serious cases have liver dysfunction and red blood cell destruction. This may be accompanied by methaemoglobinaemia, where the oxygen-carrying haemoglobin is oxidised to a compound (methaemoglobin) which does not transport oxygen. Occasionally, intravascular haemolysis with the methaemo-globinaemia can cause renal toxicity. Gyromitrin toxicity can also affect the central nervous system, as some victims have tremor, dizziness, muscle twitching and seizures which are accompanied by liver damage. These convulsions are common in gyromitrin poisoning and are probably due to reduced oxygenation of tissues, acidosis and other metabolic abnormalities. Jaundice, liver failure and hypoglycemia may lead to brain oedema as a complication.

Gyromitrin is a volatile derivative of hydrazine (see Figure 4) with the chemical formula of acetaldehyde methyl formylhydrazone and is readily converted to N-methyl-N-formyl hydrazine and N-methylhydrazine. It is primarily a liver toxin although the red blood cells and central nervous system are also involved. The

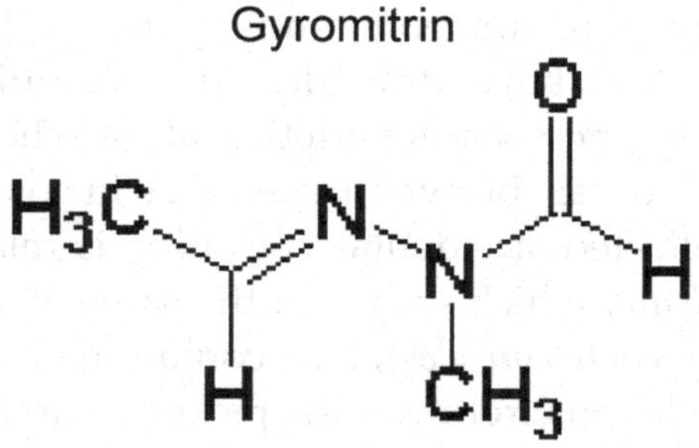

Gyromitrin

Figure 4. Gyromitrin, toxin of the False Morel mushroom.

hydrazines are irritating to mucous membranes and they also reduce pyridoxine (Vitamin B_6) levels in the CNS. This inhibits synthesis of the inhibitory neurotransmitter γ-aminobutyric acid (GABA); the neurotoxicity is particularly evident in people on isoniazid therapy as this drug also reduces pyridoxine levels in the CNS. The damage to red blood cells and hepatocytes probably reflects a reduction in levels of glutathione as the toxic metabolites of gyromitrin, which may include free radicals, are removed by combination with this thiol-containing tripeptide. The liver damage is shown by raised levels of hepatic transaminase enzymes and by alternating hyper- and hypo-glycemia. The mortality rate is relatively low (2%–4%) but the symptoms can be severe. Treatment includes decontamination of the gastrointestinal tract with oral activated charcoal, then administration of intravenous benzodiazepines and pyridoxine. Animal studies suggest that even small amounts of gyromitrin may have cumulative carcinogenic effects so that, however delicious, the mushrooms should only be eaten occasionally if at all.

Inocybe / Clitocybe Species — Muscarine-Histamine

Muscarine-histamine poisoning occurs when mushrooms from the *Inocybe* or *Clitocybe* species such as *I. geophylla,* or *C. dealbata* (Sweat mushrooms) are eaten.

Generally these mushrooms are small, brown and leathery (*Inocybe* species) or small white and leathery (*Clitocybe* species) and so are unappealing and rather unappetising. This is fortunate as even experts can have difficulty in distinguishing between poisonous and non-poisonous varieties, all of which look similar, being pale white to pale brown or even lilac in colour. As they are not normally collected, accidental poisoning is relatively unlikely although *C. dealbata*, which grows in the same area as Fairy Ring mushrooms (*Marasmius oreades*), may contaminate the collection of these choice edible mushrooms if the picker is careless.

Inocybe and *Clitocybe* species may contain up to 3%–4% of muscarine (Figure 5), a quaternary ammonium compound structurally similar to acetylcholine which can stimulate peripheral muscarinic cholinergic receptors, though not nicotinic cholinergic receptors. It cannot cross the blood-brain barrier to affect the central nervous system but causes peripheral cholinergic symptoms within 15 to 30 minutes of ingestion, the effects including drooling, perspiration, and watery eyes. Large doses can lead to urination, vomiting, nausea, abdominal pain, diarrhoea and the classic finding in cholinergic overdoses of blurred vision and pin-point pupils. Asthmatic breathing also sometimes occurs. Muscarine is not hydrolysed by plasma acetylcholinesterase so that the effects can last for several hours. Usually the major symptoms subside after two or three hours and fatalities are rare, although some cases of death from cardiac or respiratory failure have been reported.

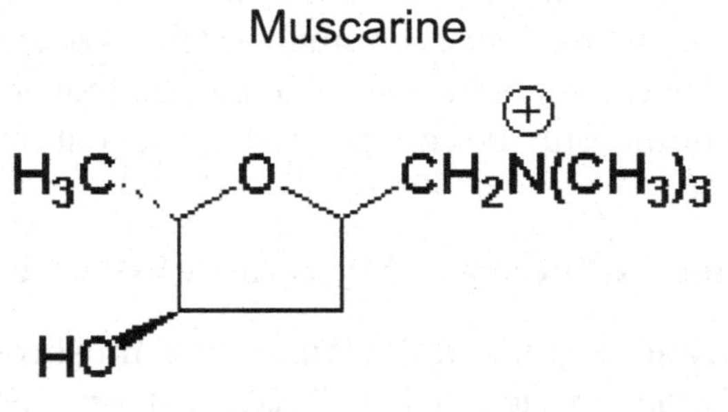

Muscarine

Figure 5. Muscarine, toxin of the 'Sweat' mushrooms.

Treatment is primarily supportive with fluid rehydration and atropine to control excessive secretions.

Psilocybe Species — Psilocybin-Psilocin Poisoning

Toxicity from these compounds (Figure 6) is caused by mushrooms of the genera *Psilocybe, Panaeolus, Copelandii, Gymnopilus, Conocybe* and *Pluteus*. Generally, these mushrooms are small, brown and nondescript; they are unlikely to be mistaken for food fungi and are usually eaten for their hallucinogenic effects which are similar to LSD intoxication. Some species such as *P. coprophila* and *P. cubensis* typically grow on cow and horse manure in fields from Florida through to Texas and also in the Pacific North-West from California up to Canada, while *P. cyanescens* grows readily in landscaped areas mulched with wood chips. In Thailand, *Psilocybe subcubensis* and *Copelandii* species occur on manure from cattle and the domesticated water buffalo and are cultivated for sale to tourists and marketed as enhancing the exotic location.

Psilocybin (O-phosphoryl-4-hydroxy-N,N-dimethyltryptamine) is metabolised *in vivo* to the non-phosphorylated compound psilocin (4-hydroxy-N,N-dimethyltryptamine); both these indole derivatives are hallucinogenic, probably because they are structurally similar

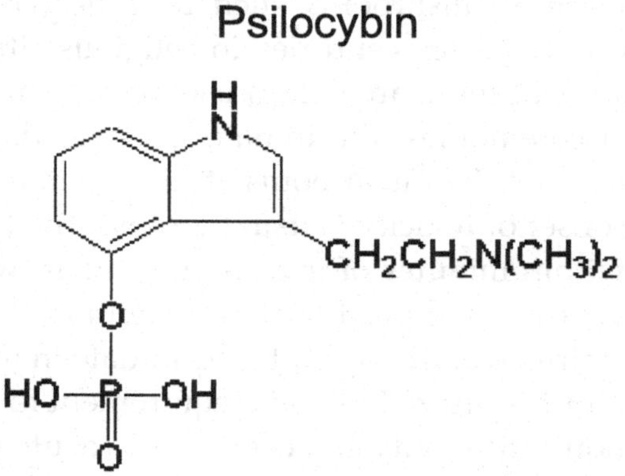

Figure 6. Psilocybin, the hallucinogen in Psilocybe ('magic') mushrooms

to the neurotransmitter serotonin (5-hydroxytryptamine) and act at the same receptors just as LSD itself does. Unlike ibotenic acid poisoning, psilocybin does not cause drowsiness or coma and those taking the mushrooms only experience hallucinations, often with alterations of colour and synesthesia ("seeing sounds" and "smelling colours"), sometimes accompanied by compulsive movements and uncontrolled laughter. Small children are more easily affected and may have hallucinations followed by fever, convulsions and coma. Some *Psilocybe* species also contain phenylethylamine and this can cause hypertension and tachycardia. Sedative benzodiazepines are used if necessary to control hyperkinesias.

Hallucinogenic species of these mushrooms commonly have a blue-staining reaction when the mushroom tissue is bruised. This is due to the presence of psilocybin itself so that the more intense the blue colour, the more hallucinogenic will be the mushroom. Each gram of dried mushroom contains about 10–15 mg psilocybin which is orally active at doses of 40 μg/kg so that the standard 70 kg man needs about a quarter of a mushroom for hallucinogenic effects and can eat the mushroom or drink an infusion to get results. The lethal dose is thought to be ~6 g psilocybin so it is unlikely that enough could be ingested to cause death. However, several instances of fatal outcomes have been reported, particularly in teenagers or students taking psilocybin mushrooms while unsupervised and then falling or jumping from buildings or bridges.

Several of these mushrooms such as *Psilocybe cubensis* and *P. mexicana* were used for centuries in religious rituals of native American tribes and these are still in use in some areas of Mexico today. In such ceremonies, the initiate fasts for days then, after purification rites, eat the mushrooms at night while resting under the stars. The onset of toxicity is usually rapid; the hallucinogenic effects occur within minutes of ingestion, peak at two hours after the dose and are over within a further two hours — it was thought that ancestral spirits would appear to the initiate in this period and give guidance for his future life. This requirement for spiritual and emotional tranquillity avoids the severe panic, acute psychosis and depressive reactions which can occur if *Psilocybe* mushrooms are eaten in a less relaxed setting. These hallucinogenic species have been used from pre-Columbian times and the finding of mushroom-shaped artefacts at archaeological sites suggests that this ritual use is an ancient practice. The Aztecs called these mushrooms "teonanacatl" and they were served at ceremonies, such as the coronation of Moctezuma II in 1502 although after the Spanish conquest, their use was forbidden. In modern times, *Psilocybe* species are often valued and cultivated to provide recreational drugs.

Coprinus Species — Coprine

Coprine poisoning occurs when the Inky Cap mushroom (*Coprinus atramentarius*) is eaten by someone who then drinks alcohol within the next three days. The mushroom is normally considered both edible and delicious and is named from the black fluid that oozes from the cap of the mature mushroom. *C. atramentarius* is a common lawn mushroom which contains coprine (see Figure 7) although other members of the genus *Coprinus* (Shaggy Mane, *C. cornatus* and Glistening Inky Cap, *C. micaceus*) do not. A coprine-type reaction may also be experienced if *Clitocybe Clavipes* is eaten. Coprine is thought to be converted to metabolites including 1-aminocyclopropanol in the human body and this compound blocks the breakdown of alcohol. Normally, alcohol is oxidised by alcohol dehydrogenase enzymes and the acetaldehyde so produced is oxidised by aldehyde oxidase enzymes. These are inhibited by the

Coprine

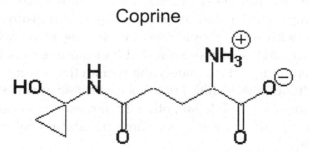

Figure 7. Coprine, the enzyme inhibitor from the Inky Cap mushroom.

coprine metabolites and high levels of toxic acetaldehyde result. This causes many of the symptoms of a "hangover" with headache, nausea, vomiting and flushing and a racing heartbeat, although chest pain and vomiting can also be present and mimic an acute heart attack. These effects last for two to three hours and resemble those produced by the drug disulfiram (Antabuse) which is given to deter alcoholics from drinking alcohol and works by the same mechanism. Although not fatal, coprine poisoning can be very unpleasant. Treatment is supportive with anti-emetics and fluid rehydration.

Gastrointestinal Irritants

Many mushrooms, including the Jack O'Lantern (*Omphalotus illudens*) and boletus species (e.g. *Boletus piperatus*, the Pepper Boletus) contain toxins which can cause gastrointestinal problems such as nausea, vomiting, diarrhoea and abdominal cramps. The main diagnostic difference from amanitin poisoning is that the onset of symptoms is very rapid, often only 15 to 30 minutes after eating the mushrooms. The responses often vary greatly within a population, some people being seriously affected while others who ate the same mushrooms are unharmed. This phenomenon is well-illustrated by a case in 1991 in Vancouver, British Columbia. During a banquet for 482 people, the morels *Morchella esculenta* and *M. elata* were prepared sliced raw in a marinade. Of the people present, 77 reported symptoms of nausea (onset 15–30 minutes), diarrhoea (20 minutes–13 hours) and vomiting (20–60 minutes) with cramps and bloated feelings. Temperature disturbances also occurred, with some people reporting feeling warm while others felt clammy. Many of those affected had numbness of the tongue and extreme thirst. Fortunately the worst effects were over in six hours. Although the white *Tricholoma matsutake* mushroom is often eaten in the US and in Japan, other *Tricholoma* species, particularly *T. pardinum*, can cause nausea, vomiting, abdominal cramps and diarrhoea.

Other Reactions

Other reactions to mushroom have also been reported. An outbreak of acute encephalopathy was described in Japanese patients with renal impairment, all of whom had eaten "Sugihiratake" mushrooms ("Angel's wing", *Pleurocybella porrigens*) about two weeks previously. Initially, the patients had difficulty in walking and were shaking; several days later they became unconscious and had epileptic fits. Of the original 32 patients, nine (27.2%) died; studies showed that there was brain oedema with lesions in the basal ganglia, the area of the brain involved in controlling movement, which would account for the problems with tremor and ataxia.

Several species, particularly *Tricholoma flavovirens* also known as *T. equestre* (Yellow Trich), and *Russula subnigricans* cause delayed rhabdomyolysis in humans and this muscle damage can be identified by its sequel of a rise in serum creatine phosphokinase (CPK) activity. Fatigue, muscle weakness and myalgia are usually seen within 24 to 72 hours of the mushroom meal; between 1992 and 2003, 12 patients in southwestern France developed these symptoms after consuming at least three consecutive cooked meals of this edible yellow mushroom. All the patients had quadriceps muscle weakness and nausea without vomiting. When they were taken to the hospital, they were found to have severe rhabdomyolysis (muscle breakdown) with mean CPK levels of 22,067 units/L in the seven female patients and 34,786 units/L in the five male patients. Three patients died despite intensive care after developing temperatures of 42°C and acute myocarditis. In the survivors, the muscle weakness persisted for several weeks and this syndrome of fatigue and muscle weakness with little gastrointestinal distress after eating large quantities of yellow mushrooms seems to be diagnostic of *T. equestre* poisoning. Similar cases, including renal damage, have been reported from Poland although it seems probable that individual susceptibility factors are involved as the mushrooms are often eaten without problems. All patients should have supportive care, CPK monitoring and forced alkaline diuresis to protect the kidneys from myoglobin released from damaged muscle. Delayed rhabdomyolysis with renal failure has been reported from Taiwan for patients who had eaten *Russula subnigricans*. Acute renal failure (mean onset ~13 hours) can also result from eating *Amanita proxima* and *A. smithiana*. Most cases of *A. proxima* toxicity occur in Europe and result from confusion with the edible *A. ovoidae*. They are characterised by initial gastrointestinal toxicity with liver damage and oliguria or anuria as signs of kidney failure. Temporary haemodialysis is required until the patients recover. *A. smithiana* is a lookalike for the edible *Tricholoma matsutake* mushroom found in the US Pacific North West and causes anorexia, nausea, vomiting, abdominal cramping, diarrhoea and dizziness at times ranging from 30 minutes to 12 hours after ingestion. An investigation on 13 patients found that they were either oliguric or anuric when they were first seen in hospital and had rising blood urea nitrogen and creatinine levels for the next seven days, with high serum alanine aminotransferase and lactic dehydrogenase although other hepatic enzymes remained at normal levels. The most seriously affected patients were treated with haemodialysis for up to four weeks. The aminoacid nephrotoxin responsible has been identified as norleucine.

Ingestion of the poisonous mushrooms *Clitocybe acromelalga* or *C. amoenolens* (Poison Dwarf Bamboo mushroom) can cause severe tactile pain in the extremities for up to a month (erythromelalgia). Patients experience numbness, intense burning pain, redness and oedema of hands and feet, made worse by heat and relieved by cold. This occurs within about 24 hours of eating the mushrooms (*C. amoenolens* is a lookalike of the edible *C. gibba* found in the high alpine meadows of France); patients show no associated gastrointestinal symptoms and the liver enzymes remain normal. Recovery usually takes from eight days to five months, although one case had pain in the feet for three years. The toxins responsible are thought to be acromelic acids (ACRO) which bind to peripheral glutamate pain receptors, including the NMDA, AMPA and kainate subtypes. ACRO has a number of isomers (A to E have been identified by Japanese toxicologists as the toxic constituents in *Clitocybe acromelalgia* mushrooms); A and B have both been shown to

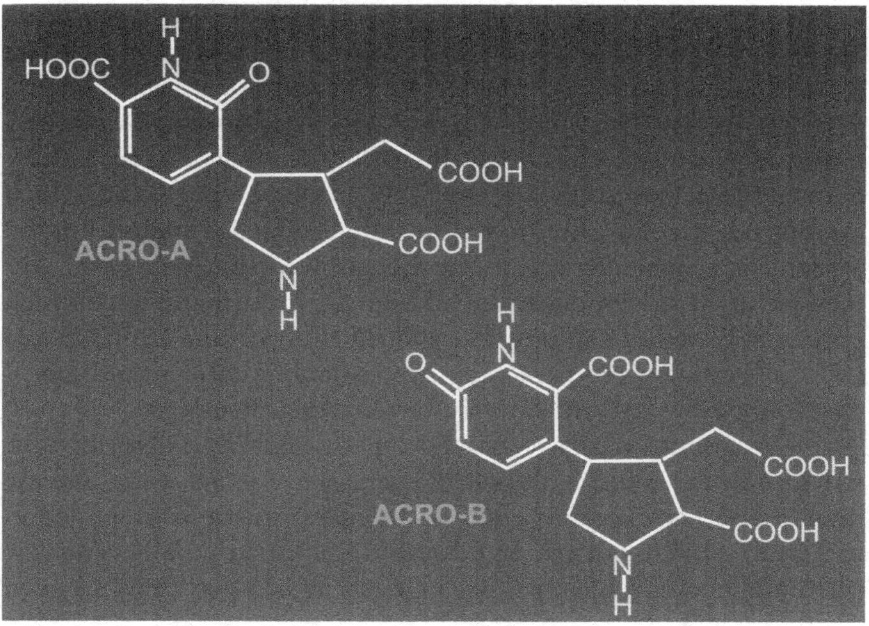

Figure 8. Acromelic acids bind to peripheral glutamate pain receptors.

damage rat lower spinal cord neurons when injected systemically at doses of 50 femtog/kg and 50 picog/kg, respectively. The situation is complex and some workers have suggested that there may be a specific ACRO receptor (Figure 8).

Other syndromes have been described including immunohaemolytic damage from eating *Paxillus involutus* (Poison Pax). The *Paxillus* syndrome usually occurs after repeated ingestion of the mushroom and is characterised by the acute onset of nausea, vomiting, gastric pain and diarrhoea 30 minutes to three hours later. This is followed by an immune complex-mediated haemolytic anaemia with loss of blood in the urine. This may proceed to acute renal failure (with oliguria or anuria) from immune complex nephritis, probably due to an allergic response to an antigenic protein constituent, involutin. Patients can be treated by haemoperfusion or haemodialysis and death is rare. Delayed central nervous system damage from eating the Purple-Dye

Polypore mushroom (*Hapalopilus rutilans*) has also been reported. Three German patients developed reduced vision, somnolence, weakness with decreased motor tone and activity at least 24 hours after the mushroom meal. Clinical biochemistry measurements found electrolyte imbalances and evidence of liver and kidney damage. These symptoms are probably due to polyporic acid, which inhibits the enzyme dihydroorotate dehydrogenase and is found in *H. rutilans* (Figure 9). Studies in rats showed that polyporic acid induced the same symptoms as had been seen in human beings which presumably reflect the reduced synthesis of uridine triphosphate (UTP). Treatment requires supportive therapy with restoration of electrolyte balances and dialysis for renal insufficiency.

Occupational allergic contact dermatitis has been described for workers in the catering industry, particularly those using *Shiitake* mushrooms, while preparing *Bunashimeji* mushrooms in Japan has been linked with a hypersensitivity pneumonitis. Allergic pneumonic syndromes also occur such as lycoperdonosis which

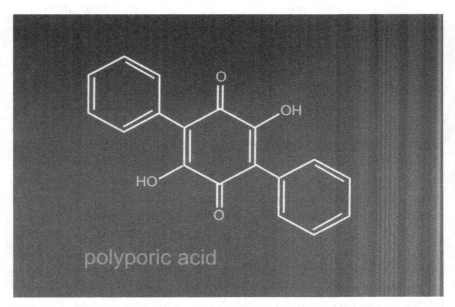

Figure 9. Polyporic acid, the toxic constituent of the mushroom, *Hapalopilus rutilans*.

is an acute bronchoalveolitis following the inhalation of spores from puffballs of the *Lycoperdon* species. These are often edible in the Autumn but dry out and decay over the winter months to release explosive clouds of spores in the Spring if the mushroom is damaged. Usually there is acute onset nausea and vomiting followed a few days later by fever, shortness of breath and inflammatory pneumonitis which responds well to steroids and the anti-fungal amphotericin B.

There are about 5000 mushroom species (Figure 10) of which about 50 to 100 are known to be poisonous to man and about 200 to 300 varieties are believed to be edible. As the consequences of eating unidentified mushrooms can be so drastic (Table 1) and as in most cases we do not have much to offer therapeutically except liver or kidney transplantation for severe cases, any toxicologist would advise prudence rather than experimental oral administration!

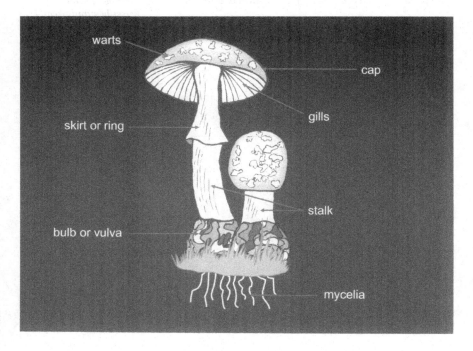

Figure 10. A typical mushroom and its component parts.

Table 1. Onset of toxicity and symptoms from ingestion of mushroom species.

Toxicity onset < 6 h	Toxicity onset 6–24 h	Toxicity onset > 24 h
Neurotoxic Cholinergic — *Clitocybe* and *Inocybe* species Glutaminergic — *Amanita muscaria* and *pantherina* Convulsive — *Gyromitra* species Hallucinogenic — *Psilocybe* species	**Hepatotoxic** Amatoxins from *Amanita, Galerina* and *Lepiota* species	**Neurotoxic** *Hapalopilus rutilans*
Allergic Immunohaemolytic — *Paxillus involutus*	**Nephrotoxic** *Amanita proxima, A. smithiana*	**Nephrotoxic** Orellanine from *Cortinarius* species
Pneumonic *Lycoperdon* species (Puffballs)	**Erythromelalgia** *Clitocybe* species	**Rhabdomyolytic** *Tricholoma equestre Russula subnigricans*
Gastrointestinal Disulfiram-type reaction — *Coprinus* Inky Caps Other reactions — *Boletus, Entoloma, Chlorophyllum* species		

12

NERVE GASES

J. Burdon

Introduction

Nerve gases are chemical warfare agents which act by affecting nerve transmission, including that in important organs like the heart and lungs. They can be lethal in very small doses — about 10 mg, and possibly as little as 1 mg, would be sufficient to kill a human being — and they are effective not only by inhalation, but also by skin absorption. They are far more toxic than the agents used in World War I — mustard gas, chlorine and phosgene — and more toxic even than cyanide.

Four nerve gases are usually described (Figure 1) but compounds which differ in minor ways from these structures would undoubtedly also be nerve gases.

TABUN (also known as GA)
Chemical name: O-ethyl N,N-dimethylphosphoramidocyanidate

$$\begin{array}{c} CN \\ | \\ (CH_3)_2N\text{-}P\text{=}O \\ | \\ OCH_2CH_3 \end{array}$$

SARIN (also known as GB)
Chemical name: O-isopropyl methylphosphonofluoridate

$$\begin{array}{c} F \\ | \\ (CH_3)_2CHO\text{-}P\text{=}O \\ | \\ CH_3 \end{array}$$

SOMAN (also known as GD)
Chemical name: O-pinacolyl methylphosphonofluoridate

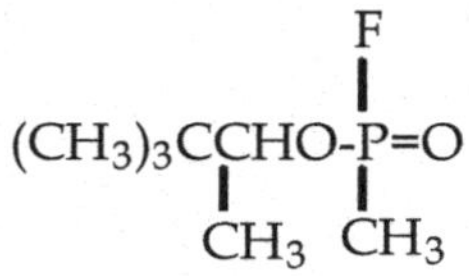

$$(CH_3)_3C\underset{CH_3}{\overset{}{C}}H\underset{CH_3}{O}-\underset{}{\overset{F}{P}}=O$$

VX
Chemical name: O-ethyl S-(2-diisopropylaminoethyl)
methylphosphonothiolate

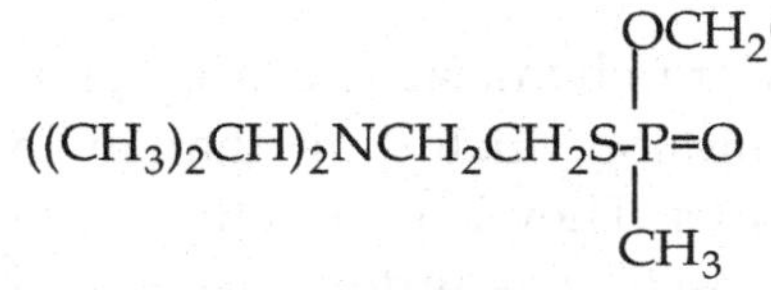

$$((CH_3)_2CH)_2NCH_2CH_2S-\underset{CH_3}{\overset{OCH_2CH_3}{P}}=O$$

Figure 1. Nerve gases (agents).

History

The first three — Tabun, Sarin and Soman — were discovered by the Germans in the late 1930s and early 1940s, and VX by the British in the 1950s. Dr. Gerhard Schrader came across the first, Tabun, through his research into organophosphorus insecticides, but the other three were deliberately invented as chemical warfare agents.

The names "Tabun" and "Soman" apparently have no particular meaning but "Sarin" is said to derive from the names of the discoverers (**S**chrader, **A**mbrose, **R**üdiger and van der L**in**de). The "G" (= "German") names were given by the Allies: GC was omitted because it was already in use as a medical abbreviation for gonorrhea.

None of the agents was used in World War II, although the Germans had begun large scale production of Tabun in 1942 and had made 12,000 tons of it by the end of the war; 105 mm shells loaded with it were found by the Allies in 1945. Sarin was being produced on a pilot plant scale, but Soman had only been made on a laboratory scale.

Chemistry

Since the compounds are so toxic, their syntheses are obviously highly dangerous, and those making them would be expected to wear full protective clothing. It has been claimed that at least ten deaths occurred in the Iraqi plant used to manufacture Tabun for the Iraq-Iran war, even though advanced containment measures were used to protect workers.

The simplest of the four to make is Tabun (Figure 2). Phosphorus oxychloride is treated with dimethylamine, and the intermediate so produced is further treated with potassium cyanide and ethanol to give Tabun.

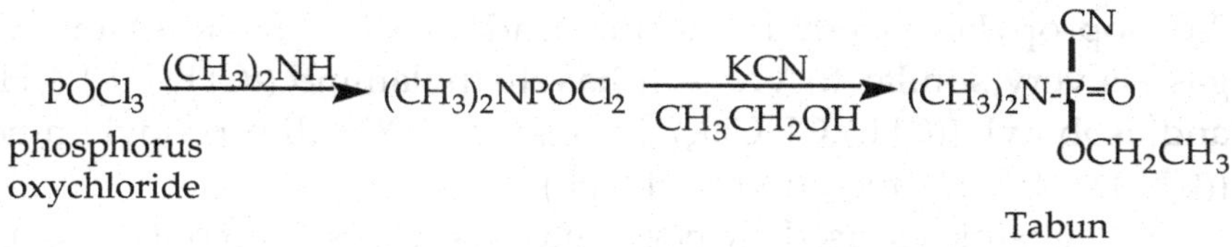

$$POCl_3 \xrightarrow{(CH_3)_2NH} (CH_3)_2NPOCl_2 \xrightarrow[CH_3CH_2OH]{KCN} (CH_3)_2N-\underset{\underset{OCH_2CH_3}{|}}{\overset{\overset{CN}{|}}{P}}=O$$

phosphorus
oxychloride

Tabun

Figure 2. The synthesis of Tabun.

Sarin can be made by more than one method, but in the usual large scale route chloromethane, phosphorus trichloride and aluminium chloride, and then water react to give methylphosphonic dichloride: this, on treatment with hydrogen fluoride (or sodium fluoride), gives a mixed chlorofluoride, which, by reaction with isopropyl alcohol, yields Sarin (Figure 3).

Soman can made in a similar way to Sarin, using pinacolyl alcohol $((CH_3)_3C(CH_3)CHOH)$ instead of isopropyl alcohol $((CH_3)_2CHOH)$.

There is also a less well-known agent, GF or Cyclosarin, which is similar to Sarin and Soman but which has a cyclohexyl group (C_6H_{11}) instead of the isopropyl $((CH_3)_2CH)$ of Sarin.

VX can also be made by more than one method. One (Figure 4) is via EMPTA (O-ethyl methylphosphonothioic acid). Chloromethane, phosphorus trichloride and aluminium chloride react to give methylphosphonous dichoride, which with sulphur followed by ethanol and potassium hydroxide gives EMPTA. EMPTA and

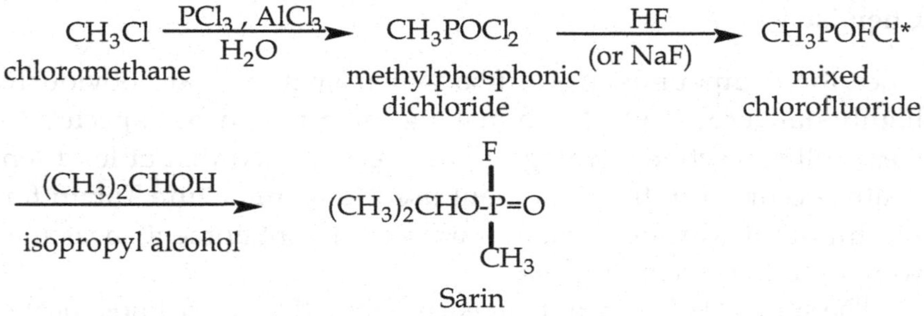

$$CH_3Cl \xrightarrow[\text{H}_2\text{O}]{\text{PCl}_3 \text{ , AlCl}_3} CH_3POCl_2 \xrightarrow[\text{(or NaF)}]{\text{HF}} CH_3POFCl*$$

chloromethane   methylphosphonic   mixed
                            dichloride   chlorofluoride

$$\xrightarrow{\text{(CH}_3\text{)}_2\text{CHOH}} \quad (CH_3)_2CHO\overset{\displaystyle F}{\underset{\displaystyle CH_3}{-P}}=O$$

isopropyl alcohol

Sarin

* CH₃POFCl may be a mixture ("di-di") of CH_3POF_2 and CH_3POCl_2.

Figure 3. The synthesis of sarin.

2-diisopropylaminoethyl chloride lead to VX. The Russian "V-gas" is very similar to VX — it has diethylamino $((CH_3CH_2)_2NH)$ and isobutyl $((CH_3)_2CHCH_2)$ instead of VX's diisopropylamino $((CH_3)_2CH)_2NH)$ and ethyl (CH_3CH_2).

Key chemicals used in nerve gas syntheses are controlled in some way in most countries. Any attempt, particularly by private individuals, to purchase them from chemical supply houses would be reported to the appropriate authorities (see also "Chemical Weapons Convention", below).

None of the four nerve "gases" are, in fact gases (nor is mustard gas): all are high boiling liquids (Table 1). For this reason, the nerve "gases" will, from now on, be referred to as nerve "agents". The two other agents well-known from World War I, chlorine and phosgene, are, however, gaseous at normal temperatures: also, hydrogen cyanide (HCN), the lethal agent in the Californian gas chamber, is a low-boiling liquid (26°C).

However, even though they are not gases, the vapour pressures of Tabun, Soman, and particularly Sarin, are sufficiently great at normal temperatures that, coupled with the compounds' great toxicity, lethal concentrations of their vapours are built up near to liquid samples (the lower the boiling point, the higher the vapour pressure at any given temperature — see Table 1). This is clear from the Japanese subway incident (see below).

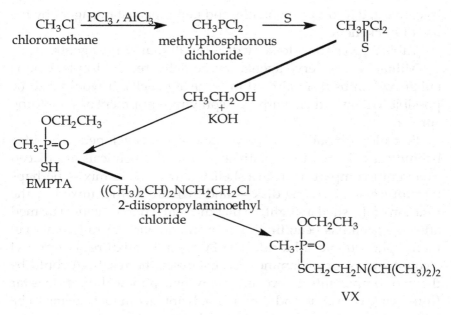

Figure 4. The synthesis of VX.

Table 1.

Compound	Boiling point (° C)
Tabun	240
Sarin	147
Soman	167
VX	over 300
Mustard gas	214

VX is much less volatile and is not dangerous in the same way. However, it can be delivered as a smoke (i.e. an aerosol), and is then even more toxic by inhalation than the other three nerve agents.

All four nerve agents are toxic by absorption through the skin. VX is particularly dangerous in this way as it is a viscous liquid, splashes of which are not easily removed from the skin. Indeed, VX has been mixed with thickening agents to provide a mixture which effectively cannot be washed or wiped off. Also, Soman can be converted into a non-volatile, and hence persistant, agent by the use of thickeners.

Table 2 gives some toxicity data for all four nerve agents.

Militarily, the nerve agents can be delivered as droplet clouds either by bombs from aircraft or by using shells. It would also be possible to deliver them with crop spraying equipment or something similar.

So-called "binary" weapons have been developed. The idea behind these is that two relatively non-toxic chemicals are placed in separate compartments in a shell. On firing these mix — perhaps by rupture of a bursting disk — and then are further mixed by the rotation of the shell in flight, so that the actual nerve agent is formed after the shell has been fired. For example, with Sarin (see above) methylphosphonic difluoride (CH_3POF_2) and a mixture of isopropyl alcohol and isopropylamine (this catalyses the reaction) could be the two components. However, methylphosphonic difluoride is far from being non-toxic and the main advantage, in fact, seems to be that the separate components are easier to store and transport than Sarin itself.

Table 2. Toxicity of nerve agents.

Compound	LCt_{50} (Inhalation)	LD_{50} (Skin absorption)*
Tabun	200	4000
Sarin	100	1700
Soman	100	300
VX	50	10

*Estimates of the doses which would be lethal to humans. LCt50 is the product of concentration and time (in mg per cubic metre for a minute) at which 50% of an exposed population would die. For example 50% would die if exposed to 50 mg per cubic metre of Sarin for two minutes. LD_{50} is the dose (in mg per person) which would kill 50% of an exposed population.

A Brief History of the Use of Nerve Agents

Far and away the most extensive use of nerve agents has been by Iraq in the 1980–1988 Iraq-Iran war, and it is probable that they were used throughout the conflict, and on many dozens of occasions. At first, mustard gas and Tabun were used, but later Sarin as well, and there are reports that VX was used in the last major battle of the war at Fao in the south. The total number of casualties is not known accurately — upwards of 50,000 with 5000 dead are the sort of numbers that are usually given.

In terms of total deaths, the Iraqi attack on the Kurdish village of Halabja in Northern Iraq on 16 March 1988 was the most serious single event in war gas history since World War I. Some 4000–7000 people of all ages are said to have died. Exactly what agents were used is not wholly clear, but it seems that several were deployed, and not just nerve agents. Mustard gas, or a similar blister agent, was certainly used, together with perhaps Tabun and Sarin, and it has been suggested that hydrogen cyanide was used as well. (Although hydrogen cyanide is lighter than air, and hence should not be suitable for use as a war gas, it is only about 7% lighter and under suitable weather conditions it can be used effectively.) A long time after the attack, workers demolishing houses reported that they suffered from itchy skin and burning eyes.

On 20 March 1995, the Japanese Aum Shinrikyo sect placed lunch boxes and soft drink containers of impure Sarin on several Tokyo subway trains. These were punctured with umbrella tips and the vapour pressure of the Sarin was enough to give a concentration in the gas phase sufficient to kill 12 people and affect 5000. There was another Sarin incident at Matsumoto City involving the same sect in 1994 in which seven people died and 600 were affected.

The only definite casualty from VX poisoning has been a member of the Aum cult who was executed by other members in 1994 when he tried to defect. He was approached in the street in Osaka and VX was sprinkled on his neck by two attackers. He chased them but quickly collapsed and then died ten days later without ever recovering consciousness. The cult also tried to kill a lawyer by injecting a VX/hair oil mixture into a keyhole, hoping that he would transfer the VX onto his skin from his key.

It has been claimed that there was some use of nerve agents by the Russians in Afghanistan in 1970s, but no clear evidence for this has been found: however, the UN investigators were unable to gain access to the area where the agents were alleged to have been used.

There have also been allegations that unspecified chemical agents were used by Renamo guerrillas in Mozambique in 1992, and by the Armenians against the Azerbaijanis, also in 1992. UN technical experts were able to investigate both of these claims, but in neither case was any evidence for the use of chemical agents found.

In an incident in Tooele county, Utah, USA in 1968, over 6000 sheep were killed by what was very probably VX when the wind shifted during a nerve agent test at the nearby Dugway proving grounds.

Symptoms of Nerve Agent Poisoning and Mode of Action

The symptoms of poisoning by all the nerve agents are very similar. However, the initial effects, particularly, depend on the type of exposure: vapours or smokes lead to respiratory, nasal, and eye effects, and skin absorption symptoms start with sweating and twitching. The onset of symptoms is much more rapid when the nerve agent enters the body via the vapour route. The order in which the symptoms appear depends to some extent on the route (skin or vapour), the dose and the duration of exposure: the usual order of appearance of symptoms when exposure is to vapours or smokes is something like: eye problems (narrowing of the pupils (miosis), and dimness and blurring of vision); runny nose (rhinorrhea); headache; tightness of chest and difficult or painful breathing (dyspnea); drooling and excessive sweating; nausea and vomiting; cramps and involuntary urination and defecation; twitching, jerking and staggering (fasciculation and ataxia); convulsions; confusion, drowsiness, and coma. Death is usually due to respiratory failure, although one of the Japanese subway victims died from brain damage. It is known that Tabun and Soman can cause brain damage and in some cases such damage was permanent: animal experiments have shown that nerve agents can cross the blood-brain barrier quite easily and so brain damage is to be expected as there are ACh dependent synapses (see below) in the brain.

High doses can lead to death within a minute, and somewhat lower doses to convulsions within one to two minutes. In the Japanese subway incident, one victim died 28 days after the attack. However most of the victims had only mild symptoms and these were mainly eye problems.

It takes at least two weeks to recover from the physical signs and symptoms, although AChE levels do not return to normal for several months. About 60% of the Japanese victims suffered from psychological problems as well — insomnia, nightmares, depression — and these continued for as long as six months; however, there is a later claim that many of them were still suffering four years after the attack. One of the victims of the Matsumoto attack had heart rate problems for 11 months afterwards. Several pregnant women were victims of the subway attack, but neither they nor their babies appear to have suffered from any long-term effects.

Hospital staff who treat nerve agent victims are at risk of nerve agent poisoning, presumably from the vapours given off by the victims' clothing: about 20% of the staff involved in the Japanese subway incident suffered in this way.

The long-term symptoms observed in survivors of the Halabja attack (see above) are not those associated with nerve agent poisoning: leukaemia, infertility, miscarriages and congenital abnormalities have all been reported at levels above those expected in the population at large.

Nerve agents act by interfering with nerve transmission in the so-called "cholinergic" system. It will be necessary to outline how this works before explaining how nerve agents act. Figure 5 is a diagrammatic representation of the brain, spinal chord and some of the nerve fibres. Nerve fibres have "breaks" in them called "synapses": synapses usually occur within or very close to the spinal cord or the brain, or within the organ that the nerve is affecting. There is also a break where a nerve connects to the muscle that it is affecting: such a break is called a "neuromuscular junction" and it

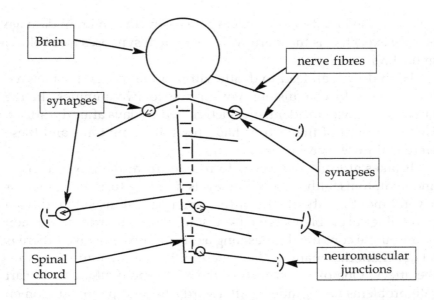

Figure 5. Brain, nerves, synapses and neuromuscular junctions (highly schematic).

operates much like a synapse. A diagrammatic representation of a synapse is shown in Figure 6. When a message comes down a nerve fibre and meets a synapse, the message is carried across the "synaptic cleft" by a "neurotransmitter" — in this case acetylcholine (ACh). The ACh is released on one side of the synaptic cleft, travels across to the other side where it encounters a "receptor". The receptor then causes the message to carry on: for example, the receptor might be connected to a muscle and receipt of a message might cause the muscle to contract. A synaptic cleft is only about 20 nm across and ACh can traverse it in about a millisecond. Once the ACh has delivered the message to the receptor, it has to be destroyed, or else it will keep on sending the same message. It is destroyed by the enzyme acetylcholinesterase (AChE) which is in the vicinity of the receptor (Figures 6 and 7), and which acts as a catalyst and hence is not destroyed in the process.

Nerve agents act by reacting with AChE and rendering it unavailable for its usual job of inactivating ACh. Figure 8 shows how this occurs. The effect of this is that the ACh continues to pass

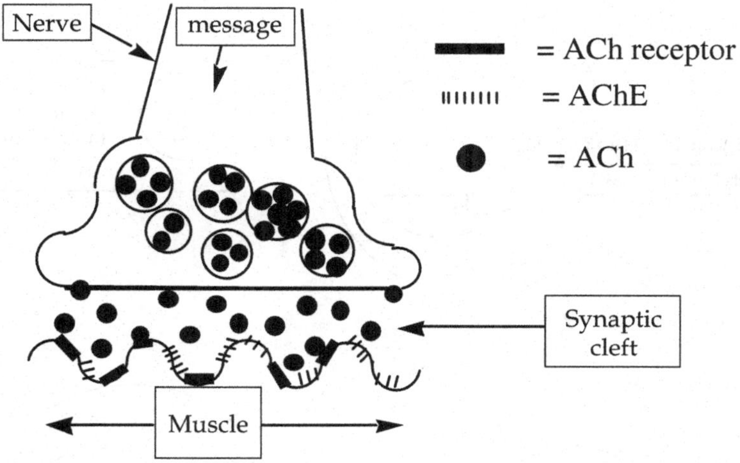

Figure 6. Neuromuscular junction (and synapse) and message transmission.

the message. The message is transmitted endlessly, or the nerve just ceases to function. If this were to occur at a neuromuscular junction, then the muscle would go into spasm.

It should not be thought that the cholinergic system is the only way messages are transmitted across synapses: there are about 40 neurotransmitter systems known but nerve agents only act against one of them — the cholinergic system. Other well-known ones are the adrenergic system, where the transmitters are adrenaline and noradrenaline, and systems in the brain which involve serotonin and dopamine as transmitters: none of these are affected by nerve agents.

Antidotes and Pre-treatments

No antidote for nerve agent poisoning, or pre-treatment when an attack is likely, is without danger, but faced with a choice between nerve agent poisoning and the side-effects of an antidote, the choice is clear. Untreated recovery can take months before the AChE levels are back to normal.

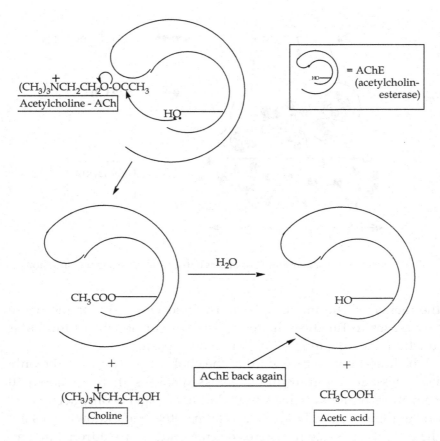

Figure 7. The breakdown of acetylcholine (Ach) by acetylcholinesterase (AChE).

There are two sorts of treatment: (i) a prophylactic one — something to be taken before a nerve agent attack; and (ii) an antidote — something that will reverse the effects of nerve agent poisoning, and which would be taken during or after an attack. These forms of treatment have been developed almost entirely with military personnel in mind, that is, fit young men.

The prophylactic now used is pyridostigmine bromide (PB) (Figure 9). This is given to troops in the form of tablets (NAP tablets — Nerve Agent Protection). PB acts by inhibiting AChE

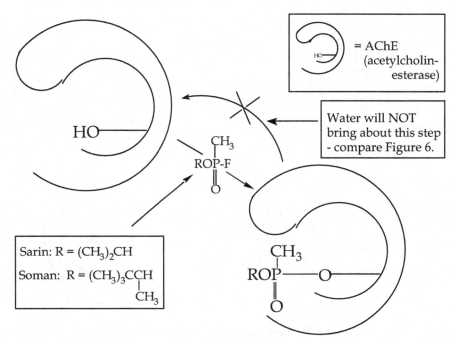

(Tabun and VX react similarly, except that with Tabun, it is the CN-group which reacts instead of the F, and with VX, the S-containing group -see Figure 1.)

Figure 8. The reaction of nerve agents with AChE.

in a reversible manner, unlike nerve agent inhibition which is, for practical purposes, irreversible. The idea is that PB inhibits some of the AChE, then the nerve agent attack comes in and the rest of the AChE is inhibited. But because the PB inhibition is reversible, it releases itself from AChE, thus liberating enough AChE for life processes to continue. Only a few percent of the body's total supply of AChE is needed for life to be possible, and so, after a nerve agent attack, even though most of the AChE will be tied up either by nerve agent or by PB, there will be enough available. However, there are clearly dangers in using PB: too much of it cannot be other than harmful. It is used clinically for the treatment of myasthenia gravis (a muscle disease) but the dose is ten to 20 times less than that used for nerve agent prophylaxis.

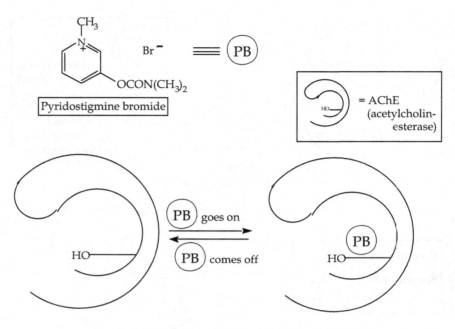

Figure 9. Pyridostigmine bromide (PB) and AChE.

The antidote to nerve agent poisoning — something which is taken during or just after an attack — is a mixture of two compounds, atropine and an oxime, which are taken by injection (diazepam may be present as a third component; it is there to control convulsions). Atropine (Figure 10) is an anticholinergic — it blocks the action of ACh. It does this by blocking the ACh receptor (see also Figure 6) so that even if AChE has been knocked out by the nerve agent, ACh cannot deliver the message because the receptor for it has been blocked. Atropine is clearly toxic — if no message is being passed across synapses, then major organs such as the heart and lungs will not function — and history records plenty of deaths from it: indeed it is the toxic principle of the deadly nightshade plant (*Atropa belladona*).

The other component of the antidote is an oxime (Figure 11) and this works by displacing the nerve agent from the AChE, thus liberating it for its usual purpose (the oxime is claimed to have

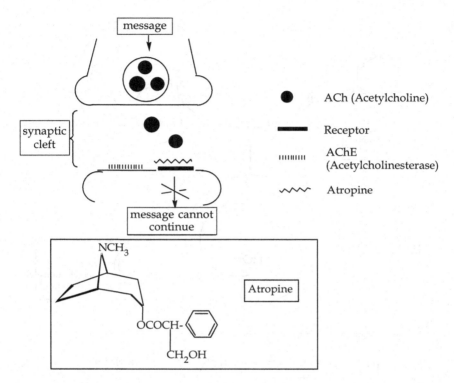

Figure 10. Atropine and the prevention of nerve transmission (see also Figure 6).

other beneficial, but less important, effects). The usual oxime is pralidoxime (also called PAM or P2S), and its mode of action is shown in Figure 11. Unfortunately it only works well with Sarin and VX. It is not very effective in reversing poisoning by Tabun or Soman. There are other oximes — dioximes — which will work against Tabun and Soman, and obidoxime (Figure 11) is such a one. It is not clear why pralidoxime is effective against Sarin and not against Tabun or Soman, or why the dioximes are effective against the latter two.

There is another complication in the use of oxime therapy against Soman poisoning. This is called "aging" and it is illustrated in Figure 12. Essentially, the pinacolyl portion of Soman comes off the Soman-AChE complex (Figure 12) and leaves a group which is

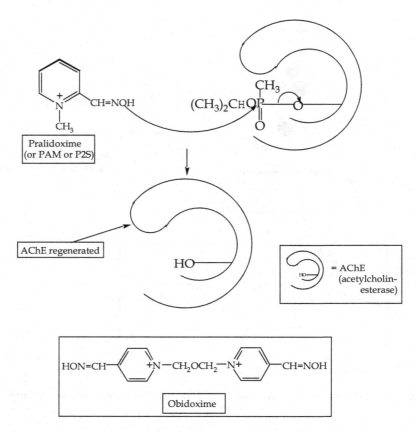

Figure 11. How the oxime antidote works.

no longer susceptible to oxime attack (it isn't susceptible because the negative charge on the oxygen repels the electrons on the oxime which are trying to carry out a nucleophilic attack on the phosphorus). Aging only takes minutes, so if obidoxime treatment is delayed for even 30 minutes after a Soman attack it is useless. Fortunately, Soman is the most difficult of the nerve agents to make and is the least stable in storage. Aging does occur with Sarin, but it is much slower — hours rather than minutes — and with VX it is slower still.

It has even been suggested that the problem of aging with Soman could be overcome by giving troops who are expecting a

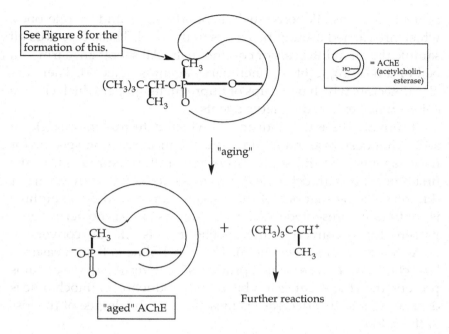

Figure 12. "Aging" of the Soman-AChE complex.

Soman attack a sub-lethal dose of Sarin. Then after the attack had passed, pralidoxime could be given and this would regenerate the AChE that had been inhibited by the Sarin. One can but wonder what state the troops would be in after such a procedure.

There are also protective suits (**M**ilitary **O**perational **P**rotective **P**osture — MOPP) that military personel can don before a nerve agent attack. These enclose the whole body and breathing is through a respirator. Wearing MOPPs for long periods of time in hot climates is, not surprisingly, very unpleasant.

Nerve Agent Detectors and Alarms

There is clearly a need for instruments which will detect a nerve agent attack in its early stages. Such detectors have been developed with military use in mind, and there are static devices — used to protect fixed installations, such as buildings — and mobile ones, which are carried around by troops in the field. These detectors all sample the air at intervals or continuously. All set off an alarm — a bell or a flashing light — when nerve agent is detected. There are also detectors which use strips of impregnated paper which change colour when exposed to nerve agents.

There are three mechanisms by which detectors may work: (i) acetylcholinesterase inhibition; (ii) a chemical reaction specific for nerve agents; and (iii) some sort of physical measurement. In the first type, butyrylthiocholine (this is much like ACh, with the major difference being that one of the oxygens is replaced by a sulphur) is continually presented to AChE: as long as there is no nerve agent present, this is converted into thiocholine, just like the conversion of ACh into choline (see Fig. 3). The thiocholine is then measured by a chemical reaction which produces a colour. If a nerve agent is present, then the AChE is inhibited, the production of thiocholine is decreased, and this decrease is measured by the degree of the loss of the colour.

In the second type, several chemical reactions have been pressed into service. One such is the reaction with butanetrione oxime $(CH_3COC(=NOH)COCH_3)$ which with G-agents releases cyanide which can be detected electrochemically This type of detector is said to operate more quickly than the first, but it could be circumvented by a new nerve agent which did not undergo the specific chemical reaction; a detector based on AChE would not have this problem — any agent, irrespective of its nature, which inhibited AChE would set off the alarm.

The third type can involve a variety of physical techniques, such as infrared or mass spectrometric detection of nerve agents, or the measurement of the amount of phosphorus in a sample of air.

The main detector used in the 1991 Gulf War was the American M8A1. It contains a very simple mass spectrometer, with the necessary ionization being achieved by the alpha particles given off by [241]Am. It dates from 1986 and over 35,000 have been manufactured and about 20 nations worldwide have purchased them. Unfortunately, as was clearly demonstrated in the Gulf War, it suffers from a very high false alarm rate — vehicle exhaust, dust, rocket propellant, Cologne, and cigarette smoke were all capable of setting it off: the alarm could also be triggered if its batteries were low. A more sophisticated detector — the M22 ACADA (**A**utomatic **C**hemical **A**gent **D**etector **A**larm) — is due to replace it. This also operates by mass spectrometry.

The most satisfactory detector deployed in the Gulf War was on the Fox reconnaissance system, which is an armoured vehicle with several sorts of detector aboard, including ones that detect radioactivity. The main nerve agent detector, the MM1, is again based on mass spectrometry but it is much more sophisticated than the M8A1, and gives rise to far fewer false alarms.

Disposal and Decontamination

These are two different problems. "Disposal" refers to the destruction of stockpiles of nerve agents under the Chemical Weapons Convention, and it is supposed to be complete by 2007 (see below). This is a difficult problem because of the sheer quantities of nerve agents held in storage (over 20 countries are believed to have stocks of nerve agents). The US, for example, has about 30,000 tons of the various war gases. There are two main contenders for disposal — incineration and chemical neutralisation (at the end of the 1939–1945 World War, half the German supplies of Tabun — about 6000 tons — were disposed of by dumping shells containing it in the Atlantic ocean at depths of about 2500 metres. This method is obviously no longer acceptable). The total cost of disposal of the US stocks alone has been estimated at US$16 billion.

Not surprisingly, incineration is not popular with people who live near incinerators — even if a small fraction of nerve agent escaped unburned, it could be very hazardous, and even the successful burning of phosphorus compounds produces highly unpleasant fumes of phosphoric acid or phosphoric anhydride. While these can be removed from the combustion fumes, this would need to be very efficient. The Americans are intending to set up about nine incineration sites, but this could be changed, such is the strength of local opposition.

All the nerve agents react with water, with the rate of reaction depending on the acidity/basicity of the water. Sarin is hydrolysed to non-toxic products quite quickly — minutes in strongly acidic or basic solutions, and a few days in neutral solutions. VX, however, is only hydrolysed at significant rates in strongly basic solutions. It follows that a generally applicable chemical method would have to employ strongly basic solutions, and then there would be the problem of disposal of the resulting solutions. This hydrolysis procedure has been used in the UK, but on relatively small quantities: after the basic hydrolysis, the resulting solution was neutralised with acid, diluted and disposed of in the sea.

The Russians are being assisted, financially and technically, by the Americans in the disposal of their stocks. They have developed another chemical method which is very effective. They treat the nerve agents with ethanolamine and water, which would hydrolyse them, and convert the products into a solid waste by mixing them with calcium hydroxide, which would hydrolyse any residual nerve agent, and bitumen. The drawback is that the final product has between three and seven times the volume of the original nerve agent.

Other chemical neutralisation methods have been suggested, as has destruction by enzymes, but the science here has not been fully developed for large scale operations. And then there is the hazard of transporting the nerve agents to the disposal plants.

"Decontamination" refers to the removal of nerve agents from ground, clothing, or machinery. As far as ground is concerned, removal is passive — it consists of letting the agents evaporate or hydrolyse. The G-agents will, over a period of days, evaporate from the ground or be hydrolysed by the moisture in it. VX is very different: it does not evaporate and hydrolyses only slowly. It therefore denies ground to attackers for considerable periods of time, and from a military point of view, this can be advantageous. Clothing and machinery can be decontaminated by washing them with solutions of bleaching powder — this brings about hydrolysis. Such treatment has to be more thorough than might be expected: liquid nerve agents have low surface tensions and can penetrate very fine cracks, including screw threads and rivet heads. Decontamination of the UK's pilot plant at Nancekuke in Cornwall was carried out by careful dismantling: all the operatives wore fully protective clothing and all equipment was washed with bleach or alkali solutions as it was dismantled. All washings were checked for the presence of traces of nerve agents and the blood of the operatives was checked throughout to make sure that their levels of AChE did not fall.

Fuller's earth (a naturally occurring clay, available in powder form) is also of value in decontamination. It does not destroy the nerve agents, but it can absorb them and render them ineffective. Steam treatment is also useful.

Organo-Phosphorus Insecticides

There are a great number of these on the market: the best known is probably malathion (Figure 13).

$$(CH_3O)_2P\overset{\displaystyle\overset{S}{\|}}{-}S\text{-}CHCO_2C_2H_5$$
$$|$$
$$CH_2CO_2C_2H_5$$

Figure 13. Malathion.

These act in exactly the same way as the nerve agents — interference with nerve transmission in the cholinergic system — but fortunately insect and mammalian AChE are sufficiently different that the organo-phosphorus insecticides are far more toxic to insects than to human beings, to whom they are also far less toxic than the nerve agents. Nevertheless, they are toxic in relatively large doses and many cases of poisoning have occurred. Also, chronic toxicity with organo-phosphorus insecticides is known, with farmers suffering long-term nerve damage from the insecticides in sheep dips.

Gulf War Syndrome

Many of the troops involved in the 1991 Gulf War have complained of symptoms of a very varied nature. Most are similar to mild Sarin poisoning, but many neurological symptoms have been reported, as well as birth defects in children not conceived at the time of the war. The cause of these is not known, and one of the many possibilities that have been put forward is that they are due to very small, sublethal, doses of nerve agents. Although there is no strong evidence for this, it cannot be entirely ruled out. US troops destroyed a large arsenal of Iraqi arms at Khamsiyah. These were thought to be entirely normal munitions — explosive bombs and shells — but it is possible that a few nerve agent shells were amongst them. The thesis is that the nerve agents were not entirely destroyed and that some escaped in a cloud and affected troops who happened to be in its path.

It has also been suggested that side effects from the pyridostigmine bromide (PB) prophylactic (see above) were responsible for Gulf War syndrome.

Another suggestion germane to this article is that the tents the troops were using were sprayed with organo-phosphorus insecticides and were then used while still wet with insecticide: on this thesis, Gulf War syndrome is due to organo-phosphorus insecticide poisoning (see above).

The Chemical Weapons Convention

There is a surprisingly long history of attempts to control the use of chemicals in warfare: so far, all have failed. The first was in 1675 when there was a Franco-German agreement not to use poison bullets. Further agreements were signed in 1874 and 1899, and then, in the light of experiences in World War I, the Geneva Protocol of 1925 came into effect.

The most recent control measure is the Chemical Weapons Convention which came into effect in 1997. Most countries have signed and ratified it, notable exceptions being Egypt, Iraq, North Korea, Sudan, Syria and Taiwan.

The two main problems in the negotiations leading to the Convention were the on-site inspection issue and the fact that many chemicals that are, or can lead too, war gases, have legitimate commercial uses. In the final draft, chemicals of possible use as war gases, or as precursors of war gases, were placed into one of three "schedules". As far as nerve agents are concerned, schedule 1 ("chemicals with no or low commercial use") contains all the nerve agents (and mustard gas) and their immediate precursors such as methylphosphonic dichloride (CH_3POF_2): schedule 2 ("dual-use chemicals with moderate commercial use and high-risk precursors") such nerve agent precursors as methylphosphonic dichloride (CH_3POCl_2) and 2-diisopropylaminoethyl chloride ($((CH_3)_2CH)_2NHCH_2CH_2Cl$): and schedule 3 ("dual-use chemicals produced in high commercial volume") is where phosphorus trichloride (PCl_3) and phosphorus oxychloride ($POCl_3$) appear. There are far more chemicals in the schedules than those just exemplified, and chemicals related to all known war gases, not just nerve agents, are listed.

The convention has a permanent Executive Council, a Technical Secretariat and inspectors, and is based in the Hague. There are complex rules relating to the amount of information that signatory countries have to declare, relating to the production, processing, consumption, and to import and export data on chemicals in all three schedules, with the "reporting thresholds" (i.e. the quantities held or made) for schedule 1 chemicals being the lowest: these rules apply to the chemical industry in all the countries and not just to government establishments. There are also rules relating to inspections, including challenge inspections. Signatories have also agreed to dispose of all stocks of all war gases, with nerve agents due to be destroyed by 2007.

13

NICOTINE AND THE TOBACCO ALKALOIDS

J. W. Gorrod and M.-C. Tsai

Occurrence, Source And Use

Nicotine is the principal alkaloid of *Nicotiana tobacum*, where it occurs throughout the plant associated with a number of structurally related alkaloids (Figure 1). Nicotine is also present in other *Nicotiana* species and a variety of other plants and fruits, principally of the Solanaceae family, including tomatoes, aubergines and potatoes (Table 1). Whilst gross nicotine exposure to humans is presently invariably associated with tobacco products for recreational smoking it was not always so. Tobacco was initially introduced into countries and their pharmacopoeias as a medicinal agent where it was used for the treatment of a variety of diseases including epilepsy, hysteria, muscle contraction, polyps, diabetes, constipation and asthma. Tobacco has also been used as an abortifacient and vermifuge. Nicotine has been used in the treatment of urticaria in the form of an ointment, and for hiccough and muscle spasms and as an insecticide in green houses.

"Tobacco" was applied to patients as compresses, infusions, decoctions, or as rectal enemas or via exposure to tobacco smoke. There is little evidence to support the efficacy of these treatments, and indeed many patients died, so that tobacco was gradually removed from the pharmacopoeias.

The introduction of tobacco for pipe smoking by Sir Walter Raleigh introduced the British population to the pleasures of tobacco smoking, which has remained; although nowadays cigarettes and

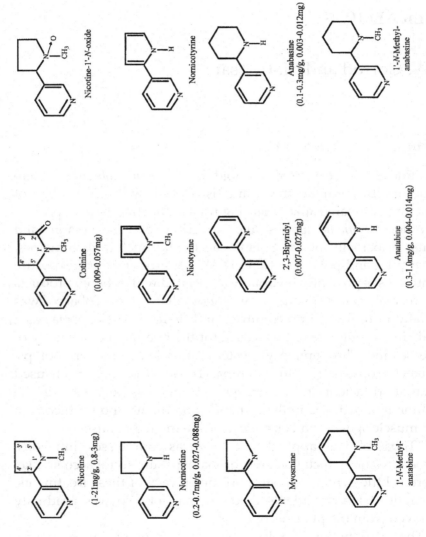

Figure 1. Structures of some tobacco alkaloids (amount in mainstream smoke per cigarette).

Table 1. Occurrence of nicotine in plant families and genus.

Family	Genus
Araceae	*Arum*
Asclepiadaceae	*Asclepias*
Compositae	*Eclipta, Zinnia*
Crassulaceae	*Sedum, Sempervivum*
Equisitaceae	*Equisetum*
Erythroxylaceae	*Erythroxylum*
Leguminoseae	*Acacia, Mucuna*
Lycopodiaceae	*Lycopodium*
Moraceae	*Cannabis*
Solanaceae	*Duboisia, Withania, Nicotiana**

* In *rustica, alata, bigelovi, gossei* and *wigandioides* of *Nicotiana* genus only nicotine occurs, without any nornicotine being present.

cigars are used more than pipes for recreational use. Smokeless tobacco, i.e. chewing and snuff tobacco is still widely used in some communities and more recently there has been increasing use of chewing gum, transdermal patches and nasal sprays as delivery systems supplying nicotine to humans as an aid to smoking cessation. Nicotine is also a powerful insecticide and concentrated solutions are widely available and used when diluted in agriculture; many gardeners still throw their cigarette ends into a bucket of water and use the solution of nicotine produced as a garden spray against green fly and black fly.

Nevertheless, it must be concluded that the principal source of chronic nicotine exposure is still via tobacco products.

Properties

Chemically the tobacco alkaloids are based on pyridine with, usually, a secondary or tertiary alicyclic base as the 3-substituent (Figure 1), in some cases the 3-substituent is aromatic or has been modified by the presence of double bonds or a carbonyl functional group. The ratio of these alkaloids in various tobacum strains varies and in certain strains the related secondary amines nornicotine or

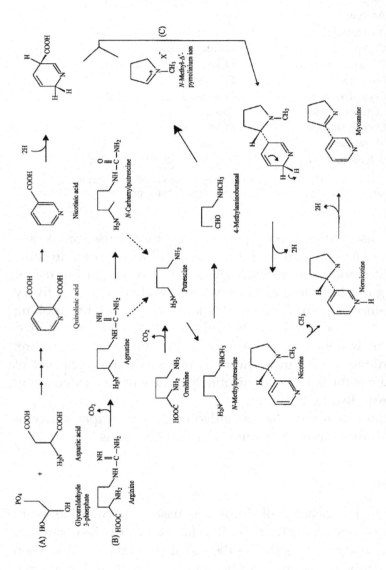

Figure 2. Biosynthesis of tobacco alkaloids. (A) Formation of nicotinic acid, (B) formation of *N*-methyl-Δ'-pyrrolinium ion, and (C) condensation to produce tobacco alkaloids. Constructed from data of Bush LP, Fannin FF, Chelvarajan RL and Burton HR (1993). In: Gorrod JW and Wahren J (eds.) *Nicotine and Related Alkaloids: Absorption, Distribution, Metabolism and Excretion.* Chapman & Hall, London, UK.

anabasine predominate. In the plant, these alkaloids are formed by complex biosynthetic pathways which initially involves the condensation of glyceraldehyde-3-phosphate with aspartic acid to yield, through a multi-step reaction sequence, quinolinic acid and thence nicotinic acid which becomes the pyridyl moiety in nicotine, nornicotine and anabasine (Figure 2).

The alicyclic 3-substituents of nicotine, nornicotine and anabasine are derived from the condensation of N-methyl-Δ′-pyrrolinium ion with 1,2-dihydropyridine to give 3,6-dihydronicotine which by loss of hydrogen yields (S)-(-)-nicotine that can undergo demethylation to nornicotine. Anabasine is produced by an analogous series of reactions from Δ′-piperidine.

Nicotine can be obtained by the steam distillation of an alkaline suspension of chopped tobacco leaves and precipitated from aqueous solution as a silicotungstic acid salt. Pure nicotine is an oily liquid having the typical fishy smell of amines: it is readily soluble in water and organic solvents. It can exist as two stereoisomers (Figure 3), but in nature the (S)-(-)-nicotine is virtually exclusively formed with a specific rotation of −168.5° for $[\alpha]^D_{25}$. However, some racemisation to the (R)-(+)-enantiomer may occur during heating, as occurs in smoking. The boiling point of nicotine is 246°C–247°C at atmospheric pressure and as it has a vapour pressure of 0.0425 mm Hg at 25°C it is volatile and easily transferred to the atmosphere. Nicotine has two basic centres capable of being ionised with pKa1 = 3.4 (pyridine) and pKa2 = 8.1 (N-methylpyrrolidine) at 25°C.

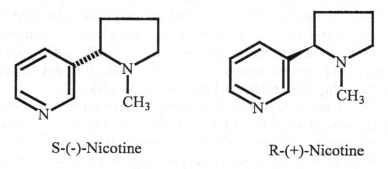

S-(-)-Nicotine   R-(+)-Nicotine

Figure 3. The stereoisomers of nicotine.

Nicotine and related tobacco alkaloids produce their effects in biological systems by reacting at nicotinic acetylcholine receptors via the protonated N'-alicyclic amino functional group. Compounds which are active tend to predominantly exist in the ionised form at physiological pH 7.4, whereas inactive compounds, such as cotinine, are virtually unionised under these conditions. Unionised nicotine can pass freely across biological membranes in contrast to the ionised form where membrane transfer is greatly inhibited; so that whilst nicotine is readily absorbed via the buccal cavity and intestine it is poorly absorbed from the acid environment of the stomach.

Pharmacology

As mentioned earlier, tobacco contains a number of alkaloids which have attracted interest as potential pharmacological agents. This has led to numerous investigations; the detailed results of which are beyond the scope of this article. The results of some comparative studies are summarised in Table 2, where it can be seen that in all the tests nicotine is the most potent agent. Similarly, comparative studies on LD_{50} values of the various tobacco alkaloids (Table 3) show that nicotine is the most toxic and, as this is usually the major alkaloid in tobacco, it is generally considered the major pharmacological agent.

The pharmacology of nicotine is complex and occurs via stereo-specific binding to nicotinic cholinergic receptors in the autonomic ganglia, adrenal medulla, neuromuscular junction and the central nervous system. These receptors are composed of various combinations of five α or β sub-units which having different characteristics produce receptors with different affinities for various ligands and act as ligand-gated channels.

As there are at least nine α-sub-units ($\alpha1$-$\alpha9$), four β-sub-units ($\beta1$-$\beta4$), as well as γ, δ, and ε subunits, the number of possible combinations is very large and this diversity of make-up probably accounts for the wide range of pharmacological activities associated with this seemingly simple molecule. Nicotine binding to a receptor

Table 2. Relative molar potency of nicotine and other tobacco alkaloids.

Alkaloid	Contraction of guinea pig ileum	Pressor action in pithed rat	Release of catecholamines from cat adrenal	Contraction of frog rectus	Blockade of contraction of diaphragm	Inhibition of cat knee jerk	Inhibition of cat flexor reflex	Inhibition of chick flexor reflex	Inhibition of chick crossed extensor reflex
Nicotine	100	100	100	100	100	100	100	100	100
Nornicotine	4.5	22	55	61	73	54	54	36	27
Metanicotine	4	3	20	/	<0.8	0.4	<0.6	/	12.5
Anabasine	17.5	20	75	28	50	17	33	33	20
Myosmine	0.2	5.5	/	3	12	3	<3	3	3
Nicotyrine	0.3	2.5	/	0.4	<0.8	17	<10	51	10
2',3-Bipyridyl	0.2	/	/	/	4	<0.1	<0.1	/	/
Dihydrometanicotine	<0.025	0.5	/	/	<0.8	<0.4	<0.6	/	/
N-methyl-anabasine	<0.023	4.6	/	3	3.5	2	<5	/	12
Cotinine	<0.001	<0.1	0.03	/	<0.8	<0.05	<0.5	/	/
Nornicotyrine	<0.028	2	/	/	<0.9	/	/	/	/

Data taken from Clark, Rand and Vanov (1965). Archives Internationales de Pharmacodynamie et de Thérapie **156**, 363-379.

produces an allosteric change and thereby modifies the functional state such that the receptor can either be resting or desensitised and have the channel open or closed. This interaction with nicotine allows the release of neurotransmitters such as noradrenaline, dopamine, serotonin and acetylcholine thereby modifying physiological processes.

Table 3. Acute toxicity of nicotine, nornicotine and anabasine in various species.

Species	Route	LD_{50} dose (mg/kg)		
		Nicotine	Nornicotine	Anabasine
Mouse	Oral	24.0–35.0		
	Subcutaneous	16.0–45.0		
	Intravenous	0.55–0.80	3.4	
	Intraperitoneal	9.99–11.77	14.66–21.7	
Rat	Oral	188.0		
	Skin	140.0		
	Subcutaneous	36.5–48.0		
	Intravenous	2.8		
	Intraperitoneal	34.0–39.0		
Rabbit	Subcutaneous	30.0		
	Intravenous	5.9–9.0	3.0	3.0
	Intraperitoneal	14.0	>13.7	
Dog	Oral	9.2		
	Intravenous	5.0		4.0
Cat	Intravenous	2.0		
Guinea pig	Subcutaneous	26.0		22.0
Pigeon	Oral	75.0		
Duck	Oral	75.0		

Compiled from data of Holmstedt B (1988). In: Rand MJ and Thurau K (eds.) *The Pharmacology of Nicotine*. IRL Press, Oxford & Washington DC, pp. 61–88.

Nicotine produces biphasic effects: at low doses it has a cholinergic effect while at higher doses initial agonist effects are followed by prolonged antagonism. The principal pharmacological effects of nicotine can be summarised as follows:

1. Central Nervous System: producing tremors and convulsions at higher doses, excitation of respiration and induction of nausea and vomiting due to stimulation of the chemoreceptor trigger zone.
2. Peripheral Nervous System: stimulating the ganglion cells and facilitating the transmission of impulses which at higher doses is changed to a blockade. Small doses of nicotine provokes a release of catecholamines from the adrenal medulla whereas larger doses prevent this release in response to splanchnic nerve stimulation.
3. Nicotine stimulates sensory receptors that respond to stretch or pressure of the skin, mesentery, tongue, lung and stomach; chemoreceptors of the carotid body; thermoreceptors of skin and tongue and pain receptors.
4. Cardiovascular System: is affected by the release of catecholamines from sympathetic nerve endings resulting in vasoconstriction, tachycardia and elevated blood pressure.
5. Gastrointestinal Tract: is affected due to parasympathetic stimulation which causes increased motor activity of the bowel, nausea, vomiting and diarrhoea.
6. Exocrine Glands: low doses of nicotine cause stimulation of salivary and bronchial secretion.
7. Protective Effects: pre-treatment of animals with low doses of nicotine has the ability to protect against the effects of much higher doses which are lethal to naive animals.

Toxicity

The acute toxicity of nicotine has long been known but despite its reputation as a dangerous toxicant, death never occurs by normal smoking where the subject is only exposed to between 0.8 and 1.1 mg nicotine/cigarette. It is claimed a heavy smoker can tolerate

40–60 mg of pure nicotine *per os* as a single dose and survived, however, 10 mg is sufficient to kill a child. Death from nicotine poisoning follows a well-defined pattern following the ingestion of large amounts of nicotine. Nicotine in large quantities is usually only obtained in solutions supplied as insecticides where they can contain up to 95% nicotine and exposure to this may occur due to suicide or as homicide or accidentally as a result of the insecticide being inadvertently stored in whisky or medicine bottles. Other sources of nicotine in high concentrations occur when it is used medicinally; particularly the use of tobacco tar condensate that collects in pipes. Death has also occurred after exposure to nicotine from cigarettes or pipe tobacco as a result of wagers to see how much tobacco could be consumed and one report indicates that a convict who had been working picking tobacco on a plantation died through rectal absorption of nicotine from tobacco leaves he was attempting to smuggle into the prison.

Non-lethal toxicity of varying severity is caused when subjects are exposed to low doses of nicotine; when a novice smoker (or a previous non-smoker) first uses tobacco or when habitual smokers consume more than normal amounts of tobacco. This latter situation can become fatal if excessive amounts of tobacco are consumed and a high dose of nicotine is absorbed.

Low doses of nicotine produce a burning sensation of the tongue which extends to the whole length of the oesophagus with higher dose. The sense of heat spreads from the stomach throughout the body to the finger tips accompanied by general excitement.

Higher doses affect the brain, causing torpor, giddiness, sleepiness and indistinct vision accompanied by an increased sensitivity to light. The subject's hearing is affected and respiration becomes laborious and rapid (dyspnoea). Even at sub-lethal doses the subjects enter a muscular debility phase at about 40 minutes after exposure and find difficulty standing and holding their head erect while their facial muscles relax. The subjects become pale and their limbs and trunk become very cold as they feel faint and lose consciousness. The subjects have a feeling of nausea, often with an extended abdomen and with a desire to defecate, eliminate flatus and urinate; all of which give some temporary relief. The patients suffer severe spasms of the whole muscular system chiefly affecting the muscles of respiration. Vomiting is common and in non-fatal nicotine poisoning sleeplessness, weakness of muscles and lethargy may persist for two to three days.

In fatal nicotine poisoning some, or all, of the above symptoms may be transiently observed but death can occur in a very short time ranging from virtually immediately after ingestion to about an hour. Rarely have subjects receiving high lethal doses of nicotine survived more than two hours.

Cause of Death and Pathology

It is generally thought that the action of nicotine in producing convulsions is via the central nervous system and is associated with the ventricle areas of the brain. Death results from peripheral paralysis of respiratory muscles. Post-mortem findings show that the blood of nicotine poisoned subjects is often dark and that excessive amounts of blood are frequently present in the heart.

Hyperaemia and congestion is a common observation for the spleen, liver, kidneys, brain, lungs and digestive tract. Oedema of brain and lungs has also been observed. Haemorrhages are thought to be caused by asphyxiation and the large haemorrhages observed in lung tissue as ruptures are caused by increased pressure in the arteries. Some changes in the histology of organs have been attributed to shock causing changes in blood distribution. These changes include thickening of the arterial walls, stasis in the capillary vessels and veins and peri-vascular haemorrhage in the cerebral areas. However, it must be emphasised that exceptions with regard to individual organs have been recorded and analysis of organs for nicotine is now considered essential in establishing the cause of death.

In most cases of death through nicotine poisoning the total nicotine content of the various organs is far in excess of the lethal dose and is distributed throughout the stomach and gastrointestinal tract, liver, heart, kidneys, as well as lungs, blood and spleen.

Chronic Toxicity

As early as 1911 it was reported that workers in the tobacco industry had high blood pressure and arrhythmia which seemed to be related to the level of nicotine and the length of exposure. During the ensuing years workers in this industry have been the subject of numerous surveys which associated them with a wide range of ills, most of which were counteracted in further studies. With the exception of the above, any causal relationships were usually with tobacco exposure rather than nicotine. As industrial hygiene has improved the reports of tobacco-induced diseases in industry have diminished. Similarly, nausea has long been associated with the harvesting of green tobacco leaves when the moist leaves are held under the arm against the skin allowing absorption of material from the leaves. Increased mechanisation of harvesting and the use of waterproof protective clothing has decreased the incidence of this illness which may have been caused through nicotine absorption.

Nicotine and minor tobacco alkaloids are present in some foodstuffs and the environment (as environmental tobacco smoke), so that virtually everyone is exposed to these alkaloids, albeit at relatively low doses, for a long period of time. Clinical symptoms associated with chronic toxicity due to exposure to nicotine alone or other minor tobacco alkaloids have not been reported in man. Until recently no long term study of low level nicotine exposure in relation to toxicity in man had been carried out and such observations that have been made were usually extrapolations from exposure to tobacco smoke or environmental tobacco smoke (ETS). This situation is changing with the widespread use of nicotine patches, chewing gum and nasal sprays in smoking cessation programmes. Eventually it may be possible to differentiate between the effects of nicotine and the numerous (over 3000) constituents of tobacco smoke.

It is now acknowledged that nicotine is an addictive drug. The complete mechanism of nicotine addiction initiation and maintenance is presently not fully understood, but probably involves receptors in the CNS. In rat brain the major receptor is $\alpha 4$, β2 which accounts for more than 90% of the high affinity binding. The density of receptors, the speed of change of receptor transition states and the length of time they stay in any active/inactive state, may well be a determinant of pharmacological, including addictive, activity. It is of interest that the brains of animals treated with nicotine and human smokers contain more receptors than their naive counterparts, which may also be associated with the psychopharmacology of nicotine.

As nicotine (and smoking) can increase both heart rate and blood pressure and increase the risk of coronary artery disease it may be thought that treatment with further nicotine in smoking cessation programmes would, at least in the early stages when subjects are still smoking, exacerbate the situation. At least this seems not to be the case when moderate doses of nicotine are given via skin patches.

It has long been known that nitrosamines derived from nicotine and certain of its metabolites are carcinogenic to experimental animals and the possibility exists that nitrosation could occur in vivo particularly in areas of high environmental nitrite/nitrate. At present, there seems to be no evidence that implicates this process in smoking cessation programmes. However, it is important that evidence of addiction, cardiovascular diseases and carcinogenesis are carefully monitored in populations exposed to "medicinal" nicotine for long periods.

Species Differences in Sensitivity to Nicotine Toxicity

The comparative toxicity of nicotine to various species has been the subject of many studies with many conflicting results depending upon the route of administration, pH of solution, nature of the salt, volume of solution administered, time taken to administer the drug and, as mentioned earlier whether the animals were naive or had been pretreated with a low dose of nicotine. An additional variable in reports of acute nicotine toxicity is the time that elapses between administration and death of the animal. These factors have made any absolute comparison of sensitivity towards nicotine acute toxicity very difficult. Comparing routes of administration in any one species, it is clear that intravenous administration requires a much lower dose of nicotine to produce a lethal effect and that intramuscular, intraperitoneal, subcutaneous and oral dosing requires between ten and twenty times the intravenous dose to produce the same effect. The LD_{50} doses (mg/kg) of nicotine for various species are found to be the highest following the oral route of administration.

In general, the degrees of toxicity of nicotine (LD_{50}) vary in different species comparing the same route of administration. The LD_{50} values of nicotine, nornicotine and anabasine for various species are shown in Table 3. For oral administration of nicotine to various species, the increasing order of LD_{50} values are as follows: dog < mouse < pigeon = duck < rat, whereas for intravenous administration of nicotine, the order of LD_{50} doses are: mouse < cat < rat < dog < rabbit. With the intraperitoneal and subcutaneous routes of administration, the increasing order of LD_{50} values are: mouse < rabbit < rat, and mouse < guinea pig < rabbit < rat, respectively. Rat is found to be the species that can withstand the highest doses of nicotine by various routes of administration, with the exception of the intravenous route where the rabbit is the most resilient species. Among the species studied, the mouse is the most sensitive species towards nicotine.

With nornicotine, using intravenous administration, LD_{50} doses of more than 3.0 mg/kg have been observed for mouse, rabbit and dog.

These differences in sensitivity are probably due to variation in transfer across bio-membranes, pharmacodynamics and receptor composition and affinity.

Metabolism and Reactive Intermediates

Despite the apparent simplicity of the chemical structure of nicotine, its metabolism is complex and involves a variety of intermediates and enzyme systems (Figure 4). The metabolism of nicotine involves

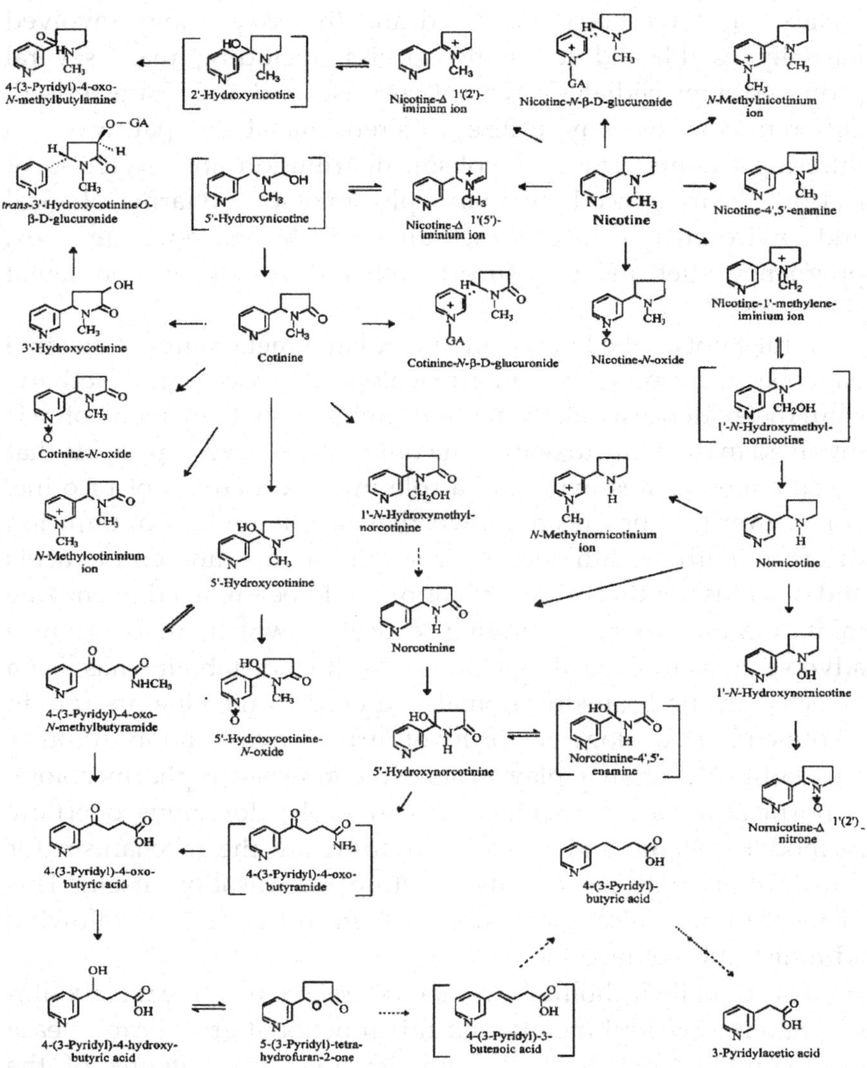

Figure 4. Metabolic pathways of nicotine in mammals. Solid arrows show established pathways; broken arrows indicate unconfirmed reactions.

both phase 1 (oxidative, reductive and hydrolytic) and Phase 2 (conjugation) reactions. Whilst the major pathways of nicotine metabolism have been elucidated and the enzymology involved partially established for some species, including man, several proposed intermediates have not been isolated and characterised. Different species may utilise different metabolic pathways to differing extents. The metabolism, distribution and excretion of nicotine is influenced by many physiological, pharmacological and environmental factors including genetic make-up, age, sex, pregnancy, diet and exposure to other drugs or environmental chemicals.

Whilst not all the known and postulated metabolites of nicotine have been examined for pharmacological/toxicological activity, from the evidence available it seems unlikely that any metabolite is involved in the acute toxicity of nicotine. However it is possible that specific metabolites may play a role in certain effects of nicotine. For example, it has been shown that nicotine-$\Delta^{1'(5')}$-iminium ion can react with glutathione, other sulphur-containing amino acids and proteins. By this process nicotine could be involved in enzyme inhibition and covalent binding to protein which could produce adverse immunological reactions. As this metabolite has been detected in the brains of animals exposed to nicotine, its role in nicotine pharmacology is worthy of further investigation. Another metabolite, which may play a role in nicotine neuro-pharmacology is nornicotine, as this metabolite can evoke dopamine overflow in model systems and *in vivo* can stimulate the mechanism for sensitisation without overtly eliciting sensitisation itself. This metabolite has also been detected in brain tissue following administration of nicotine.

Additionally it should be remembered that nicotine metabolites still retain a pyridyl moiety and this functional group can release nicotinamide from NADPH and generate an analogue of the coenzyme via a glycohydrolase. As these analogues may not be able to participate in the normal oxido/reduction reactions of intermediary metabolism certain pathways may be inhibited leading to accumulation of substrates e.g. glucose-6-phosphate and diminution of availability of products e.g. ribose, and thereby affect purine, pyrimidine and nucleic acid biosynthesis.

Sporadic reports of effects of other nicotine metabolites on biological systems have appeared in the literature but these seem unlikely to produce major pharmacological or toxicological effects when metabolically derived from nicotine, due to the low levels formed.

Minor Tobacco Alkaloids

The most important natural sources of minor tobacco alkaloids are from *Nicotiana* species, and at least eight minor tobacco alkaloids are shown in Figure 1. Since the chemical structures and physical properties of these minor tobacco alkaloids are similar to that of nicotine, some of them are shown to exhibit similar pharmacological activities as those of nicotine, although with a much lower potency. Table 2 shows their relative molar potency in some pharmacological systems. When nornicotine or anabasine was applied to the cat cervical ganglion, initial stimulation was followed by paralysis. On the autonomic ganglion and neuromuscular junction, nornicotine is only one-fifth to one-tenth as active as nicotine. Both nornicotine and anabasine have vaso-depressor action and affect the respiratory system.

The acute toxicity of nicotine, nornicotine and anabasine in various species, using different routes of administration, is presented in Table 3. Very few studies have been carried out on the toxicity and metabolic fates of the other minor tobacco alkaloids. *In vitro* studies of anabasine have demonstrated its conversion to the corresponding N-hydroxy derivative, which is sequentially oxidised to the nitrone, in a similar manner to nornicotine. Anabasine also undergoes α-carbon oxidation to form 2'-oxoanabasine *in vitro*. N-Methylanabasine is reported to undergo N-oxidation to form diastereoisomeric N'-oxides. Both cytochrome P450s and flavin-containing monooxygenases are involved in the formation of diastereoisomeric N'-oxides of N-methylanabasine. Furthermore, N'-demethylation of N-methylanabasine to anabasine and α-carbon oxidation have also been reported *in vitro*.

Treatment

From the foregoing it can be seen that death occurs very quickly after exposure to massive doses of nicotine and in these cases little can be done in the way of treatment. In cases of accidental exposure, treatment has to be supportive and artificial respiration should be initiated immediately and continued even if the heart has stopped. The latter used to be treated by intra-cardiac adrenaline but electro-stimulation may currently be more appropriate. Other treatments are aimed at removing unabsorbed nicotine. These include gastric lavage, purgative enemas or the induction of vomiting. Other measures include trying to absorb the toxicant on to a slurry of medicinal charcoal. We believe that gastrointestinal washing with dilute hydrogen peroxide to convert any residual nicotine to the much less toxic nicotine-1'-*N*-oxide should be tried. Brandy and ammonia have been used as reflex stimulants, and coffee, caffeine, nikethamide, atropine and strychnine as stimulants.

In the case of surface skin contact, washing with large volumes of water or preferably dilute acid solution (acetic, tartaric or citrate) should help minimise absorption.

Specific treatments recommended include gastric lavage with 1:5,000 or 1:10,000 potassium permanganate followed by medicinal charcoal, with convulsions being controlled by phenobarbitone or anaesthesia. An alternative adsorbent in place of charcoal is a mixture of magnesium oxide, tannic acid and charcoal (1:1:2).

For mild nicotine toxicity such as the nausea, weakness and vomiting that occurs during the harvesting of tobacco leaves, prochlorperazine is recommended.

Summary

Nicotine and the related tobacco alkaloids are simple chemical moieties with diverse pharmacological activities. Under normal conditions of human exposure acute toxicity with these compounds would not be expected. However, chronic exposure to nicotine in smoking cessation programmes may induce behavioural changes and pathological conditions which call for long term monitoring.

Suggested Further Reading

Benowitz NL (ed.) (1998). *Nicotine Safety and Toxicity*. Oxford University Press, New York, USA.

Gorrod JW and Jacob P (eds.) (1999). *Analytical Determination of Nicotine and Related Compounds and Their Metabolites*. Elsevier, Amsterdam, The Netherlands.

Gorrod JW and Jenner P (1975). The metabolism of tobacco alkaloids. In: *Essays in Toxicology*. Academic Press, USA, Vol. 6, pp. 33–78.

Gorrod JW and Wahren J (eds). (1993). *Nicotine and Related Alkaloids: Absorption, Distribution, Metabolism, Excretion*. Chapman & Hall, London, UK.

Larson PS, Haag HB and Silvette H (1961). *Tobacco: Experimental and Clinical Studies*. Williams & Wilkins, Baltimore, USA and Supplements by Larson and Silvette in 1968, 1971 and 1975.

14

PARACETAMOL (ACETAMINOPHEN)

G. Steventon and A. Hutt

Introduction

Paracetamol (acetaminophen, 4-acetaminophenol, 4-hydroxy acetanilide) is one of the most widely used non-prescription drugs in the world today. In the late 19th century, a new class of synthetic analgesics and antipyretics was discovered based on a coal tar derivative used in the dye industry. The first of these to be marketed commercially was Antifebrin (acetanilide) by Kalle and Company. In Germany, during this period at the laboratories of Farbenfabriken Bayer the organic chemists under the guidance of Carl Duisberg began to investigate the utilisation of 4-aminophenol, a by-product of its synthetic dye production. The simple chemical process of O-acetylation converted the waste product into 4-ethoxyacetanilide (also known as phenacetin). Phenacetin was found to possess both antipyretic and analgesic properties on subsequent clinical investigations. Phenacetin was marketed in 1888 and it could be stated that the development of phenacetin (from concept to market) should be regarded as the beginning of the pharmaceutical industry in its modern form.

Paracetamol (acetaminophen) was synthesised first in 1888 as an intermediate in the synthesis of phenacetin. Unfortunately, the organic chemists at Bayer never tested this compound for its medicinal properties. The prevailing thought at the time was that phenols were too toxic for clinical use. It was the Swedish physiological chemist, Karl Morner in 1891 who discovered that paracetamol (acetaminophen) was a urinary metabolite of phenacetin. This is a result of the O-de-ethylation of phenacetin by the cytochrome P450 monooxygenase system located in the liver. The first recorded investigation of the medicinal properties of paracetamol (acetaminophen) was in 1893 when the German physiologist, J. F. von Mering reported that the compound showed both antipyretic and analgesic properties. Since phenacetin was already a successful product on the market, paracetamol (acetaminophen) was not developed commercially as a drug. During the 1950s Bernard Brodie in the US and Boreus and Sandberg in Sweden provided experimental data that indicated that paracetamol (acetaminophen) may be a useful drug in its own right. Finally two pharmaceutical companies, Squibb Pharmaceuticals and McNeil Laboratories marketed paracetamol (acetaminophen).

The drug has been effectively and safely used by large numbers of individuals (adults and children) since its introduction to the market place in the 1950s. However, paracetamol (acetaminophen) hepatotoxicity has occurred in individuals following either accidental or deliberate overdose with the drug. It was in 1966 that hepatotoxicity due to paracetamol (acetaminophen) was first reported in man. Since paracetamol (acetaminophen) was on the market in the 1950s it did not undergo the stringent toxicity testing that modern drugs undergo today. It would without question fail to reach the market today.

Toxicity Profile

Acute exposure

In humans, exposure to paracetamol (acetaminophen) is primarily by the oral route from a wide variety of formulations (tablets, capsules and liquids) either on its own or in combination with other drugs. An estimate of paracetamol (acetaminophen) consumption [prescribed paracetamol (acetaminophen), combination tablets and paracetamol (acetaminophen) without a prescription] was 3500 million 500 mg tablets in the year 2000! The acute toxicity of paracetamol (acetaminophen) in man and experimental animals is due to its dose-dependent metabolic profile.

The biotransformation of paracetamol (acetaminophen) at therapeutic concentrations in man can be seen in Figure 1. The major route of metabolism is via conjugation of the phenolic –OH group. The hepatic sulphotransferases have a high affinity for paracetamol (acetaminophen) but are rapidly saturated (either due to cofactor depletion or limited sulphotransferase availability), as were the uridine diphosphate glucuronosyl transferases (UDPGTs) which

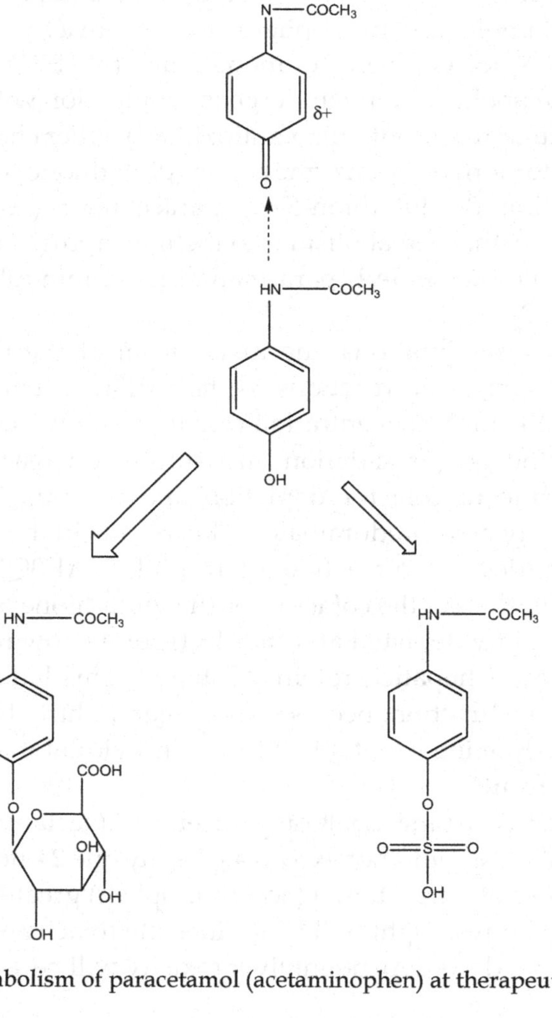

Figure 1. Metabolism of paracetamol (acetaminophen) at therapeutic doses.

have a lower affinity for paracetamol (acetaminophen) but a higher capacity to produce the glucuronide metabolite. Approximately 60%–70% of a therapeutic dose of the drug appears in the urine as the glucuronide metabolite, 20%–25% as the sulphate conjugate, 1%–3% as the mercapturic acid conjugate (a metabolite of the glutathione conjugate) and approximately 1%–2% as unchanged drug. A very minor metabolic pathway is the oxidation of the drug by the cytochrome P450 system. The isoenzymes CYP1A2, 2A6, 2D6, 2E1 and 3A4 have all been implicated in the oxidative metabolism of paracetamol (acetaminophen) to the reactive electrophilic metabolite N-acetyl-p-benzoquinone imine (NAPBQI). This highly reactive metabolite readily undergoes conjugation with glutathione (major cellular thiol containing antioxidant) either chemically (via a Michael addition) or by enzyme action (glutathione S-transferases). The resulting 3-(glutathion-S-yl) paracetamol (acetaminophen) undergoes further metabolism into the mercapturic acid conjugate (3-(N-acetyl-cysteine-S-yl) paracetamol (acetaminophen)) metabolite (Figure 2).

In overdose situations, the metabolism of the drug switches from the conjugation reactions to the oxidative pathway (Figure 3). Due to the high concentration of paracetamol (acetaminophen) achieved, the glucuronidation and sulphation reactions become saturated due to cofactor depletion and the minor cytochrome P450 pathway now predominates. This results in the production of large quantities of the reactive electrophile NAPBQI. The level of hepatic glutathione (the cofactor for the glutathione S-transferases) becomes rapidly depleted and the electrophile now reacts with the thiol groups of hepatic proteins (Table 1). This latter event leads to cellular dysfunction, necrosis and organ failure. If the antidote, N-acetyl-L-cysteine is not given to the individual in question then death will result.

In the UK, warning labels on paracetamol (acetaminophen) state that the patient should not exceed 4 g in any one 24-hour period but an acute dose of paracetamol (acetaminophen) greater than 7.5 g or 150 mg/kg body weight will result in acute toxicity. Once the dose reaches 10 g a dose-dependent liver toxicity will be seen. In the UK,

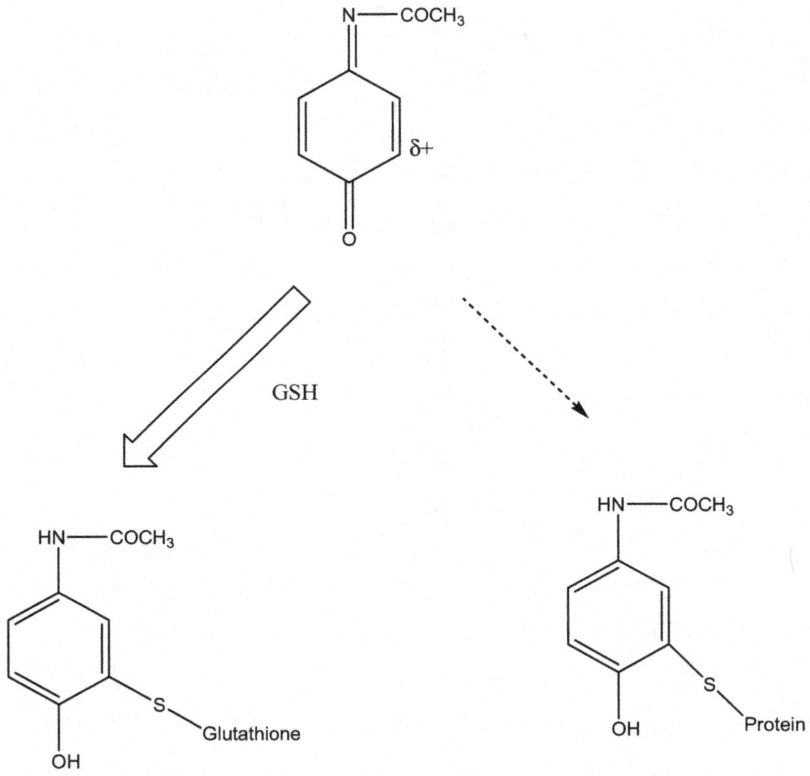

Figure 2. Metabolism of reactive paracetamol (acetaminophen) metabolite.

approximately 500 deaths annually were attributed to paracetamol (acetaminophen) which accounted for 15% of fatal poisonings.

Chronic exposure

Hepatotoxicity from chronic paracetamol (acetaminophen) therapy in adults have been reported in the literature, but compared to the acute toxicity seen with deliberate or accidental overdoses, these are very few. The majority of these are associated with chronic ethanol intake, smoking, isoniazid, rifampin, phenobarbital and phenytoin (inducers of the cytochrome P450 system). Short-term fasting and

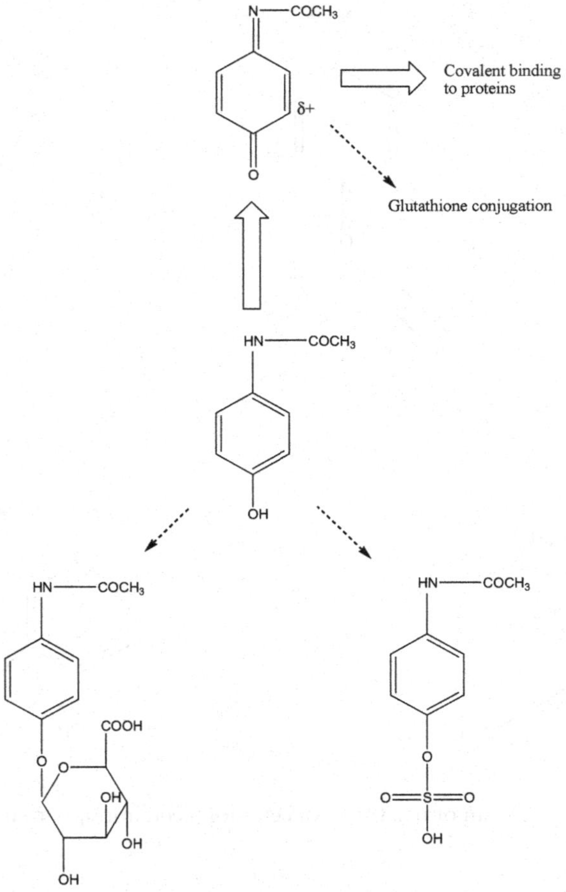

Figure 3. Paracetamol (acetaminophen) metabolism in overdose.

malnutrition have also been associated with an increased risk of hepatotoxicity in individuals taking excessive chronic amounts of paracetamol (acetaminophen). A case study of chronic paracetamol (acetaminophen) poisoning will be discussed latter.

Mechanisms of Biological Interaction

Two mechanisms of biological interactions of the reactive

Table 1. Hepatic proteins that form adducts with paracetamol (acetominophen) in the mouse.

Protein	Cellular location
Aldehyde dehydrogenase	Cytosol, mitochondria
ATP synthetase α-subunit	Mitochondria
Glutamine synthetase	Endoplasmic reticulum
Glutathione peroxidase	Cytosol, mitochondria
Glutathione S-transferase	Cytosol, mitochondria
Glycine N-methyltransferase	Cytosol
3-hydroxyanthranilate-3,4-dioxygenase	Cytosol
Tropomyosin-5	Cytoskeleton
Urate oxidase	Peroxisomes

Adapted from James *et al.* (2003).

paracetamol (acetaminophen) metabolite, NAPBQI, have been hypothesised:

(1) Covalent binding to cellular macromolecules
(2) Oxidative damage to cellular macromolecules

In 1973, at the Laboratory of Chemical Pharmacology, National Institute of Health, USA, James Gillete and colleagues published a series of papers on the covalent binding of paracetamol (acetaminophen) to mouse hepatic proteins. The binding was grossly biased to the liver compared to other organs, and the protein adducts were found in all the subcellular fractions investigated (endoplasmic reticulum, cytosol, mitochondria and nucleus). Pre-treatment of mice with classical cytochrome P450 inhibitors reduced the protein adduction, while pre-treatment of the mice with classical cytochrome P450 inducers increased hepatic protein adduct formation. Some of the proteins covalently modified by adduct formation with NAPBQI can be seen in Table 1. The outcome of adduct formation is to perturb the biochemical function of the enzyme and reduce or inhibit its mode of action. This will result in extensive biochemical/cellular dysfunction and cellular necrosis leading to organ failure. Paracetamol (acetaminophen) hepatotoxicity shows characteristic centrilobular (Zone 3) necrosis. This is of interest since the highest concentration of hepatic cytochrome P450 isoenzymes is to be found in Zone 3 and the highest concentration of reduced glutathione in the liver is located in Zone 1.

NAPBQI is an electrophile and as such will cause oxidative damage to cellular nucleophiles. This in turn will initiate a cascade of free radicals to be formed with the consequential oxidative damage resulting from these events. In addition, inhibition of ATP synthetase will result in a decrease in cellular ATP concentrations, mitochondrial dysfunction and more free radical formation. Other workers have put forward alternative hypotheses to the above two, and these include disruption of calcium homeostasis, altered mitochondrial function, Kupffer cell activation, protein nitrosylation, activation of inflammatory mediators, damage to DNA and apoptosis to name but a few!

Treatment

This is based on the result of a serum paracetamol (acetaminophen) concentration determined after four hours of an acute single oral ingestion of the drug. Prior to this the drug may not have been completely absorbed from the gastro-intestinal tract. The result to treat the patient will be made from a paracetamol (acetaminophen) nomogram. This nomogram was developed using serum concentrations (four to 24 hours post-ingestion) and known times of ingestion. It cannot be used to interpret the results of chronic drug intake.

USA

N-acetyl-L-cysteine is administered orally as follows. A loading dose of 140 mg/kg followed by 17 doses of 70 mg/kg every four hours. N-acetyl-L-cysteine is diluted to the correct concentration in fruit juice or a soft drink.

UK

N-acetyl-L-cysteine is administered by intravenous infusion as follows. Initially 150 mg/kg over 15 minutes followed by 50 mg/kg over four hours and then 100 mg/kg over 16 hours. N-acetyl-L-cysteine is diluted to the correct concentration in 5% glucose intravenous infusion fluid.

The antidote therapy is highly successful between four to 14 hours post-overdose. Between 14 to 24 hours the treatment is less successful, and after 24 hours post-overdose the success rate falls dramatically. If the antidote therapy fails, a liver transplant is the only option available to the patient. If a liver cannot be found, then death occurs seven to ten days post-overdose.

The object of both treatment protocols is to provide L-cysteine, the thiol-containing amino acid used in the synthesis of reduced glutathione. Thus, the antidote provides the thiol precursor for glutathione synthesis. Once hepatic glutathione levels are restored, the NAPBQI can be removed from the liver by conjugation with glutathione and any oxidative damage caused can be reversed. Glutathione is one of the major antioxidants in the liver.

Case Histories of Paracetamol (Acetaminophen) Poisoning

Case studies in the literature concerning acute paracetamol (acetaminophen) poisoning are no longer reported due to the high incidence of such poisonings seen in hospital accident and emergency departments and the routine treatment of such cases. Here is a report of a rarer chronic paracetamol (acetaminophen) poisoning.

A 54-year-old woman was admitted to the accident and emergency department with altered mental status and decreased appetite. She was lethargic and responded poorly to questions.

She did however complain of lower abdominal pain, weakness and drowsiness for one weeks' duration. Physical examination revealed a well developed and well nourished lethargic woman in acute distress. Laboratory investigations found that she had hyponatremia, renal insufficiency and hepatic damage. Serum salicylates and ethanol were undetectable but the paracetamol (acetaminophen) concentration was 66 µg/ml (normal 0–20 µg/ml). Her hepatitis profile was negative and her thyroid profile was normal.

Treatment was initiated for paracetamol (acetaminophen) overdose. N-acetyl-L-cysteine is administered orally as follows. A loading dose of 140 mg/kg followed by 17 doses of 70 mg/kg every four hours. The patient was transferred to the intensive care unit. Her liver function tests and serum electrolytes returned to normal values during the next eight days (Table 2). The patient's

Table 2. Laboratory investigation of a chronic paracetamol (acetaminophen) overdose.

Test	Reference range	Day 1	Day 2	Day 3	Day 4	Day 5	Day 6	Day 7	Day 8
Paracetamol (acetaminophen)	0–20 µg/ml	66	30	4	ND	ND	ND	ND	ND
AST	15–38 U/L	7816	6691	2836	1212	371	187	119	94
ALT	5–37 U/L	2918	3066	2508	1771	1344	993	761	563
INR	<1	10.4	5.6	4	1.9	1.5	1.5	1.3	1.3
Bilirubin (T)	0.2–1.3 mg/dL	2.9	2.7	3.8	5.9	6.5	5.6	3.6	3.3
Bilirubin (C)	0–0.2 mg/dL	1.3	1.3	2.1	3.0	3.5	2.8	1.7	1.5
AP	30–95 U/L	162	136	153	141	162	173	183	183
BUN	5–22 mg/dL	32	29	13	10	6	7	11	15
Creatinine	0.5–1.2mg/dL	5.4	2.4	1.0	0.9	0.6	0.5	0.7	0.7

Table was adapted from Lane et al. (2002).

ND: Not detected, AST: aspartate aminotransferase, ALT: alanine aminotransferase, INR: international normalised ratio (blood cloting indicator), Bilirubin (T): total bilirubin, Bilirubin (C): conjugated bilirubin, AP: alkaline phosphatase, BUN: blood urea nitrogen.

mental status was restored to and the scleral icterus and cutaneous jaundice disappeared. She was discharged from hospital on day 8.

The patient revealed on questioning that six to eight weeks prior to admission she had taken one 500 mg paracetamol (acetaminophen) capsule every three to four hours per day and Lortab 10 [an analgesic containing 500 mg paracetamol (acetaminophen)] four to five times a day. This is approximately 5.0–6.5 g of paracetamol (acetaminophen) per day. The recommended daily maximum is 4.0 g per day. Her last dose of paracetamol (acetaminophen) was taken ten hours prior to admission. The patient was taking the paracetamol (acetaminophen) for chronic leg, back and neck pain.

Suggested Further Reading

Bartlett D (2004). Acetaminophen toxicity. *Journal of Emergency Nursing* **30**, 281–283.

James LP, Mayeux PR and Hinson JA *et al.* (2003). Acetaminophen-induced hepatotoxicity. *Drug Metabolism and Disposition* **31**, 1499–1506.

Josephy PD (2005). The molecular toxicology of acetaminophen. *Drug Metabolism and Disposition* **37**, 581–594.

Lane JE, Belson MG, Brown DK and Scheetz A (2002). Chronic acetaminophen toxicity: a case report and review of the literature. *Journal of Emergency Medicine* **23**, 253–256.

Sheen CL, Dillon JF, Bateman DN, Simpson KJ and McDonald TM (2002). Paracetamol toxicity: epidemiology, prevention and costs to the health-care system. *Quarterly Journal of Medicine* **95**, 609–619.

15

Paraquat and Diquat

R. M. Harris

Description

Paraquat and diquat are both extremely effective non-selective contact herbicides which have been used globally for the last 40 to 50 years. Unfortunately, the very property which makes them so efficient at destroying weeds also lends itself to the destruction of animal tissues and both are extremely toxic to humans. Despite their rather different structures (Figure 1), both are members of the bipyridilium (sometimes referred to as dipyridilium) family of compounds. Moreover, their structural differences mean that their toxicity manifests itself in different ways even though the underlying mechanism is the same for both.

Paraquat is usually manufactured as the dichloride salt but is also available as the dimethylsulphate. In the pure form, the salts are white, odourless, crystalline powders which are hygroscopic and hence readily absorb moisture from the air. However, the grade of material used for herbicides tends to contain yellowish (usually highly toxic) impurities. It is extremely soluble in cold water (about 700 g/l at 20°C) but only slightly soluble in ethanol and virtually insoluble in the more oily types of organic solvent. When dissolved in water, the solution has a fairly neutral pH and is stable under these or acid conditions but not under alkaline. By contrast, whereas paraquat is made as the dichloride, diquat is marketed as the dibromide which is pale yellow. However, apart from the difference in colour, the physical properties of diquat are virtually identical to those of paraquat. Although their high solubility in water is useful in the manufacture of highly concentrated solutions

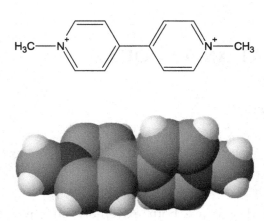

Paraquat

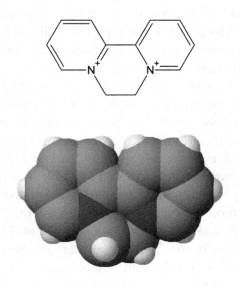

Diquat

Figure 1. Line- and space-filling structures of paraquat and diquat.

for retail purposes, it does cause problems when the herbicides are applied as they are repelled by the waxy surface of the plants. To counteract this, many commercial preparations also contain surfactants which allow the herbicides to cling to and penetrate plant tissues.

Various laboratory techniques have been used to detect paraquat depending on the nature of the sample. Relatively large amounts can be reacted with an alkaline solution of sodium dithionite to produce a blue colour (diquat gives green) which can be measured spectrophotometrically at 600 nm. For trace amounts (below 1 ppm), a different wavelength of 396 nm (diquat, 379 nm) is used which can allow the detection of concentrations of around 10 ppb. Comparable levels of sensitivity can be achieved by gas chromatography which can be coupled with a mass spectrometer for improved identification. Similar techniques can be used for the detection of diquat, but there also exist immuno- and bioassays capable of detecting diquat at less than 1 ppb.

Source

Paraquat and diquat are not naturally occurring substances. Paraquat was first synthesised in 1882 but the herbicidal properties were only discovered in 1955 in the ICI (now Syngenta) laboratories; since the 1960s it has become one of the most globally used herbicides and is marketed as formulations of 23%–44% for commercial use and around 2.5% active ingredient for the domestic market. Likewise, diquat was also developed by ICI in the mid-1950s and is usually available domestically as a 0.1%–0.25% solution or commercially at concentrations of up to 37.45%. Despite long-running campaigns to ban them, their cost-effectiveness means that they are still widely used, particularly in developing nations.

Uses

In the 1930s, paraquat was known as methyl viologen and, due to its ability to undergo a single electron reduction, was used as a redox indicator dye. However, following the discovery of its herbicidal properties, both paraquat and diquat have been used as highly effective herbicides and defoliants. They destroy plant tissues very rapidly and, as they require photosynthesis and hence sunlight to work, they should be applied late in the day so that they have time to become dispersed within the plant tissues and destroy the whole weed rather than isolated patches on the leaves. Diquat is also licensed for use against aquatic plants, where it is mainly used for the destruction of weeds clogging drainage channels, streams and estuaries. As their mechanism of action causes a massive loss of water from terrestrial plants, paraquat and diquat are also used as desiccants and sprayed on cereal crops prior to harvest in order to speed up the drying process.

Mechanism of Toxicity

Paraquat and diquat exert their toxic effects in plants mainly by interacting with photosynthesis, and in animals (and, but probably to a much lesser extent, also in plants) by interacting with mitochondrial respiration. A detailed description of the processes of photosynthesis and mitochondrial respiration would require many pages and is beyond the scope of this book. However, as they are key to the discussion, the following paragraphs have been included to provide a brief overview which is summarised in Figure 2. If the reader is interested in a more detailed account, this can be obtained from any standard biochemistry textbook.

In green plants, photosynthesis takes place in chloroplasts. These are intracellular organelles containing the pigment chlorophyll which is incorporated into two photosystems (imaginatively called photosystem I and photosystem II). Perversely, the process begins with photosystem II where water molecules are split into molecular oxygen, protons and electrons. The oxygen is released into the atmosphere while light interacting with chlorophyll is used to excite the electrons into a higher energy state. A series of intermediate complexes transfers the electrons to photosystem I where their energy is increased further to the point where they

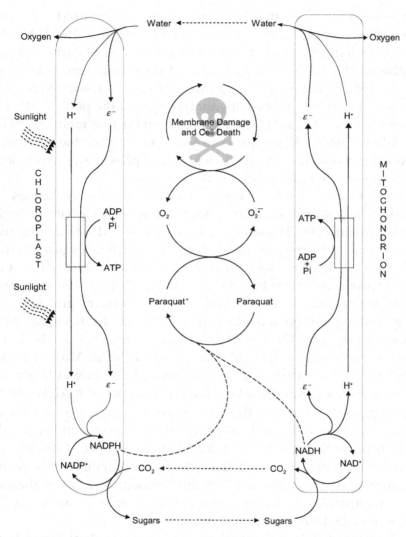

Figure 2. Cycles of life and death. A brief overview showing how paraquat (and similarly diquat) interacts with two of the most fundamental processes of life; photosynthesis in the chloroplast and respiration in the mitochondrion. Abbreviations: ADP, adenosine diphosphate; ATP, adenosine triphosphate; e⁻, electron; H⁺, proton; NAD⁺ and NADH, oxidised and reduced forms of nicotinamide adenine dinucleotide; NADP⁺ and NADPH, oxidised and reduced forms of nicotinamide adenine dinucleotide phosphate; Paraquat•, paraquat radical; Pi, inorganic phosphate; $O_2^{•-}$, superoxide radical.

can be used to reduce the chemical $NADP^+$ (nicotinamide adenine dinucleotide phosphate) to form NADPH. In turn, the NADPH is used to reduce carbon dioxide in the first step of carbohydrate synthesis. This leads to the formation of sugars and starches which serve as intermediate and long-term energy stores, respectively. The protons, which are released at various stages of the photosynthetic process, are used to set up a proton gradient across membranes in the chloroplast and this, plus some of the energy from the electrons, drives the production of ATP (adenosine triphosphate) which acts as an energy donor for other cellular processes.

Most of the chemical energy stored in the carbohydrates is released in organelles called mitochondria which occur in both plants and animals. The sugars are first converted into an organic acid, pyruvate, by the process of glycolysis which takes place in the cell cytosol. The pyruvate then enters the mitochondria where it is further processed in the tricarboxylic acid cycle (also known as the Krebs or citric acid cycle). Here its hydrogen atoms are stripped away leaving carbon dioxide which is eventually released back to the atmosphere. The hydrogen atoms are removed in the form of separate protons and highly energised electrons. Most of these electrons are captured and used to reduce the molecule NAD^+ (nicotinamide adenine dinucleotide) converting it into NADH which is analogous to NADPH in the chloroplast. The electrons are passed from NADH through a series of different molecular complexes and, in the process, their energy is bled away in a controlled fashion and used to pump protons across the mitochondrial membranes. As in the chloroplast, the proton gradient thus formed is used to synthesise ATP. Meanwhile, the de-energised electrons are recombined with the protons and oxygen atoms to make water.

The processes of photosynthesis and mitochondrial respiration share several common features and it is via one of these, the production of high-energy electrons, that paraquat and diquat act. As previously noted, paraquat is capable of undergoing single electron reductions and was originally used as a redox indicator dye. This property is shared by diquat and the molecules are able to steal electrons, from suitable donor molecules such as NADPH and NADH, converting themselves into free radicals. In a normal covalent molecule, the electrons orbit the atomic nuclei in pairs. This stabilises the electrons and makes the molecule less chemically reactive because the electrons must unpair before a reaction can occur. However, in a free radical, one of the electrons is already unpaired and this greatly increases the reactivity of an otherwise stable molecule. Hence, rather than releasing the energy in a controlled fashion and reducing oxygen to water, the paraquat and diquat radicals are able to transfer electrons directly to oxygen molecules to produce not water but superoxide radicals ($O_2^{\cdot-}$, see Figure 3). Transference of the electron returns the paraquat/diquat molecules to their normal state in which they are free to take up further electrons in a succession of reduction and oxidation steps known as a redox cycle. Meanwhile, the superoxide radicals unleash a chain of events culminating in the destruction of the cellular membrane. The overall effect is rather like exploding a pin cushion inside a beach ball with multiple perforations of the outer membrane allowing the cell cytosol to leak out. Although it is less obvious in animal tissues, in the case of plants the membrane damage causes a massive loss of water resulting in wilting and rapid desiccation. It is the rapid loss of water which kills the plants rather than the disruption of their energy supply and enables paraquat and diquat to be used as both herbicides and pre-harvest desiccants for crops such as grain.

Although they share similar mechanisms of toxicity, through an evolutionary fluke, the structure of paraquat makes it more toxic than diquat. Both chemicals are capable of damaging many of the major organs, with the kidneys and nervous system being especially vulnerable due to their high energy requirements. However, in most cases of paraquat poisoning, it is damage to the lungs which proves fatal. The lungs are particularly vulnerable for two reasons. Firstly, they are continuously exposed to oxygen which, as we have seen, is converted into superoxide radicals in a redox cycle. Secondly, the lungs are a surprisingly hostile environment and the lung cells undergo a continuous cycle of replacement. By necessity, this involves the synthesis of large amounts of DNA. Although

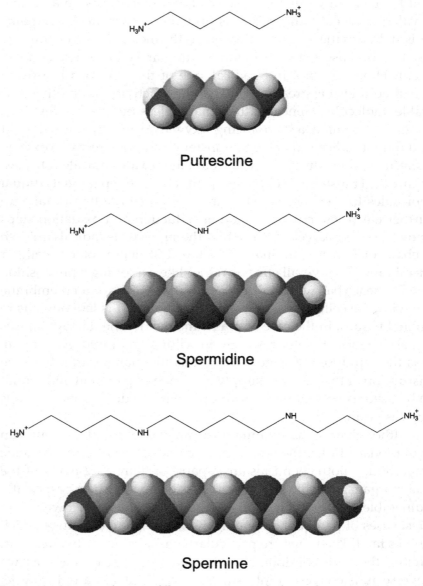

Figure 3. Line- and space-filling structures of the polyamines putrescine, spermidine and spermine. With a little imagination, it is possible to envisage how the essentially linear paraquat molecule might fit into a transport system that has evolved to convey these chemicals into lung cells.

DNA molecules are intrinsically stable structures, they are further stabilised by the binding of molecules called polyamines. As their name suggests, these carry several positively charged amine groups (see Figure 3) which can bind to the negatively charged phosphate groups on the DNA molecules. Unfortunately, the lung cells cannot make polyamines and have to use an active-transport mechanism to take them up from the blood. Paraquat, but not diquat, molecules are sufficiently similar in structure to the polyamines that they too can be taken up by the transport system. Hence, paraquat rapidly becomes concentrated in the lung tissue which is quickly destroyed by the cascade of free radicals that result.

Metabolism and Detoxification

In humans, the oral route is the major route of exposure to paraquat. However, paraquat is relatively poorly absorbed by this route and the majority of the dose is usually excreted unchanged in the faeces. Paraquat can be absorbed across the skin; the concentrate is corrosive and can cause severe burns if it remains in prolonged contact with the skin which accelerates absorption. Nevertheless, although the diluted "working-strength" solution can be absorbed through the skin it is a lengthy process unless the skin is already damaged. Splashes of the concentrate into the eyes can also cause severe damage, but this is mainly due to its corrosive rather than toxic properties. Inhalation of paraquat spray is not considered a high risk as the droplets are large but workers have reported skin problems including burns, ulceration, dermatitis and nail loss and nose bleeds due to local irritation of the upper respiratory tract. Modern formulations of paraquat now contain an emetic, a stench agent and are coloured blue to make accidental ingestion less likely so that most cases are now from patients who have tried to commit suicide or from accidental exposure to people in developing countries who either cannot afford or are forced to work without the proper protective clothing required.

Once absorbed, paraquat is distributed to all the major organs and, although there is selective uptake by the lungs, there is no evidence that bioaccumulation occurs. Likewise, there is no evidence for metabolism and the compound is largely excreted unchanged in the urine. Since the kidneys play such an important role in paraquat excretion, early kidney failure provides a poor prognosis for the victim. The minimum lethal dose by oral ingestion for humans is ~35 mg/kg body weight (10–15 ml of a 20% solution); generally patients with mild poisoning (<20 mg paraquat/kg body weight; less than 7.5 ml of 20% w/v paraquat concentrate) recover after severe gastrointestinal symptoms while those with 20–40 mg paraquat/kg body weight (7.5–15 ml 20% w/v paraquat) develop severe lesions in the gastrointestinal tract, acute renal failure and progressive pulmonary fibrosis; if the latter occurs it leads to death two to three weeks after the initial ingestion. Patients with more than 40 mg paraquat/kg/bw (>15 ml 20% w/v paraquat) die within a few hours to a few days from multiple organ failure; a paraquat blood level of 1.6 µg/ml at six hours after dose is always fatal.

Like paraquat, diquat is also poorly absorbed from the gastrointestinal tract but any of the compounds reaching the liver is partly metabolised to the less toxic mono- and dipyridones. Probably for most species, less than 20% of the dose is metabolised, with the gastrointestinal flora thought to be responsible for some of the chemical transformation. Although diquat is not concentrated in the lungs, it does accumulate in the kidney and is also thought to undergo more enterohepatic circulation than paraquat with around 5% being excreted in bile. Faecal elimination accounts for roughly 90% of the dose, with about 10% in urine. Subcutaneous injection in rats gave 87% of the dose in urine, with approximately 3% converted to the monopyridone and 6% to the dipyridone. Diquat is less corrosive and toxic than paraquat, although delayed wound healing, nail changes and nose bleeding have all been described in chronic users. Acute poisoning leads to gastrointestinal problems, with vomiting, mucosal ulceration and grass-green diarrhoea (due to reduction of diquat by gut bacteria in the same way that reduction by dithionite produces a green colour in laboratory tests). Abdominal bloating occurs from an accumulation of water in the lumen of the intestines but the lung damage seen with paraquat is not found. Progressive haemoconcentration is a feature of diquat poisoning and it has been suggested that the reduction in renal function induced by diquat is secondary to water redistribution although necrosis of proximal and distal tubules has been seen in rats. The dehydration induced by diquat poisoning can cause collapse of the circulatory function with hypotension and tachycardia leading to shock and death. Higher doses give some CNS effects with lethargy followed in serious cases by coma, convulsions and death. Fatal cases of diquat poisoning in man have shown consistent pathological brain changes involving brain stem infarction, particularly involving the area of the pons. In human cases of diquat poisoning, clinical signs of neurotoxicity are often seen. These include irritability, restlessness, aggressiveness, disorientation, nervousness, an inability to recognise familiar faces and confusion. Long-term use of both paraquat and diquat has been linked to an increased incidence of Parkinson's disease, possibly due to dysregulation of the oxidative phosphorylation pathway or to generation of toxic free radicals in brain tissue. A review of a series of cases suggests that the lethal dose of diquat dibromide in man is around 6–12 g. However, even survivors of diquat and particularly paraquat poisoning will have sustained serious systemic damage to many organs and will experience considerable morbidity.

Suicides and Homicides

Ironically, although they can destroy plants in a few hours, paraquat and diquat are not quick-acting mammalian poisons and can take several days or even weeks to kill a human. However, the use of a substance for murder or suicide tends to depend more on its availability than to its speed as a killing agent. Hence, both paraquat and diquat have been widely used by those with suicidal intent. The lingering, unpleasant nature of the death inflicted by these chemicals also makes them attractive options for those whose culture makes a "macho" form of suicide more socially acceptable. In addition, without proper laboratory tests, the symptoms of paraquat poisoning could be mistaken for a lung infection. There is evidence to suggest that paraquat was considered and may have been used as a chemical weapon for assassination in South Africa during the apartheid years. The reputation of these chemicals has also been tarnished by amateur gardeners persuading their agricultural friends to let them have some of the industrial-strength solution which they then store in old soft drink bottles. Inevitably, children playing in the garden find Daddy's bottle of lemonade in the shed and take a swig with predictable consequences.

To combat both intentional and accidental ingestion, modern formulations of paraquat now contains a blue dye, stenching and vomiting agents. However, prior to this, there have been several high profile cases where paraquat has been used as the murder "weapon". One of these, "the case of the poisoned pie", occurred in Westcliffe-on-Sea a quiet town in southeast England. Due to inclement weather, a Mr. Barber returned home unexpectedly from an aborted Saturday fishing trip to find his wife in a compromising situation with a Mr. Collins who lived a few doors down the road. The man fled, but although the Barbers appeared to have patched up their relationship, 12 days later Mr. Barber was given tablets by his company doctor for a severe headache. That weekend, his wife called their doctor because he had developed a sore throat, abdominal pains and was being violently sick. Although the doctor prescribed antibiotics and a syrup for the sore throat, his condition deteriorated and he was taken to Southend Hospital where, three days later, he was placed in intensive care. He was diagnosed as having Goodpasture's syndrome which is a relatively rare condition characterised by rapidly progressing kidney failure and haemorrhaging in the lungs which cause the patient to literally spit blood. However, despite being placed on a ventilator, his condition continued to worsen and a few days later he was transferred to a specialised renal unit at London's Hammersmith Hospital. Meanwhile, his wife calmly received the news that he was unlikely to survive. At Hammersmiths, the doctors were unable to confirm the diagnosis of Goodpasture's syndrome and suspicions of paraquat poisoning were raised. Unfortunately, due to a series of administrative errors, samples intended for the National Poisons Reference Centre were never sent and the doctors were told that the results were negative in an attempt to disguise the blunder. Mr. Barber finally died 23 days after the initial headache and was cremated six days later. The ceremony was attended by Mrs. Barber and Mr. Collins who promptly left his wife and moved in with Mrs. Barber that night. However, the relationship was somewhat transitory and he was soon replaced by the next in a series of lovers as Mrs. Barber took advantage of the financial rewards of her husband's death. Her enjoyment was to be short-lived as the suspicions of doctors at Hammersmith's hardened into facts and the police, aided by Mr. Collins who had been privy to some of the events, accumulated evidence against her. Nine months after her husband's death, she confessed to putting weed-killer into a steak and kidney pie before giving it to her husband. This was followed by another dose and then still more in his throat medicine when nothing appeared to happen. Both she and Mr. Collins were arrested and charged with conspiracy to commit murder. At their trial they pleaded guilty and were sentenced to life and two years' imprisonment, respectively. When sentencing, Mr. Justice Woolfe told Mrs. Barber, "I cannot think of a more evil way of disposing of a human being".

Antidotes

There is no known antidote to paraquat or diquat poisoning. Some clinicians have used antioxidants such as vitamins C or E, N-acetyl cysteine, nitric oxide donors, and a combination of cyclophosphamide and corticosteroids to prevent inflammation and pulmonary fibrosis in severe cases. However, the effectiveness of these treatments appears to be marginal. If the chemical has been ingested, the main treatment is to prevent further absorption from the gastrointestinal tract. This is accomplished by the standard methods: gastric lavage with saline solution to wash the substance from the stomach or ingestion of activated charcoal or Fuller's earth to act as absorbents. Since one of the effects of both paraquat and diquat is to inhibit peristalsis in the gut, the use of cathartics to speed elimination from the bowel is not advised. However, the time frame for this type of treatment is relatively short and the effectiveness diminishes rapidly beyond 60 minutes post-ingestion. Once the herbicide has been absorbed, haemodialysis may also be used to remove it from the blood, but only a relatively small amount is likely to be eliminated by this method and it is unlikely to make a great deal of difference to the outcome. On rare occasions paraquat poisoning has been successfully treated by a complete lung transplant, but this can only be performed when there is no more poison in the blood stream, otherwise the new lungs will be destroyed by absorbing the residual material. Hence, success is more likely if the operation can be delayed until the poison clears although this may take some weeks if the kidney proximal tubules are severely damaged which is too late for most patients.

Case Histories

It was a hot day when the boy came in from the garden. All the time he was cutting the grass, he'd been looking forward to that bottle of lemonade in the shed. His father had told him not to touch it, but he reckoned it was just what he needed and new that his father wouldn't begrudge him a drink. He put the bottle to his lips and took a gulp. The liquid burned his throat as it went down and the boy felt sick, then started to vomit. He staggered to the house and was immediately rushed to the hospital by his frantic parents, where doctors used activated charcoal and Fuller's earth to prevent further absorption of what they now knew was the herbicide paraquat. They gave analgesics to help control the pain from the boy's oesophagus, which had started to ulcerate, hoping that it would not rupture and cause death. They waited to see whether their patient would survive, realising that he might be dead within a few days from lung damage or kidney failure. Fortunately, the initial vomiting episode had cleared most of the chemical from his body and, after further treatment for the oesophageal damage, he was released from the hospital a few days later. Meanwhile, both he and his father had received stern lectures on, "not touching bottles in your Dad's shed" and "the responsible use and storage of garden chemicals", respectively.

Abandoned by her husband for his much younger secretary, a woman in her early forties decided that she'd had enough. Nursing a hangover from drinking heavily over the weekend, she took an empty medicine bottle with her when she set off to her job at an agricultural warehouse. During her lunch break, she slipped down to the chemical section and, knowing how dangerous it was from the company safety talks, filled the bottle with paraquat-based weed-killer. That evening, back in her flat, she put her favourite CD on the stereo, lay down on the sofa and tried to drink the contents of the bottle. Her mouth and throat felt like they were on fire but she managed to consume most of it. A while later, she began to feel queasy but was able to stop herself from vomiting for some time. However, eventually she was violently sick several times which further aggravated the pain in her throat. The following morning, after an almost sleepless night, the pain was getting steadily worse and she tried to drink some milk but had difficulty in swallowing it. Meanwhile at work her absence was noted and one of her friends decided to call round at lunchtime to see if she was ill. Although they rang the doorbell several times and thought they could hear movements inside there was no answer. Worried, the friend went to find the caretaker who, after also ringing the bell and receiving no reply opened the door with his master key.

Paramedics were called and took the woman to the accident and emergency department of a nearby hospital. Her urine gave a positive dithionite test for paraquat and her blood plasma was later found to contain 1.7 $\mu g/ml$ of the chemical. She was immediately placed on haemodialysis to remove paraquat from her bloodstream but it was decided that it was too late to try using activated charcoal to prevent further absorption. In an attempt to combat the free radical damage, she was given intravenous ascorbic acid (vitamin C), α-tocopherol (vitamin E) and N-acetylcysteine. A chest X-ray showed that the lower lobes of both lungs were beginning to fill with fluid. Her blood chemistry showed abnormally high levels of urea, creatinine and liver enzymes. The next day she was having increasing difficulty in breathing and a second X-ray showed that air was leaking from the lungs into the surrounding tissues. The decision was made to intubate and provide mechanical ventilation. Meanwhile, her kidney function was deteriorating causing a further increase in blood urea and creatinine concentrations. Despite further supportive therapy, she died at 1.30am the following day from multiple-organ failure.

Suggested Further Reading

International Programme on Chemical Safety (1984). *Environmental Health Criteria 39: Paraquat and Diquat*. Published by the World Health Organization, Geneva, Switzerland (ISBN 92-4-154099-4).

McKeag D, Maini R and Taylor HR (2002). The ocular surface toxicity of paraquat. *British Journal of Ophthalmology* 86, 350–351.

Smith P and Heath D (1975). The pathology of the lung in paraquat poisoning. *Journal of Clinical Pathology* 28(Suppl 9), 81–93.

Case Reports

Erickson T, Brown KM, Wigder H and Gillespie M (1997). A case of paraquat poisoning and subsequent fatality presenting to an emergency department. *The Journal of Emergency Medicine* 15(5), 649–652.

Sittipunt C (2005). Paraquat poisoning. *Respiratory Care* 50(3), 383–385.

Walder B, Bründler M-A, Spiliopoulos A and Romand JA (1997). Successful single-lung transplantation after paraquat intoxication. *Transplantation* 64(5), 789–791.

16

PHOSPHORUS

S. C. Mitchell

Introduction

"In mere size and strength it was a terrible creature which was lying stretched before us. It was not a pure bloodhound and it was not a pure mastiff; but it appeared to be a combination of the two — gaunt, savage, and as large as a small lioness. Even now, in the stillness of death, the huge jaws seemed to be dripping with bluish flame, and the small, deep-set, cruel eyes were ringed with fire. I placed my hand upon the glowing muzzle, and as I held them up my own fingers smouldered and gleamed in the darkness.

'Phosphorus', I said".

The Hound of the Baskervilles, Sir Arthur Conan Doyle, The Strand Magazine, London, 1901–1902.

It was this ability of phosphorus to shine in the dark, to emit an eerie greenish glow, to be luminous without visible combustion, that led to a fascination with this substance, and also to its name. Phosphor, the morning star; the planet Venus when appearing before the sunrise — bringing a glow of light to a premature dawn ("the bright morning Star, Dayes harbinger", Milton).

The element we now know as phosphorus was discovered by Hennig Brand in 1669. Brand, a somewhat uncouth and dubious physician and merchant from Hamburg, had acquired wealth through marriage but sought to augment this, like so many at the time, by converting cheap base metals, such as lead, into gold. During his many alchemic experiments, which usually involved roasting a mixture of substances together with *"vital"* concoctions containing a mystical *"life-giving force"*, Brand had fortuitously isolated crude phosphorus. Once discovered, this process remained a closely guarded secret. It was sold for profit to J. D. Kraft on the understanding that a rival experimenter, Johann von Kunckel, would not be told. However, the latter individual did learn that Brand's phosphorus was obtained from human urine and soon afterwards discovered its mode of preparation. The properties of this new element were made public in 1676 and the preparation unequivocally unravelled by Robert Boyle (1680) but only published after his death (1693) − phosphorus prepared by evaporating urine to dryness and distilling the residue with sand (Figure 1).

Confusion still reigned at this time. During the 17th century the term "phosphorus" was loosely applied to more than one "light-bearing" substance. In fact, the term was suffixed to any substance or mixture that was capable of becoming luminous in the dark. Particularly popular was Baldwin's phosphorus (Bologna phosphorus), the name given to barium sulphide produced by roasting the sulphate ore (barytes) with charcoal which, on cooling, glowed with a red-orange light (the phosphorescence was due to heavy metal impurities, e.g. bismuth). To distinguish the "new" phosphorus from other phosphorescent minerals, Ambrose Godfrey Hanckewitz, R. Boyle's assistant who had for a time the monopoly of its sale, called it English phosphorus, Boyle's phosphorus as well as Kunckel's phosphorus, Brand's phosphorus and Kraft's phosphorus or just simply *"phosphorus mirabilis"*.

Phosphorus is one of those elements that can exist in several forms (allotropy). White phosphorus is a colourless (tending toward white) transparent crystalline solid with a waxy appearance which darkens on exposure to light. Often, impurities give a yellowish tinge to this allotrope. White phosphorus is always the form obtained by chemical preparation. It is volatile, unstable and extremely inflammable, burning with an intensely bright light and generating great heat. It spontaneously catches fire when dry in air at around 35°C and should therefore be kept under water. In air at ordinary temperatures it undergoes slow combustion (oxidation)

Figure 1. "The Discovery of Phosphorus" by Joseph Wright (1771). The full title of this work was, "The Alchemist in Search of the Philosopher's Stone Discovers Phosphorus and Prays for the Successful Conclusion of His Operation, as was the Custom of the Ancient Chemical Astrologers". Reproduced with permission of the Derby Museums and Art Gallery, Derby, England.

and hence appears luminous in the dark. Red phosphorus is made by heating white phosphorus to 260°C in the absence of oxygen. This form is more stable and reacts only at high temperatures. Other allotropic forms exist. Violet phosphorus may be formed by heating a solution of red phosphorus in molten lead to 500°C and black phosphorus resembling graphite in texture may be produced by heating white phophorus under extreme pressures. Such forms are of limited practical importance.

In nature, phosphorus is not found free but only in combination with oxygen in the form of phosphates. The commonest mineral is phosphorite (tricalcium phosphate) which occurs in Russia, North Africa, parts of the Southern United States and in smaller quantities around the globe. There are many other native phosphates including coprolite and the apatites (fluorapatite, chlorapatite and hydroxyapatite). Industrial preparation involves the heating of calcium phosphate from such mineral deposits or bone ash with silica chippings or sand and coke or anthracite in an electric furnace to about 1500°C. Phosphorus distils over as a vapour and is condensed under water to produce the white allotrope. This can be further purified and converted into red phosphorus if required, although the bulk is now used in the production of phosphoric acid.

The main use of elemental phosphorus was in the match industry where its inherent combustible nature permitted a vast improvement in the production of a portable means of ignition. In 1807, an individual had to carry around "chemical tinder" — primitive matches containing a crude mixture of potassium chlorate and sulphur which were set alight by dipping them into sulphuric acid. Alternatively, a "fire-box" could be carried in the pocket. This was a small glass phial internally coated with white phosphorus and kept sealed until a small piece was removed onto the end of a common match whereupon, exposed to the air, it instantly caught fire. In a chemistry lecture delivered at the Royal Institution of Great Britain in 1828, a Professor Brande stated that this latter "fire-box" was "... *very simple, and, indeed, the best mode of obtaining light*". (*Lancet*, Sat. 12th Jan. 1828, Vol. 13i, pp. 552–554).

The actual incorporation of white phosphorus into the match mixture permitted the construction of "friction matches" which were introduced in 1833 (by Friedrich Moldenhauer of Darmstadt, Germany) and became known as "congreve matches" (after Sir William Congreve). However, this name was relatively short-lived and the previous term of "lucifer matches" applied generally to the earlier non-phosphorus-containing versions ("sulphur matches") continued to be used.

The relatively non-poisonous red phosphorus was discovered in 1845 (Anton Schrötter in Vienna) and had been used since 1851–1852 (some sources say the Paris Exhibition of 1855), either in the match itself or within the material of the surface upon which the match was rubbed ("safety matches"). However, "friction matches" incorporating white phosphorus continued to be made for a least another half-century and it wasn't until the realisation, in 1898 (Sévène and Cahen in France), that the non-poisonous phosphorus sesquisulphide (P_4S_3; previously discovered by Berzelius) could be used successfully as a substitute, that an agreement (eventually in Europe, the "Berne Convention 1906"; in USA, the "Esch" law and "Match Act" of 1912) finally put an end to the use of white phosphorus in the match industry.

Phosphorus itself also finds use in chemical syntheses (mainly via phosphoric acid), the firework (pyrotechnic) industry and in the manufacture of incendiary devices, smoke bombs and tracer bullets. Phosphorus is added to bronze (copper and tin) in order to make the extremely strong, tough and fatigue-resisting alloy, phosphor-bronze, which was at one time the choice for ship's propellers. The potential for incorporation of phosphorus into more modern alloys is also of increasing importance. White phosphorus, spread as a paste (2%–5% by weight) on bread, cheese or other foods has been, and in some areas still is, employed as (an extremely dangerous) rat poison. Its odour and taste are apparently attractive to rats. It has been anecdotally reported that in the 19th century, when white phophorus was common in matches, rats often caused serious fires by gnawing at the heads and igniting the whole box.

Toxicity Profile

White phosphorus is the poisonous allotrope and causes most concern. Red phosphorus is nonvolatile, insoluble in water and most solvents and is practically unabsorbable. Consequently, red phosphorus is relatively non-toxic unless it contains traces of white phosphorus as an impurity, although it may provoke dermatitis and inhalation of dust may initiate acute pneumonia.

The external application of white phosphorus to the skin gives rise to severe burns. Since the body temperature is about 37°C the handling of white phosphorus may ignite it. If this occurs the phosphorus adheres to the flesh being difficult to remove, causing an extremely painful and intractable burn. Indeed, an abstract from a French source reported in the "Foreign Department" section of the weekly London medical journal, the *Lancet*, advocated the use of burning phosphorus in medical practice (moxibustion).

> "A small piece of phosphorus, of the size of a lentil-seed, being placed on the skin, is set fire to: the burning causes considerable pain, and is followed by much ... intense inflammation ... This method M. Paillard asserts to have been successful in several cases of asthenic ulcers, chronic bronchitis, rheumatism, and several other inveterate diseases of an asthenic character ..."
>
> *Lancet, Sat. 30th May 1829, Vol. ii, p. 261.*

White phosphorus is thus always handled with tongs and manipulated, where possible, under water. Perhaps unbelievably, continued small doses of white phosphorus have been prescribed to stimulate blood-making and the growth of bone, to act as a tonic for the nervous system and even to offset sexual impotence — benefits apparently not achieved from either organic or inorganic phosphates. In 1678, Johann von Kunckel published a public pamphlet entitled, *"Oeffentliche Zuschrift vom Phosphor Mirabile und dessen leuchtenden Wunderpilulen"*. It is not known if Kunckel's "shining wonder-pills" were meant to be taken internally or simply "magic-pills" for entertainment purposes. If ingestion was intended, the dose would have to have been calculated extremely carefully so as not to result in spectacularly fatal results!

Exposure to white phosphorus may be acute — a single relatively large dose over a short period of time, or chronic — repeated minute doses over long time periods. Acute exposure usually occurs through the accidental or intentional ingestion of rat poison. The chewing of small fireworks containing phosphorus or the soaking of "lucifer" match heads in water or an alcoholic beverage were also popular methods of preparing phosphorus "tonics" or poisons. Chronic exposure typically occurs in industrial situations, notoriously in the old white phosphorus match industry, but more recently in firework factories and chemical processing plants.

Acute exposure

The approximate fatal dose of white phosphorus is 50–100 mg, but as little as 15 mg may produce poisoning. A reasonable estimate would be about 1 mg phosphorus per kilogram of body weight. Prompt medical intervention may prevent fatalities, but those who have ingested larger doses (1 g and over) will probably not recover.

It has been suggested that there are different types of phosphorus poisoning. Patients have been collected together into various groups dependent upon their overriding signs and symptoms, although the boundaries of these groups appear to be blurred and overlap. This may be correct, with certain individuals having inherent susceptibilities and reponding in distinct and discreet fashions. However, differences such as the actual amount and physical state of phosphorus taken, the way in which it has been administered and physiological factors of the patients (e.g. age, nutritional status) may help create these apparent groupings.

Following ingestion, a burning sensation develops immediately in the upper gastrointestinal tract — the mouth, throat and oesophagus. The individual starts to belch gas which usually has an odour described as resembling that of garlic. A feeling of nausea overcomes the patient which will be accompanied by severe abdominal and epigastric pain. Vomiting commences anything from half to six hours after intake, but once initiated it becomes persistent and exhausting for the patient who usually develops a raging thirst. The vomit is copious and blood-streaked, may have a characteristic garlic-like odour and glow if observed in the dark — a diagnostic feature. Constipation is usual, but diarrhoea may occur later when the watery faeces may contain blood and be luminescent. An asymptomatic period occurs during the second or third day and may last for 24–48 hours. During this time there is a remission of symptoms and the patient may appear and feel much better, so much so that they may be discharged from hospital. In other individuals, the symptoms recede but do not disappear. However, the gastrointestinal problems commonly reappear accompanied by organ involvement. The liver is usually enlarged and may be readily palpable below the lower rib margin. This enlargement continues and is typically accompanied by jaundice, first detected in the eyes and then in the skin generally, which may take on a yellowish, or even greenish, appearance. The earlier jaundice appears, the worse the prognosis. Subcutaneous bleeding may be apparent on the body surface and such events may be extensive both externally and internally, leading to severe haemorrhagic complications. Circulatory failure may precipitate shock and hypotension. Central nervous system involvement may include lethargy, restlessness and stupor, progressing to convulsions and delerium, through coma to death.

This sequence of events typically take up to a week to unfold and involves patients who have ingested moderate amounts of phosphorus (400–800 mg). Those individuals who have taken larger amounts (1 g and over) may die more quickly (six to 48 hours) and develop a "fulminating acute poisoning" where cardiovascular collapse, especially in the peripheral vessels, is the cardinal feature.

Chronic exposure

Chronic poisoning resulting from exposures exceeding five years involves a classic condition known as "phossy jaw" as well as an increased tendency for spontaneous bone fractures throughout the body, anaemia and weight loss.

"Phossy jaw", first reported in Vienna in 1839 (first case in Britain 1846), is a progressive condition in which degenerative changes take place in the jaw bone. A thickening and inflammation of the periosteum (fibrous membrane covering the bone) occurs together with ulceration, infection and the development of an offensive discharge. The teeth may decay and become loose with the sockets failing to heal after extraction. The jaw bone is exposed and entry of bacteria cause necrosis with the formation of pieces of dead bone which become detached from the surrounding healthy tissues (sequestrum). Disability occurs together with gross distortion and disfigurement of the jaw. There is direct damage to the walls of the blood vessels with subsequent continuous exudation and the linings of the mouth may become a dull red in appearance. A lowered immune response to local infection occurs owing to the decreased blood supply as the channels within the bone which convey the vasculature (Haversian canals, Leeuwenhoek's canals) are gradually filled in with dense bone. Some individuals are more susceptible to this condition than others (time to development after exposure to phosphorus, ten months to 18 years) and poor dental hygiene has been shown to contribute to susceptibility. Fortunately, this condition is now rare in industrialised countries.

Phosphine gas poisoning

Completely pure phosphine (PH_3) is odourless, but impurities (higher phosphines and organic alkyl derivatives) usually impart a strong fishy or garlic odour. The gas is only slightly soluble in water and denser than air therefore lying close to the ground. It is spontaneously inflammable in the atmosphere if, as customarily, there are traces of diphosphine (phosphorus dihydride, P_2H_4) present, when it burns with a luminous flame.

The faint luminosities occasionally seen hovering or flitting over marshy ground (*ignus fatuus* — foolish fire, Will-o'-the-wisp, Jack-a-lantern) and those noctilucous spectres allegedly dancing over graveyards and believed to forewarn the observer of funereal doom (corpse-candle, corpse-fire) are purported to be due to the spontaneous combustion of phosphine derived from decaying organic matter. It is doubtful if sufficient phosphine to cause such a phenomenon could be produced by decay alone, but a small quantity would provide the ignition for an admixture of hydrogen and methane (marsh gas) also produced by microbial decay.

Exposure to phosphine arises from the action of moisture on phosphide salts. Aluminium phosphide is used in this way to produce large amounts of phosphine gas which acts as a "denser-than-air" fumigant throughout the holds of grain freighters. Calcium phosphide is employed in the production of flare mines and emergency flares that ignite spontaneously on contact with, and are not extinguished by, sea water. Zinc phosphide is incorporated into a variety of rodenticides. Acetylene, a gas used for welding (and previously for lighting, etc.), is usually contaminated with traces of phosphine (and hydrogen sulphide) when produced by the addition of water to calcium carbide. The offending phosphide is derived from impurities within the coke used to make the calcium carbide.

Acute inhalation of phosphine causes a feeling of restlessness and nausea followed by tremors, fatigue, slight drowsiness and increased thirst. Severe abdominal pain in the region of the diaphragm occurs together with a feeling of coldness, nausea, vomiting and diarrhoea. There is often a headache with dizziness, double vision, paresthesia (burning, pricking, numbness sensation) and distressing disturbances of balance. An awareness of chest pressure and burning substernal pain may follow, and the patient may find it difficult to breath (dyspnoea), developing a cough and sputum. Bronchitis, lung damage and oedema may ensue and convulsions and coma often precede death. If the exposure is overwhelming, rapid death from pulmonary oedema may result. Chronic poisoning, continued exposure to very low concentrations of phosphine, is characterised by necrosis of the nasal septum, gastrointestinal disturbances, anaemia, bronchitis, and neurological events such as visual, speech and motor function problems. Damage to the nervous system has been shown to persist for at least 18 months after exposure has ceased and may continue to do so indefinitely.

Mechanisms of Action of Phosphorus Compounds

The mode of action of phosphorus on living tissues is unknown. It is, as yet, impossible to correlate the observed clinical or pathological features of phosphorus poisoning with the disruption of function of any particular enzyme or group of enzymes, although some appear to be inhibited. There are a great many measurable biochemical effects, but no definite insights into the underlying biochemical lesion.

It is fashionable to refer to phosphorus as a general protoplasmic poison and it has been suggested that it uncouples oxidative phosphorylation in mitochondria, thereby decreasing the availability of high-energy phosphate compounds (e.g. ATP; adenosine triphosphate) for general metabolic processes and the overall maintenance of cellular integrity. Phosphorus certainly appears to interfere with fat metabolism, resulting in its accumulation in the liver and other organs (e.g. fatty degeneration of the heart), but there are a number of mechanisms which may underly this complicated pathological response. In addition, fatty liver (steatosis) is not unique to phosphorus ingestion but occurs following exposure to many other toxic compounds.

The clinical pictures of both phosphorus and phosphine poisoning are similar, despite the differing routes of intake. It has been intimated that phosphorus toxicity may be due to phosphine, and that phosphorus is converted to phosphine in the liver, or more probably that this reduction would occur via microbially-assisted metabolism within the gastrointestinal tract before absorption. Phosphine has been detected in the flatus of cows. However, although interesting, this merely avoids the issue and does not provide any further understanding of the molecular mechanisms of toxicity.

It has also been commented upon that although phosphorus has a distinctive odour when burnt in limited oxygen supplies, it does not resemble that of garlic, and that the garlic-like odour repeatedly referred to in the literature concerning phosphorus poisoning may actually be that of an arsenic contaminant. Such arsenic may have been derived from the coal within the furnaces when phosphorus was extracted. It is true that certain parts of the phosphorus toxicity profile do resemble those of arsenic poisoning and this potential co-ingestion would certainly complicate the overall picture.

Treatment

There is no specific therapy for phosphorus poisoning and there are no known antidotes.

The removal of phosphorus as rapidly as possible after ingestion by the induction of vomiting or gastric lavage is of utmost importance. Dilute potassium permanganate or hydrogen peroxide solutions are usually used in preference to water as they may oxidise some of the phosphorus to harmless phosphate. Mineral oil (liquid paraffin) should also be given via stomach tube to act as a purgative, thereby hastening the gastrointestinal transit of the ingested phosphorus and also helping to decrease its absorption. Additional doses of mineral oil may be given regularly by mouth. The use of blood transfusions to enhance elimination are of questionable benefit and should only be undertaken if renal failure develops. A diet high in carbohydrates and protein but low in fats, accompanied by large amounts of crude liver extract and B vitamins, is usually recommended. It has been suggested, though not proven, that supplements of the sulphur-containing amino acids methionine and cysteine/cystine, may offset or reduce liver necrosis. The use of cortisone acetate has also had beneficial effects in one case of severe poisoning.

General supportive therapy involves closely monitoring the patient in an "intensive care" environment with special regard for heart dysrhythmias, whilst replacing necessary fluids and electrolytes and measuring blood chemistry and liver and kidney functions on a regular basis. Owing to the corrosive nature of phosphorus, the patient's clothes, as well as ejected material (stomach washings, vomit, faeces) should be handled with care by workers wearing suitable protective clothing.

Case Histories

An interesting case was reported in 1829 amidst the "Foreign Department" section of the *Lancet* medical journal. This must be one of the earliest recorded instances of known phosphorus poisoning.

"M.Ch.E. Dieffenbach, chemist at Biel, has lately fallen a victim to his zeal for science. He had been for some time engaged in making experiments on several powerful remedies, and at the end of last year began to try the effect of phosphorus, first in a dose of one grain, which was eventually increased to three grains. On the evening of the day on which he had taken the latter dose, he felt very ill, and a violent pain in the stomach, which he unfortunately attributed to a cold, and took no notice of it. After a few days, the pain in the abdomen having increased, he began to vomit a great quantity of greenish matter of a garlick-like smell. A physician was at last called in, and everything done to allay the irritation of the stomach, but without any effect; convulsions, and a paralytic affection of the left arm succeeded, and the patient died on the 12th day after the experiment".

Lancet, Sat. 20th June 1829, Vol. ii, p. 357

A grain, one of the smallest apothecaries' weights (originally the weight of a grain of wheat taken out of the middle of the ear), is equal to one-60th of a drachm, and is the equivalent of 0.0648 g. An ingested dose of 0.194 g (three grains) by an average man of 70 kg weight gives a dose rate of 2.78 mg/kg — well within the fatal range. This illustrates the potency of phosphorus — a lethal dose was almost below the limit of accurate weighing during those times.

A second illustration of phosphorus poisoning has been taken from the turn of the century and appeared in the *Lancet* on 22 September 1900.

"On August 11th I was called to see a man who was suffering from vomiting and intense burning pain in the stomach and bowels. His history was that on the 7th, when 'in drink', he had swallowed three-pennyworth of rat-killer. Immediately after swallowing it he complained of violent pain in the stomach and sickness. He was given salt and hot water, after which he was very sick and vomited freely and expressed himself relieved. On the next day he was a little better, although he felt far from well; he, however, dressed himself and walked about. On the 11th he was taken much worse and I was sent for. When I saw him his temperature was normal, he was quite conscious and coherent, and his pulse was 90. He complained of thirst, constant vomiting, and great pain in the stomach and abdomen. An examination of the vomit proved it to consist wholly of altered blood of a very dark colour. His stools also were dark and pitchy in character. He was given the usual remedies, but they were of no avail, and he quietly sank and died on the following Tuesday, the 14th, having lived exactly a week after swallowing the poison. The vomiting of dark coloured blood continued up to the time of his death".

Newey WE , A case of phosphorus poisoning, Lancet, 1900, Vol. ii, pp. 875-876.

A post-mortem carried out the next day revealed a blue staining of the neck and the superficial veins of the limbs. The gastrointestinal tract showed signs of severe irritation — inflammation and ulceration — with leakage of blood and haemorrhages were evident in the greater omentum. The liver, heart and kidneys showed signs of commencing fatty degeneration.

A final example has been taken from the report of a paper read at the Therapeutical and Pharmacological Section of the Royal Society of Medicine on 7 December 1909. It concerns a young girl who ingested rat-poison containing phosphorus in the attempt to procure a miscarriage. This practice, although quite common in Germany, was relatively unusual in Britain.

"An unmarried girl, aged 19 years, on her mistress becoming aware that she was about two months pregnant, was summarily dismissed from her situation and went into lodgings. Towards noon on the following day she took a quantity, afterwards estimated at a drachm, of a rat poison containing about 4 per cent of phosphorus (about 156 mg phosphorus), and within a short time complained of abdominal pain and a feeling of distension. On the next day she was unable to get up until the afternoon, and while dressing she vomited some clear fluid. The sickness recurred and was accompanied by great thirst. The bowels were not moved. She took no food whatever on this day. On the third day, after drinking tea she rose at 10 am and during the morning vomited repeatedly. In the afternoon, finding her left foot swollen, evidently owing to the first subcutaneous haemorrhage, she walked to the hospital, but due to some irregularity on her part was not attended, and returned to her lodgings and had tea. She went to bed early and had a supper of bread and milk. There was some abatement of the symptoms on this day, as was shown by the fact that she was able to get out and also take food. At 2am on the next day she felt so ill that she aroused the people of the house. Vomiting frequently, blood was present in considerable quantity, abdominal pain was severe, and a profuse uterine haemorrhage appeared. Thirst was continuous and unrelieved by the large quantities of water she drank.

The patient first came under observation at 11 am on the fourth day of her illness. She was semiconscious, extremely restless, perpetually turning from side to side in bed, unable to answer questions, the only reply to attempts to rouse her being an appeal for water. Haematemesis was frequent and the uterine haemorrhage copious. On examination a remarkable condition was disclosed. Below the waist-line more than half of the total skin area was the seat of enormous subcutaneous haemorrhages. The front and back of each thigh, both buttocks, the greater part of the left leg and the right foot presented continuous patches, dark purple in colour, with sharply-defined and often crescentic margins. The intervening skin, as elsewhere throughout the body, was normal. There was no jaundice. No satisfactory examination of the abdomen could be made owing to the pain and the girl's restless condition. There was incontinence of urine. The bowels were not moved. The temperature was subnormal, respiration 48, pulse 120 and very feeble. She died at 7.40pm — less than three (? four) and a half days after taking the poison".

> Veale RA and Hann RG, A fatal case of poisoning by phosphorus, with unusual subcutaneous haemorrhages, Lancet, 1910, Vol. i, pp. 163-164.

A post-mortem carried out 19 hours later showed substantial subcutaneous haemorrhaging and bruising around a needle injection site. Numerous haemorrhages were seen in the mediastinal, omental and mesenteric tissues and in the retroperitoneal area, especially in the neighbourhood of the left kidney. The liver was enlarged, bulging from its capsule, and was of a brilliant canary-yellow colour. It presented the appearance of acute fatty degeneration, confirmed by microscopic analysis. The kidneys, on section, were also pale yellow in colour suggesting fatty infiltration.

Suggested Further Reading

For a general history of phosphorus

Threlfall RE (1951). The story of 100 Years of Phosphorus Making 1851-1951. Albright and Wilson, Oldbury, England.

For detailed descriptions of "phossy jaw"

Hughes JPW, Baron R, Buckland DH, Cooke MA, Craig JD, Duffield DP, Grosart AW, Parkes PWJ, Porter A, Frazer AC, Hallam JW, Snawden JWE and Tavenner RWH (1962). Phosphorus necrosis of the jaw: a present-day study. British Journal of Industrial Medicine 19, 83–99.

For reports of nefarious homicidal uses

Polson CJ, Green MA and Lee MR (1983). Clinical Toxicology, 3rd ed. Pitman Books Ltd, London, Chap. 27, pp. 522-539.

17

RADON

J. Woodhouse

Description

Radon (chemical symbol Rn) is a naturally occurring radioactive material which originates from the decay series of the primordial radionuclides uranium and thorium, present at the formation of the earth. It is a member of the noble (or inert) gas family which comprises helium, neon, argon, krypton, xenon and radon, in order of increasing atomic weight. There are several different isotopes, depending on the route of origin. The isotope ^{220}Rn arises from the thorium-232 decay chain (and was historically known as thoron). ^{222}Rn arises from the uranium-238 decay chain (see Figure 1 and Table 1) and is the isotope of main concern to man, being the most prevalent radon isotope in the environment. It is ^{222}Rn that we are mainly concerned with here.

Occurrence and Source

The ultimate source material of ^{222}Rn, i.e. the most abundant uranium isotope ^{238}U, is present in the soil and rock of the earth's crust in varying amounts. A typical range of uranium activity found in ordinary soil is from 7 to 40 Bq kg^{-1}. As the decay chain of the parent progresses, the first few daughter products are retained *in situ*, being solids, until the decay of ^{226}Ra to ^{222}Rn occurs. This daughter product is both a gas and chemically very unreactive. It may therefore percolate through the rocks and soil to escape into the atmosphere or into ground water. The level of release to atmosphere depends *inter alia* on the nature of the underlying rock

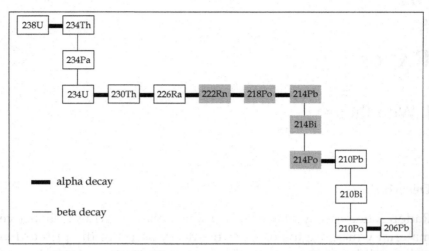

Figure 1. The decay chain of ^{238}U leading to production of ^{222}Rn and the important radon daughters (boxes highlighted). The half-lives of the chain members are listed in Table 1 below. The relatively high abundance of ^{238}U in rocks and soils means that radon gas is, and always has been, present to some degree in the atmosphere.

Table 1. Elements in the uranium-238 decay chain and their half-lives. Note the very long half-life of ^{238}U (approximately equal to the age of the earth) which provides the continuing source of all other chain members. The half-lives of the radon daughters (highlighted in Figure 1) are all less than 30 minutes and so may reach varying degrees of equilibrium with radon in air depending on the factors which act on them.

Key to nuclides and their half-lives:

Uranium	^{238}U	4.5×10^9 years				
	^{234}U		2.7×10^5 years			
Thorium	^{234}Th			25 years		
	^{230}Th			8×10^4 years		
Protactinium	^{234}Pa				7 hours	
Radium	^{226}Ra		2×10^3 years			
Radon	^{222}Ra			4 days		
Polonium	^{218}Po				3 minutes	
	^{214}Po					1.6×10^{-4} seconds
	^{210}Po			138 days		
Lead	^{214}Pb				27 minutes	
	^{210}Pb		22 years			
	^{206}Pb	**Stable**				
Bismuth	^{214}Bi				20 minutes	
	^{210}Bi			5 days		

and can vary greatly, being highest in regions underlain by igneous rocks high in uranium such as granite and much lower in regions underlain by sedimentary rocks such as chalk.

The radon concentration in air is subject to diurnal and seasonal variations and depends on local meteorological conditions but a level of some 4 Bq m^{-3} is considered typical over average soil. Clearly variations around this value are large. It has been estimated that some 75% of radon escaping from soil is formed in the top 2 m. Not only does the rate of radon escape vary greatly between different places but also at one place at different times.

An atlas of the radon levels present in the British Isles has been compiled by the National Radiological Protection Board (NRPB) which clearly illustrates these variations and the links to the underlying geology.

Obviously this radioactive gas has always been present in the environment and humans (as well as all other species of plants and animals) have evolved in this environment. Indeed the radiation levels must have been rather higher in the past. However, some of man's activities have increased his exposure levels in the very recent past. In particular, underground mining activities result in occupational exposure of miners as radon gas is released from the enveloping rocks, especially where ventilation levels are limited. The severity of this situation again depends on the nature of the rocks being mined and is particularly problematical in uranium mining — as you would expect. More recently still the habit in western civilisation of building dwellings which are well insulated and draught-proofed has led to a situation where radon gas released under the house, or from the materials of which the house is built, is effectively trapped within the house leading to build-up of radon concentration indoors. Again the severity of the problem depends on location. Of course the same considerations may lead to build-up of radon gas in workplaces also.

Thus radon presents an interesting case of a naturally occurring toxic material (i.e. a hazard) which has been present in the human environment throughout evolution but which has only recently come to be considered as presenting a risk due to human activities which have produced scenarios of potential for increased levels of exposure.

Mechanism of Toxicity

The mechanism of toxicity associated with radon is via cellular damage produced by the radioactive particles that its immediate daughter products emit following inhalation and deposition in the airways. The degree of hazard from radon itself is relatively small in comparison with that from the daughter products as most radon inhaled is immediately exhaled again. The exact mechanisms by which damage is produced (i.e. tumours occur following a latent period of some years) are uncertain but include direct damage to chemical bonds in DNA and production of free radicals in the environment of DNA.

As can be seen in Figure 1, radon itself and its polonium daughter products are alpha emitting nuclides, while the isotopes of lead and bismuth produced are beta/gamma emitters. The short half-lives of the daughter products prior to ^{210}Pb (Table 2) result in the rapid production of a mixture of airborne radioactive materials which may attain equilibrium concentrations within a relatively short time. The half-life of ^{210}Pb is 22 years and at this point in the decay chain any activity inhaled is largely removed from airways in which it is deposited before any appreciable decay occurs.

Table 2. Half-lives and main emissions of the "Radon daughters". The alpha emissions are considered to be the source of biological damage to the respiratory system, the beta and gamma emissions depositing, relatively, a very small amount of energy in the target tissues.

Radionuclide	Half-life	Main radioactive emissions
^{222}Rn	4 days	5.5 Mev alpha particle
^{218}Po	3 minutes	6.0 Mev alpha particle
^{214}Pb	27 minutes	Beta and gamma emissions
^{214}Bi	20 minutes	Beta and gamma emissions
^{214}Po	<1 millisecond	7.69 Mev alpha particle

The radon daughters are solid elements and are formed in a charged state. As a result they readily attach to aerosols (very small particles in the air) and are therefore breathed in, along with the parent gas. While the majority of radon gas is immediately breathed out again the solid daughters deposit in the airways where they can irradiate the cells of the bronchial region. The main consequence is an increased risk of lung cancer although other lung disorders may also be produced at very high doses (e.g. emphysema).

The degree to which radon daughters attach to aerosol particles is an important factor in determining the radiation dose to which bronchial cells are exposed. The unattached fraction (f) remain charged and are deposited in the airways more efficiently than the fraction attached to aerosol dusts. They therefore produce a higher resultant radiation dose to sensitive tissue. The particle size distribution of the aerosol-attached activity will also have a major effect on the airway deposition pattern and hence the radiation dose received in various regions of the respiratory system.

A further consideration of importance in determining the radiation dose resulting from inhalation of a particular mixture of radon and its daughter radionuclides is the degree of equilibrium reached in the decay process. To take account of the wide variations in degree of equilibrium reached in a particular situation, a somewhat complicated (but operationally convenient) system of quantifying exposure level was introduced in the uranium mining industry and is now widely used in quantifying radon exposure levels. A working level (WL) is defined as any combination of radon progeny (i.e. ^{218}Po, ^{214}Pb, ^{214}Bi and ^{214}Po) in one litre of air that results in the emission of 1.3×10^5 Mev of alpha particle energy. This quantity (which can be measured by air sampling), in combination with duration of exposure for a particular worker allows his exposure to be expressed in terms of working level months (WLM), where 1 WLM is equivalent to exposure at 1 WL for 170 hours (the hours worked in one month). This unit is also now used in the domestic situation. For example, exposure at 1 WL for 30 days at an occupancy of 12 hours per day would produce a cumulative exposure of 2.12 WLM (i.e. 360 hours × 1 WL/170). There is clearly scope for confusion in this system of exposure units.

In addition to the above physical factors which govern toxicity in respect of a particular radon and daughters mixture, a wide variety of biological factors will affect the eventual dose received by an individual exposed to the mixture. In particular, the breathing characteristics of the exposed individual — tidal volume and breathing frequency — will clearly determine the total activity entering the body while nose or mouth breathing pattern will affect deposition in the upper respiratory passages. The pattern of nose or mouth breathing is frequently dependant on degree of physical activity and is an obvious factor that will differ between domestic and working situations.

Finally, the particular characteristics of the lung of an exposed individual will affect dose. The bronchial morphometry will determine deposition pattern following which the efficiency of the mucociliary clearance mechanism for that individual will determine the time of exposure. Mucus thickness and the depth of the target cells will also be important and may be affected by smoking habits. As most of the mining workers studied to assess radon effects were smokers this introduces another point at which extrapolation of epidemiological results from the mining populations to domestic situations is not straightforward.

The various factors that have an influence on the dose received by respiratory tract cells following inhalation exposure to radon and its daughters are summarised in Table 3.

Health Effects of Radon and Daughters

Many studies have established that cancers can be caused by exposure to ionising radiation, including exposure to alpha-emitting radionuclides. In the case of radon the exposure route of relevance is inhalation and the effect seen is an increase in levels of lung cancer in the exposed population. This effect has been seen in many epidemiological studies of miners working in both the uranium mining industry and in mining for other metals. Because lung cancer also occurs as a result of many other causes, for example smoking, the establishment of a quantitative dose-risk relationship

Table 3. Main factors influencing the radiation dose received by an individual following exposure to radon and daughters. Clearly, different individuals exposed to the same air would receive different doses. Broad models, based on the above parameters have been devised to assess average doses to average individuals for dosimetry purposes. Results from such models vary by factors of about 2 or 3, but this represents only one part of the sources of uncertainty in estimating risks from Radon exposure.

Physical characteristics of air:
    Size of unattached fraction
    Particle size of aerosol particles to which daughters attach
    Degree of radioactive equilibrium achieved

Breathing pattern of exposed individual:
    Volume of air breathed in (tidal volume)
    Rate of breathing
    Whether mouth or nose breather

Lung characteristics of exposed individual:
    Morphometry of airways
    Clearance rate of mucus lining airways
    Distance to target tissue - mucus thickness

---

is problematical. To date there is no clear evidence from studies of radon in homes that such exposure leads to excess lung cancer risk and the assessment of such risk depends on extrapolation of data from mining scenarios to the domestic scenario — a process fraught with uncertainties. Nonetheless, several studies have attempted to quantify the risk and have produced figures for lifetime risk of lung cancer mortality due to a lifetime exposure to radon progeny — i.e. the domestic situation. Values of from about 100 to about 800 excess lung cancer deaths per 10^6 person WLM have been suggested, but to date there is no way of validating such estimates.

Experimental work with animals (mainly rats and dogs) has been carried out in France and in the US. Early work considered only acute exposure conditions but more recent work has involved chronic exposure and has demonstrated lung cancer induction in both species. Extrapolation of these results to humans is not straightforward however given significant differences in, for example, lung morphometry and the fact that the location and histopathology of the experimentally produced tumours in animals differ from the normal pattern seen in humans.

As would be expected the efficiency of tumour production was found to increase with increased cumulative exposure, the increase being approximately linear. However, a more unexpected finding from this animal work was that, for a given total exposure level, tumour incidence increased with decreasing rate of exposure. Efficiency of tumour production was also found to increase with an increase in the unattached fraction and with an increase in the degree of disequilibrium of the daughter isotopes. These results serve to indicate the complexity of the problem.

In addition to the induction of lung cancer, the animal studies indicated an association between radon exposure and other non malignant respiratory diseases. Exposed animals developed emphysema and interstitial fibrosis, but only at very high levels of exposure. Some epidemiological studies on miners have also reported an increase in mortality from non-malignant respiratory diseases. However, these studies cannot distinguish between the effects of radon daughter exposure and the effects of, for example, other mining dusts, diesel-engine exhaust, etc. found in a mining scenario and a causative relationship with radon exposure has not been established. Levels of radon exposure in domestic situations are such that these types of effect are not relevant.

Radon in Buildings — Exposure of the Public

For many years the potential problem of radon exposure in ordinary buildings — either domestic homes or workplaces — was unrecognised. This was largely due to the inappropriate treatment of exposure from the short lived daughters in early assessments of the radon exposure situation. Only when the major contribution to total dose presented by the daughter product decay was properly accounted for was the full impact of radon on the natural background radiation level, to which everyone is exposed, appreciated. Current assessments of natural background levels now place radon exposure as producing in excess of 50% of all background dose.

The air pressure inside buildings is normally a little below outdoor atmospheric pressure as a result of warm internal air rising and the effect of wind blowing across openings. Air is therefore drawn into a building. Where ventilation is limited the indoor radon concentration can become rather higher than the outdoor concentration. Internal radon concentrations have been found to be log-normally distributed, with a few results being very much higher than the mean.

Values differ between countries due to differences in climate, rocks, construction methods and materials and living habits. A worldwide figure of 25 Bq^{-3} for the geometric mean level of radon has been estimated by the United Nations Scientific Committee on the Effects of Ionising Radiations (UNSCEAR). Surveys in the UK suggest a national average of about 20 Bq m^{-3}. In some areas − e.g. Finland, Sweden − concentrations of many thousands of Bq m^{-3} have been measured in large numbers of houses.

Studies of the behaviour of radon in dwellings have now been carried out in many countries. These studies show that overall the rate of radon entry is log-normally distributed, with a wide range between maximum and minimum rates. Thus, in one study in the UK the range of entry rate was from <1 Bq m^{-3}h^{-1} to >150 Bq m^{-3}h^{-1}. In Sweden the range measured was from about 3 to >600 Bq m^{-3}h^{-1}.

Radon enters buildings from several sources:
- The soil/rock underlying/surrounding the building.
- The building materials.
- Water supplies.
- Natural gas supplies.

Underlying soil and rock

Radon originating in underlying soil may enter a house by either diffusion or in response to the pressure driven flow of air through the house structure and openings. The second of these two mechanisms is normally the most important contributor. Pressure driven flows arise from differences in indoor-outdoor temperatures and from the wind. In any particular case, the situation will be dependent on, *inter alia*, construction methods, weather conditions, living habits of occupants. On average the contribution of inflow of radon from soil and rocks has been estimated at about 80% of the total.

Building material

Radon in building materials may enter a building by diffusion. Concrete and brick are the most common building materials. The level of ^{226}Ra in building materials, in combination with the effective porosity of the material to radon arising from the ^{226}Ra, will govern the level at which radon enters the air of the living space. Levels of ^{226}Ra in building materials range from as much as 1500 Bq kg^{-1} for Swedish aerated concrete based on alum shale to as little as 10 to 20 Bq kg^{-1} for most other forms of concrete. For most forms of brick ^{226}Ra concentrations are in the range 20 to 60 Bq kg^{-1}. The average contribution to indoor radon from building materials has been estimated at about 10% to 15% of the total, although deviations around this figure may be very large.

Water supplies

Levels of radon in water vary widely, normally being much greater in deep water supplies than in surface water. As a general rule levels in surface water will be similar to ^{226}Ra levels — i.e. about 10 Bq m^{-3}. For ground water extracted from deep wells drilled in granitic areas much higher levels are found. For example, levels of up to 77 MBq m^{-3} have been found in Finland and of around 20 MBq m^{-3} in the US. It has been estimated that from 1% to 10% of the population of the world drink water derived from deep wells with a radon concentration in excess of some 100 kBq m^{-3} while the remainder drink water from surface sources with an average radon concentration of below about 1 kBq m^{-3}.

Radon in water enters the air as a result of degassing. Concentrations will vary being highest in rooms to which water is delivered — kitchen and bathroom. Assuming a radon concentration in tap water of 1000 Bq m^{-3} and reasonable factors for degassing efficiency and water use rate, a radon entry rate of about 0.1 Bq m^{-3}h^{-1} has been estimated, indicating this to be a very minor source in comparison with the contribution from underlying soil and rocks. It can be important, however, in those regions having exceptionally high concentrations of radon in water.

Natural gas supplies

Radon levels in natural gas vary over a wide range — from undetectable to levels of some 50 kBq m^{-3}. Where storage occurs prior to usage some decay of activity occurs but domestic burning of gas for purposes such as heating and cooking will lead to some enhancement of indoor radon levels. Few measurements have been made in this area but in the US it has been estimated that average radon concentrations in natural gas are about 1000 Bq m^{-3} and that for a reference house this would lead to a radon entry rate of about 0.3 Bq m^{-3}h^{-1}, a small contribution in comparison with that from soil and building materials.

Preventative Measures

Several measures are available to reduce indoor radon levels. Their application would depend on a cost-benefit analysis approach in accordance with the recommendations of the International Commission on Radiological Protection (ICRP). The measures available include:

Soil depressurisation. Reduction/reversal of the pressure differential between the building interior and the radon source (soil). This is normally achieved by drawing air from under the building with a small fan and discharging it to the external atmosphere. The cost of this approach is not great and the effectiveness is usually high.

Sealing of surfaces. This works both by reducing ingress of radon through structural cracks and by limiting emanation of radon from building materials. In practice it is difficult to achieve. Costs are moderate but effectiveness is also moderate.

Radon source removal. This involves water treatment, which may be worthwhile where radon levels in water are a major contributor to overall levels, or removal of underlying soil. In the latter case, costs can be high while the costs of water treatment are more moderate. In both cases effectiveness is high.

Increasing ventilation. The objective is to dilute the radon and progeny. However the level of ventilation required will often be unacceptable to occupants and may result in high costs for heating (or cooling). Furthermore, a decrease in building pressure may occur which will tend to increase radon concentration and negate the intended effect.

Increased air movement. The objective here is to increase levels of radon daughter deposition on surfaces, so removing them from breathable air. Alternatively some form of filtration could be employed. The costs of this option are generally low, but so is the effectiveness.

Estimation of Doses from Indoor Radon

For a known level of radon in air a variety of other factors come into play in estimating the radiation dose resulting from domestic exposure. For example an occupancy factor is required. Clearly this will vary depending on individual lifestyles and habits. A global figure of 80% indoor occupancy has been derived by UNSCEAR while work in the UK suggests a slightly higher figure of 90% is applicable here. Dose will vary according to whether mouth breathing or nose breathing is involved. State of health might affect levels of bronchial mucus present and effectiveness of clearance of mucus from the airways. Smoking habits will be important.

Several different dosimetric models have been developed to calculate radiation dose due to radon exposure taking into account all the various parameters involved. Such models are generally complex and beyond the scope of this text. In most epidemiological studies, dose figures for the exposed population are not available and such studies are carried out using exposure level — in terms of working level months (WLM) — rather than radiation dose to target tissues. As is apparent from the above the relationship between exposure and dose is far from direct.

Mining and Lung Cancer — How Radon was First Identified as a Problem

The fact that miners showed an unusually high death rate from lung disease was noted as early as the 16th century in relation to silver mining activities in the Schneeberg region of Germany. The observation was reported by Paracelsus (1493–1541) in a book printed in 1567, after his death. As mining activity increased through the 17th and 18th centuries the level of disease also increased but it was not clearly identified as lung cancer until 1879. At this time as many as 75% of miners in this area died of this disease.

The cause of the disease was initially assumed to be inhalation of metallic ore dust and accordingly Paracelsus called the disease "Mala Metallorum". Following the discovery of radon by Pierre and Marie Curie in the late 1890s from ore obtained from the region, a high concentration of radon was demonstrated in the mines in the early 1900s. At about the same time the first cases of cancer resulting from radiation were reported and a relationship between the lung cancer rates and the high levels of radon in the mines was assumed in some quarters. The causal link was still not proved, however, and there was dispute as to whether it was the radon that was responsible for the lung cancers or inhalation of ore dusts, or arsenic or other mine contaminants. The poor health of the workers was also put forward as a possible cause. Research carried out in the 1930s by Rajewsky, which involved radon measurements in the mines, measurement of alpha activity in tissue samples and histopathology of lung tissues from lung cancer victims, provided much evidence in support of the theory that radon inhalation was a causal mechanism of the cancers. However, the relationship could not be quantified and the role of the radon daughter products was not appreciated at this time.

In the 1940s a great deal of effort was directed to uranium mining by several countries for military purposes. Little attention was paid to radon levels and few measurements were made. In general it was thought that radon levels in the new mines were much lower than in the old Schneeberg mines. Furthermore, research into the radiobiology of radon gas did not seem to support the idea that this gas could be producing lung cancer. The problem was that the impact of the radon daughters was not appreciated at this time. It was not until the 1950s that the role of the radon daughters was understood and it was realised that the radiation dose to the lung from these daughters would be far greater than that from the parent radon gas. As a consequence the concept of the Working Level Month (WLM) was devised and introduced to quantify exposure levels.

In the 1960s and 1970s health status studies were carried out on various uranium miner populations. One such study on the uranium miners in Colorado suggested a significant excess of lung cancer in the group and resulted in improved control of radiation hazards in the mines. Further studies in the US and in Czechoslovakia found an increasing risk with increasing exposure to radon progeny, although the rate of increase differed significantly between the two studies. Many other studies have now been carried out, including studies in non-uranium mines, which confirm the general finding but quantification remains elusive given the many confounding factors which may arise.

Suggested Further Reading

ICRP (1987). Lung cancer risk from indoor exposures to radon daughters. ICRP Publication 50, *Annals of the ICRP* **17**(1).

ICRP (1993). Protection against Radon-222 at home and at work. ICRP Publication 65, *Annals of the ICRP* **23**(2).

NRC (1988). *Health Risks of Radon and other Internally Deposited Alpha-Emitters.* US National Research Council Report BEIR IV, National Academy Press, Washington, DC.

NRC (1990). *Health Effects of Exposure to Low Levels of Ionising Radiation.* US National Research Council Report BEIR V, National Academy Press, Washington, DC.

Stannard JN (1988). *Radioactivity and Health. A History.* Office of Scientific and Technical Information. US Department of Energy.

UNSCEAR (1988). *Sources, Effects and Risks of Ionising Radiation.* United Nations Scientific Committee on the Effects of Atomic Radiation, 1988 Report to the General Assembly, United Nations, New York.

ANNEX

Radioactive Decay

The radiations produced

All atoms can be conceptualised as consisting of a small central nucleus which contains positive charges, plus a number negatively charged of electrons orbiting outside the nucleus rather like the planets orbiting the sun. The number of orbiting electrons equals the number of positive charges in the nucleus so that the atom overall is electrically neutral. In some atoms the nucleus is unstable and radioactive decay involves loss of energy by such a nucleus as it moves towards a more stable state. This energy can be lost through the ejection of a small particle from the nucleus or the emission of electromagnetic energy from the nucleus, or commonly both events occur together. The two main types of particle emitted are termed alpha and beta particles while the electromagnetic energy released is termed gamma rays.

Alpha particles are identical to the nuclei of the element helium, the smallest of the noble gas family (of which radon is also a member — see above). Each alpha particle carries a positive charge of 2 and has an atomic mass of 4 (note: radon has a nuclear charge of 86 and an atomic mass of 222 — so the alpha particle — mass 4 — represents a very small piece of radon nucleus.). These properties mean that alpha particles strongly interact with whatever material they pass through and despite being ejected from the nucleus at high speed they are stopped very quickly — i.e. they have a very short path length (a few cm in air and much less than a mm in tissue). Thus they deposit all the energy they take away from the nucleus in a very small amount of surrounding material. It is this intense energy deposition that leads to biological damage. Furthermore, the remaining nucleus now has two less positive charges than before the decay and is now a different element, which may decay further in its turn, leading to a chain of decays as, for example, shown in Figure 1.

Beta particles are identical to the electrons orbiting the nucleus and are created in the nucleus at the time they are ejected. They carry a negative charge but very little mass (they are some 7000 times lighter than the alpha particles) and interact with matter over a rather longer distance than alpha particles — from a few mm in tissue to many cm in air, depending on the starting energy of the particle. They therefore deposit their energy over a much larger volume of material than do alpha particles, with correspondingly less biological damage at the points of energy deposition. The loss of one negative charge from the nucleus changes the net positive nuclear charge — increasing it by 1 — and the remaining nucleus is again a different element from that which underwent decay.

Gamma rays are a form of electromagnetic radiation — just as are radio waves, microwaves and light rays. The difference between these various types of electromagnetic radiation arises purely from the amount of energy they carry — gamma rays having very high energy, microwaves relatively low energy and light rays being somewhere between the other two. They have no charge and no mass and interact with matter much less than do alpha and beta particles. Gamma rays therefore travel much further than do the particle forms of radiation having the same energy and they deposit this energy over a correspondingly larger volume. As they have no mass or charge, emission of a gamma ray does not change the elemental identity of the nucleus that emitted it.

Half-life

The loss of a particle or gamma ray from a nucleus constitutes radioactive decay. For each type of isotope the rate at which the decay occurs is characteristic of that isotope. The **decay constant** (λ) is the fraction of nuclei which decay in unit time and the **half-life** ($T\frac{1}{2}$) is the time taken for half the nuclei present to decay. Thus, if at time 0 the number of atoms of an isotope present is $N_0 = 100$, then after one half-life the number present will be $N_0/2 = 50$ and after a further half-life will be $(N_0/2)/2 = 25$, etc. This leads to the familiar exponential pattern of radioactive decay.

18

Ricin

M. J. Ruse

Introduction

Occurrence and historical perspectives

Ricin, one of the most toxic compounds in the Plant Kingdom, is found in the seeds of the castor oil plant, *Ricinus communis*. The toxin is synthesised as the seed matures and is concentrated in storage granules. The castor bean plant, which belongs to the *Euphorbiaceae* (spurge) family, is a large bushy shrub which can grow up to eight to ten metres in height in warm climates. The palmate leaves, alternating from the upright trunk, have a deep green/red colour, a prominant central vein and are divided into seven or nine lobes. The grey-brown or mottled red-brown beans are contained within a brown globular capsule bearing soft spines. Each capsule houses three beans and are clustered with other capsules around a central stalk (Figures 1 and 2). Owing to the attractive nature of these beans, they are sometimes used to make necklaces and other ornaments. The plant is found across the globe but is thought to have its origins in tropical Africa or possibly Egypt or Abyssinia. Scholars have suggested that the large leaves of the plant may be those which sheltered Jonah:

> "And the Lord God prepared a gourd, and made it come up over Jonah, that it might be a shadow over his head, and to deliver him from his grief. So Jonah was exceedingly glad of the gourd".
>
> *(Jonah 4: 6-7. Bible, King James version).*

Figure 1.

Figure 2.

In the Hebrew language the word for gourd is also the name of the castor bean plant. Castor bean seeds are oval shaped varying in size from 0.8 to 2.2 cm and have been found in ancient Egyptian sarcophagi surrounding mostly the graves of priests. These date back 4000 years to when the seeds were worshipped in this area of the world. From Egypt the plant was introduced to Greece and then to the Latin peoples who were aware of its medicinal qualities. In 1764 the plant was mentioned in England in a published paper by Cancane entitled, "A dissertation on the Oleum Palmae Christi, sive Oleum Ricine or Castor Oil".

Today, the castor bean plant is widely cultivated around the world in tropical and temperate climates for the production of castor oil, which reaches about 800,000 tonnes a year. The oil has technical application particularly in the motor industry and in rubber and plastic foam production. As the beans contain nearly a quarter of their weight as protein, the mash left over after oil extraction is often used as animal feed and as fertiliser. The castor bean plant and the oil have had during the course of history, and still continue to have, widespread use in herbal preparations for therapeutic applications throughout the world. The oil is used in Haiti against bronchitis. In tenth century Iran it was used for the treatment of apoplexy. The bean has been used to treat many types of digestive disorders. In China it is thought to reduce swelling of the tongue, and in Mexico the crushed beans are used to counteract gastralgia. Both the bean and the oil are considered to have anthelmintic properties in Brazil, Italy and China. However, the commonest use throughout the world is as a purgative. In India and Mexico both raw and roasted castor beans are taken as cathartics, emetics and as treatments for leprosy, and in combination with honey for curing syphilis. They have also found application in relieving problems associated with the bones, urogenital system, infectious diseases, skin problems, venereal diseases, eye disorders, nervous system complications, and in obstetrics and gynaecology. The fabled panacea indeed!

Medical uses

The purgative properties of the oil are chiefly related to the triglyceride of ricinoleic acid (12-hydroxyoleic acid). The poisonous component of the plant is a substance called ricin. Although ricin is highly toxic it has been examined for various therapeutic applications and as a tool in cell surface property studies as well as other experimental situations. When conjugated with monoclonal and polyclonal antibodies, ricin has been investigated for use in cancer treatment and AIDS therapy. However, a problem resulting from the use of ricin-immunotoxins is the "vascular leak syndrome" where fluids escape from the circulatory and lymphatic systems causing hypoalbuminuria, weight gain and pulmonary oedema. Ricinimmunotoxins have been successfully used to destroy T-lymphocytes in bone marrow during transplant procedures thereby reducing the chance of rejection of marrow by the recipient ("graft-versus-host" disease). With increasing knowledge of the movement of ricin around the cell and the targeting of the poison to specific cellular components, the use of ricin-immunotoxins holds great promise for the future treatment of cancer. In addition, neuroscientists use ricin to selectively kill neurones and induce neuronal lesions. This permits the detailed investigation of many degenerative neuronal diseases and nerve damage scenarios, leading to a greater understanding of the processes involved.

Toxicity Profile

Although spread throughout the plant, it is the castor bean itself that contains significant quantities of ricin. Fortunately, the ricin remains in the fibrous residues of the seeds after castor oil has been extracted from them. Safety is ensured by heating the oil as ricin is temperature sensitive and irreversibly decomposes. If the bean is swallowed without chewing then the testa or seed coat prevents the release of ricin and poisoning does not occur. Chewing prior to ingestion breaks the testa and ricin is released.

Initially, the first symptoms are related to gastrointestinal problems including burning sensations in the alimentary tract, thirst, vomiting, diarrhoea, gastrointestinal tract bleeding, nausea and abdominal pain. Other symptoms include fluid and electrolyte depletion, haemolysis, hypoglycaemia, shivering and fever, and dilation of the pupils. Serum enzyme levels, aspartate and alanine aminotransferase, and extracellular lactate dehydrogenase are also raised. The total bilirubin levels are increased and blood sugar levels decreased. Where poisoning has been severe, haemorrhagic gastritis and dehydration can result. The primary organs where cellular damage occurs are the liver, kidneys and pancreas and post-mortem reports document necrotic lesions in these organs. Death resulting from ricin poisoning is due to hypovolaemic shock, caused by a reduction in blood volume.

There is normally a delay of several hours between the exposure to ricin and the onset of clinical symptoms, but the cytotoxic effects may not occur until two to five days after ingestion. In contrast, certain individuals may display allergic reactions almost immediately after exposure, and this is presumed to be related to the glycoproteins that the beans contain. Episodes of asthma have been linked to the inhalation of ricin dust at castor bean mills — in one incident in 1952 at Bauru in Brazil, 150 people living close to a bean mill were affected by sudden asthma attacks.

The ingestion of eight beans in an adult is thought to lead to death as this quantity contains the lethal dose of 1 to 10 mg/kg body weight. However, doses as low as three beans have been cited as fatal. It should be remembered that because of the possibility of anaphylaxis following ricin ingestion, the consumption of one bean or less could be lethal and this may certainly be sufficient to kill a child. Contrariwise, evidence accumulated from the literature shows that fatalities appear to be low (less than 10%) in individuals who have ingested a supposedly "fatal dose". This may be because of the variable release of ricin from the seed coat and matrix, the incomplete absorption of ricin, or its potential deactivation by, and binding to, components within the alimentary tract.

However, haemorrhage of the gastrointestinal lining and rupture of the capillary vessels may permit ready access directly into the circulatory system.

The toxicity of ricin is increased several 100-fold when administered parenterally (not via the gut). Fatal levels in rats given ricin by intravenous injection have been shown to be as low as 0.3 µg/kg body weight (about 60 ng/animal), with mice being less sensitive (LD_{50} 2.7 µg/kg). Weight for weight, ricin is twice as poisonous as cobra venom.

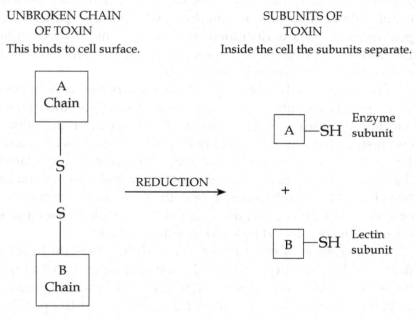

UNBROKEN CHAIN
OF TOXIN
This binds to cell surface.

SUBUNITS OF
TOXIN
Inside the cell the subunits separate.

Diagram 1. Representation of the ricin toxin subunit structure. The A-subunit is an enzyme and the B-subunit a lectin. The B-subunit plays a crucial role in binding to the cell surface thereby allowing the entry of the A-subunit into the cell. Once inside the cell the A-subunit separates and exerts its toxic effects.

Mechanism of Action

Ricin, the chief poison in the castor bean, is a glycoprotein and consists of two non-identical subunits, the A-chain or effectomer

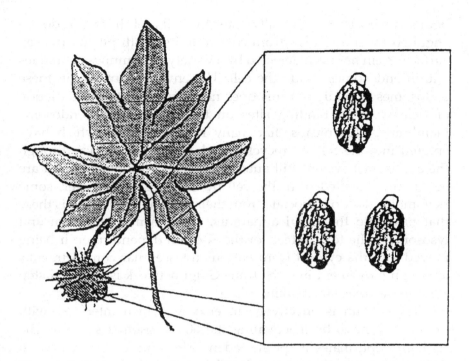

Diagram 2. Leaf, capsule and beans of *Ricinus communis*. The palmate leaves have a deep deep green colour. The capsule is surrounded by soft spines and contains brownish red beans.

and the B-chain or haptomer. The two glycoprotein subunits are joined together by a single disulphide bridge which can be broken by reduction to give two reactive sulphydryl groups, one attached to each subunit.

The B-chain is a lectin which binds to galactose residues present in cell surface receptors. There are two binding sites with different affinities for galactose. Additionally, the B-chain also contains mannose-rich oligosaccharides which can become bound to mannose receptors on the surface of reticulo-endothelial cells. The binding of the B-chain to the cell surface acts as an anchor and facilitates the entry of the A-chain into the cell.

About 100 molecules of ricin can bind to a single cell at the galactose-containing receptors. However, only a fraction of these bound molecules will actually enter the cell and those that do are engulfed by endocytosis from coated and smooth pits on the cell surface. Ricin has been detected by a variety of immuno-techniques within endosomes inside the cell. The environment within these endosomes is acidic but this does not encourage the dissociation of ricin from its binding sites on the inside of the endosome membrane. This means that many ricin molecules which have entered into the cell are recycled back to the cell surface through the endosomal system still attached to their binding sites and are eventually expelled from the cell by exocytosis. However, some ricin molecules do dissociate from their binding sites and it is these that pass into the Golgi apparatus, endoplasmic reticulum and lysosomes. The toxic effects of the A-chain depend upon it being moved into the cytosol from various intracellular compartments, although movement into the trans-Golgi network is a crucial step prior to the toxic effect taking place.

The A-chain is effectively an enzyme which interferes with protein synthesis by inactivating the 60S ribosomal subunit. The A-chain depurinates a specific adenine residue of the ribosomal RNA; the adenine ring is hydrolysed, by the N-glycosidase action of the A-chain, when it becomes situated between two tyrosine rings in the enzyme's active site. As the ribosome is modified it can no longer act as a site of protein synthesis and this leads to the eventual death of the cell. The A-chain enzyme then moves on to deactivate another ribosome. Although only a very small proportion of the ricin molecules that enter the cell are actually moved into the cytosol, one A-chain is sufficient to destroy it. A single ricin molecule is able to deactivate more than 1500 ribosomes per minute.

Treatment

Many of the accidental poisonings involving castor beans occur in children who should have been prevented from having access to the beans. If a plant is present in the garden then it should not be allowed to flower and produce seeds. Necklaces containing castor beans should be avoided.

If a suspected ingestion of the beans has occurred then it is vital to determine whether or not the beans were chewed before swallowing or if the outer coat had been removed in some other way. Anyone handling the beans should wear protective gloves. It should be remembered that the castor beans contain a number of allergenic glycoproteins which can produce skin reations and even anaphylaxis.

Treatment of a person poisoned by ricin involves alimentary canal decontamination procedures so as to prevent absorption of the toxin. These include the use of syrup of ipecac to induce vomiting, activated charcoal to adsorb the toxin and cathartics to accelerate expulsion. Where a suspected poisoning has occurred but the patient remains asymptomatic, alimentary canal decontamination should still be undertaken and hospital observation for at least six hours after suspected poisoning should take place. The patient should be told to return immediately if symptoms begin. Where more severe poisoning has occurred treatment with intravenous fluids, monitoring for haemolysis and hypoglycaemia, supportive care and the possibility of hypovolaemia should be considered.

There is no vaccine or antidote available. Ricin can be inactivated with dilute (0.5%) hypochlorite solution. Experiments have demonstrated that antibodies raised against ricin can be effective in protecting rabbits from ricin poisoning when administered quickly after intoxication and work is underway to develop this for cases of human exposure.

Case Histories

Accidents and suicide

Many instances of ricin poisoning have involved children as the castor beans often have an attractive appearance and allegedly taste like hazelnuts. Owing to their colour they are sometimes used in necklaces which requires that a hole is bored through the bean. This breach in the bean's seed-coat allows toxins to reach the skin where they can enter the vascular system through superficial scratches. In 1993 the charity Oxfam had to recall over 1100 castor bean necklaces sold to customers. Fortunately, an observant botany student in Birmingham (UK) spotted the potentially poisonous beans and informed the Trading Standards Authority before any accidents occurred.

A case has been reported of intoxication in a four-year-old boy who had eaten parts of four castor beans which he had taken from a necklace. The boy's mother induced vomiting at home with the use of syrup of ipecac, and later activated charcoal and cathartics were used at the hospital. The child was released after three days and one week later was found to be well.

Castor beans were distributed amongst a group of children in Georgia (USA) after they had been given as a present to one child by his grandfather. Subsequent investigation revealed that 23 children, ranging from seven to 12 years of age, had taken the beans and nine were found to have eaten them. Vomiting was induced in these nine by using syrup of ipecac and they were treated with activated charcoal. No symptoms were related to the bean ingestion (presumed whole) but three had minor diarrhoea. This case received high media coverage and some months later there was an attempted suicide by a 38-year-old woman who had ingested at least 24 castor beans. She had chopped the beans into pieces as she had learnt of the protective nature of the bean's outer coat against poisoning. She was treated in hospital and, surprisingly, no symptoms of poisoning were observed.

In 1985, an English woman who was on holiday in the Murcia region of Spain ate a "nut" taken from a castor bean plant. Afterwards she began to vomit and her mouth and tongue became irritated. Owing to a history of severe asthma she thought that an asthma attack had begun. On the way to the hospital she collapsed and died. The Coroner's report recorded the cause of death as "accidental death by castor oil poisoning".

An interesting case has been described of a laboratory chemist in England who injected himself intramuscularly with ricin which he had extracted from castor beans. The extraction was carried out using cut beans left in water for a week. Apparently, he had not intended to commit suicide but was interested in the effects of ricin after reading a review article on the toxin.

Murder and bioterrorism

Probably the most well known case of suspected ricin poisoning was that of Georgi Markov, a Bulgarian dissident journalist who was a well known playwright and novelist. From London (UK), where he lived, he was involved in disseminating propaganda against the then Communist government in his country of origin. According to the wife of Mr. Markov, who gave evidence at the inquest after his death, on the 7 September 1978 her husband was waiting for a bus on Waterloo Bridge when he felt a sharp prick in the back of his right thigh. He looked around and noticed a man picking up a dropped umbrella who apologised and left immediately in a taxi. Only a few hours after returning home Mr. Markov felt unwell and was running a high fever. The following day he was admitted to St. James Hospital in Balham, South London, as the fever was continuing. An examination revealed an area of circular inflammation on the back of his right thigh which had a central puncture mark. His condition worsened, his blood pressure and temperature fell, and a tentative diagnosis of septicaemia was made. On the third day after the incident at the bus stop Mr. Markov died.

The post-mortem investigation discovered a metallic sphere, 1.52 mm in diameter (about the size of a pin's head), just below the surface of the skin at the site of the puncture mark. The sphere, made of an alloy of platinum (90%) and iridium (10%), had two holes in it, of diameter 0.35 mm, that extended across the entire sphere. This would allow about 0.28mm^3 of a substance to be stored. The pellet was checked by the Government Chemical Defense Establishment at Porton Down but no traces of any substance could be found. Despite this, owing to the symptomatology and the very high toxicity required for so small a dose, the Coroner was satisfied that the cause of the poisoning was ricin. The pellet found in Mr. Markov was placed in the Black Museum at New Scotland Yard but has since been removed to serve as a piece of evidence in a law suit brought by Mrs. Markov against the Bulgarian government.

Shortly before the death of Mr. Markov, another Bulgarian dissident living in Paris, Vladimir Kostov, was shot at by something that sounded like an air pistol. For 12 days after the incident he was hospitalised for a fever. Examination by X-ray revealed a pellet in Mr. Kostov's back which was later shown to be identical with the one found in the thigh of Mr. Markov.

The potential for terrorist attacks involving ricin has been recognised by the Organisation for the Prohibition of Chemical Weapons and ricin is now included in Schedule 1 of the Chemical Weapons Convention. During the commercial production of castor oil the mash left over after oil extraction can contain up to 5% ricin and this is viewed as a potential source of the toxin for terrorists. Ricin could be used in a number of ways to attack civilian and military populations. These include the contamination of water supplies or food, and the use of aerosols to release the toxin into the air. The latter is expected to lead to necrosis of the lungs, resulting in pulmonary oedema and hypoxic respiratory failure. The Department of the United States Navy dealing with defence predicts that ricin could be released as a toxic cloud as well as being injected. The Detroit News reported (Wednesday, 26 November 1997) that the United Nations Special Commission in Iraq had stopped a professor leaving Baghdad University who allegedly was carrying a file containing detailed information about the extraction of ricin from castor beans.

Table 1. Properties of ricin.

CAS (Chemical Abstract Service Registry Number	9009-86-3	
Physical properties	Colour	White
	Form	Powder
	Isoelectric point	7.1
	Solubility	Soluble in water and glycerin
Synonyms	Ricin A, Ricin B, RCA (Ricin Communis Agglutin), RCA 60, RCA 120, Ricin D, RCL 111, RCA 11.	

Suggested Further Reading

Challoner KR and McCarron MM (1990). Castor bean intoxication. *Annals of Emergency Medicine* **19**, 1177–1183.

Crompton R and Gall D (1980). Georgi Markov — Death in a pellet, *Medico-Legal Journal* **48**, 51–62.

Ellenhorn MJ, Schonwald S, Ordog G and Wasserberger J (1997). Ornamental "beans" and seeds — castor beans. In: *Ellenhorn's Medical Toxicology: Diagnosis and Treatment of Human Poisoning*, 2nd ed. Williams and Wilkins, Baltimore, pp. 1847–1849.

Knight B (1979). Ricin — a potent homicidal poison. *British Medical Journal* **1**, 350–351.

Scarpe A and Guerci A (1982). Various uses of the castor oil plant (*Ricinus communis* L.). *Journal of Ethnopharmacology* **5**, 117–137.

USN (May 1996). *Biological Warfare Defense Information Sheet — Ricin.* United States Navy Nuclear, Biological and Chemical Defense (http://nmimc-web1.med.navy.mil/MED-02/med-02C/ricin.htm).

Wiley RG and Oeltmann TN (1991). Ricin and related plant toxins: mechanisms of action and neurobiological applications. In: Keeler RF and Tu AT (eds.). *Handbook of Natural Toxins*. Marcel Dekker, Inc., New York.

19

SNAKE TOXINS

I. C. Shaw

Introduction

Snakes are amongst the most feared of all creatures simply because a handful of species are lethal to humans. It is perhaps a fortunate quirk of evolution that snakes strike fear into us, because this means that we leave them alone. On the other hand some harmless snakes, for example the Common Water Snake (*Nerodia sipedon*) from north America, are killed because they look like dangerous species (in this case the dreaded Water Moccasin or Cottonmouth (*Agkistrodon piscivorus*)). This is an unfortunate quirk of evolution, because the Brown Water Snake has evolved to mimic the dangerous Water Moccasin in the hope of frightening away potential predators. The outcome, however, as far as humans are concerned, is quite the opposite.

Of 3200 species of snakes, nearly 1300 are venomous. The venomous snakes fall into only three families. Of these families the best known is the Viperidae (or the true vipers) which includes the infamous Cobras (Genus Naja). The venomous snakes produce a wide array of toxic chemicals to kill their prey and to help the snake to protect itself from attackers. For example, the Spitting Cobras (e.g. Ringhals Cobra (*Hemachatus hemachatus*)) from south east Asia spit venom very accurately into the eyes of their attacker rendering the latter blind for a few minutes, just enough time for the Cobra to escape to freedom. The Cobra has secured an important place in history, because it is thought that Cleopatra committed suicide by allowing a deadly Egyptian Cobra (*Naja haie*) to bite her.

Why Do Snakes Produce Venoms?

The production of venom is energy-expensive and therefore snakes will only evenomate (produce venom when they bite) if there is no other way around the situation. For example it might be sufficient for the snake to simply bite its attacker, so frightening or injuring it sufficiently to prevent a sustained attack. In fact, in studies on the bite of the Malaysian Cobras it was found that only 13% of bites resulted in evenomation. Perhaps the most likely, reason for evenomation is to kill prey. What better reason to expend energy than in the pursuit of food?

We must not lose sight of the reasons for snakes producing venoms, namely to kill prey and deter or kill attackers. However, from a human perspective the potential for a snake to bite and kill a person is perhaps the first thought that springs to mind when snakes are encountered. Even if a particular snake produces a lethal toxin, in order for the toxin to kill it must be injected into the person at a dose that is sufficiently high to result in death. Snakes fall into two anatomical classes, those with hinged jaws and those without. The former are able to open their mouths fully and puncture an object of large diameter with their fangs. Those without hinged jaws are only able to bite objects with small diameters. If we set this into a human perspective, snakes with hinged jaws would be able to bite an arm or a leg (e.g. the Mambas, Genus Dendroapsis) and those without would perhaps only be able to bite a finger or a toe (e.g. the European Adder or Viper (*Viper berus*)). At the extreme end of this biting scenario are many of the sea snakes. The sea snakes have some of the most toxic venoms known, but they have extremely small jaws and so rarely kill people. An Indonesian fisherman told me recently that he had only ever heard of fishermen being fatally injured by sea snakes when they were taking their catch out of their nets and a sea snake had been caught inadvertently. The snake was then able to bite the fingers of the unsuspecting fisherman and inject enough venom to kill.

Fangs and Venom Glands

Most people can visualise a fang and realise that it is the fang that delivers the snake's venom to its prey or attacker. Snakes have venom glands which produce and secrete the venom. These glands fall into three types which can be recognised easily when they are cut and prepared for microscopical examination. Their classification is based upon histology and anatomical structure, but conveniently coincides with snake families or genus:

1. *Elapidae* (including the Cobras and the Sea Snakes)
2. *Astractaspis* (Mole Vipers)
3. *Viperidae* (True Vipers)

The venom glands secrete and store venom, they are situated at the base of the fangs. When the snake bites and evenomates, the venom is conducted along a groove in the fangs into the wound, so injecting it deep into the prey or attacker's tissue. The venom is then easily taken up by capillaries in the tissues from where it is transported to its site of action quickly via the blood stream. This speed of transport explains why some snake venoms kill very quickly indeed (Sea Snake bites can kill a person within minutes). Some snakes (e.g. Cobras) have particularly well evolved injecting fangs. Here the groove in their teeth is enclosed to form a channel (almost a tube within the fang) which is particularly effective at delivering venom very efficiently.

The Nerve Poisons

Venom is a complex mixture of biological molecules in a water-based liquid containing inorganic ions. Perhaps the most important components of venom are the toxins. Often the venom from a particular snake will contain several different toxins of completely different type and mode of action. It is likely to be this fact that makes the venoms so incredibly toxic, often far more toxic than their individual components.

The nerve poisons, or neurotoxins, are wonderfully well designed to act quickly to immobilise the snake's prey or attacker. They inhibit the formation or conduction of nerve impulses and so prevent movement. This is a particularly sinister approach to securing prey, because the unfortunate animal is immobilised while not being killed. Indeed, it is likely that in some cases the prey is conscious and aware, but simply unable to move because the venom acts as a motor nerve block (a similar approach is used in anaesthesia to facilitate muscle relaxation, particularly in abdominal surgery and one hears of horror stories of immobilised, helpless patients being subjected to surgery while not properly anaesthetised — this is how the snake's prey might feel!).

Before we can understand how the neurotoxic venoms work, it is essential to understand how a nervous impulse is formed and transmitted. For this I will use a specific example, wiggling the big toe. If you want to move your big toe, you first think that you would like to do so (this might not be a conscious thought). This occurs in the frontal region of the brain. This message of desire is passed to the region of the brain which controls movement (the Motor Cortex in the Central Sulcus) where an impulse is generated. The impulse passes from the Motor Cortex, via the spinal cord, to peripheral nerves in the leg and eventually to the toe. The toe then moves. This whole process occurs within a second.

Transmission of the impulse along the nerves (neurones) *en route* to the toe is akin to an electrical current passing down a wire. The nerve is the wire. Nerves are composed of very large cells; their membranes have specialised ion pumps which allow Na^+, K^+ and Cl^- to be pumped in and out fuelled by energy from adenosine triphosphate (ATP — a high energy molecule that is used as a source of energy by cells). The passage of the impulse along the nerve is facilitated by an exchange of + and − charges across the membrane (depolarisation). This process would be exceptionally fast (around the speed of light) if it were not for gaps between individual nerve cells. These gaps are the junctions between two individual nerve cells and are called synapses. It is necessary for the impulse to traverse this gap. This is done by the cell on the brain

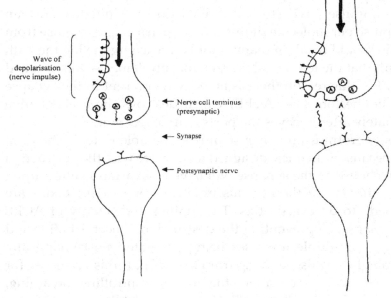

Figure 1. The beginning of a nerve impulse. An exchange of charged atoms (ions) across the nerve cell membrane sends a wave of depolarisation towards the nerve cell terminus. The neurotransmitter, acetyl choline is held in vesicles ready for release.

Figure 2. When the wave of depolarisation arrives at the nerve cell terminus, acetylcholine is released into the space (synapse) between the two nerves. It moves across the synapse to the postsynaptic receptors.

side (i.e. the presynaptic neurone) producing a chemical which can carry the charge across the synapse (i.e. the neurotransmitter), one such chemical is acetylcholine. Acetylcholine, produced in the presynaptic neurone, is secreted across the presynaptic membrane into the synapse in response to a nervous impulse arriving from the brain. The acetylcholine, crosses the synapse and fits like a key in a lock into a receptor on the postsynaptic membrane. When the acetylcholine docks into the receptor it causes membrane depolarisation which restarts the nervous impulse on its way to the toe. This will happen many hundreds of times *en route* to the big toe.

It is important that as soon as the acetylcholine has restarted the nervous impulse, it is destroyed so that it does not initiate more than one impulse per molecule (if this did happen a single impulse from the brain would result in many impulses reaching the big toe with the result that the toe would vibrate uncontrollably — this is called tetany). For this reason there is an enzyme present in the synapse (acetylcholinesterase — AChE) which destroys the acetylcholine immediately after it leaves the postsynaptic receptor.

This neurotransmission system is vulnerable to toxic chemical attack, because a chemical designed to interfere with the destruction of acetylcholine in the synapse will have a devastating effect upon the transmission of the nervous impulse. Some snake toxins are "designed" to do exactly this. They inhibit the activity of AChE in the synapse, so preventing the destruction of acetylcholine and resulting in multiple impulses being generated postsynaptically from a single impulse arriving from the brain. If this occurs in, for example, the nerves controlling breathing, death will occur very quickly.

Several of the most poisonous snakes in the world use inhibition of AChE as their means of killing. For example the Black Mamba (*Dendroapsis polylepis*) from Africa secretes fasciculins into its venom. The fasciculins are potent AChE inhibitors and make the Black Mamba's venom amongst the most potent venoms known; only 21 mg (equivalent to about three grains of salt) would be needed to kill an average sized person. Indeed, a Hospital in South Africa reported treating seven patients for Mamba bites — all of whom died within 24 hours. The symptoms associated with a Mamba bite are tetany,

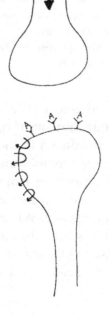

Figure 3. Acetylcholine docks with the postsynaptic receptors which initiates a new wave of depolarisation and another nerve impulse.

associated with spasm of the pulmonary muscles, usually resulting in death by asphyxiation within about 15 minutes of being bitten.

In order to maximise the toxic potential of their venoms, many snakes have several toxins in their venoms which act by different biochemical mechanisms. This is an ingenious ploy which means that more than one of the body's vital systems is hit by the venom so making death more certain than if only one were hit. The Black Mamba is an excellent example of a snake with multiple toxic components in its venom. In addition to the fasciculins, Mamba venom has dendrotoxins which inhibit neurotransmission by blocking the exchange of + and − ions across the neuronal membrane. This prevents passage of the nerve impulse. If the impulse is *en route* to the big toe the toe will be paralysed — this is certainly not life-threatening. However, if the impulse is to the pulmonary muscles, respiratory failure and death will result. The dendrotoxins from the Black Mamba are very much less toxic than

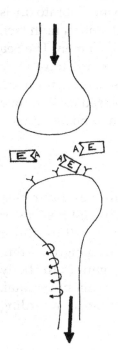

the fasciculins (it would take 1.6 g to kill a person), however the combined effect of the two toxins is far more toxic than the toxicities of the individual components (this is termed *synergy*) which is why the Black Mamba is lethal to humans.

There is one final way in which snake toxins can interfere with neurotransmission. As explained above, when the impulse is carried across the synapse as a charge on the neurotransmitter molecule, to re-initiate the impulse postsynaptically the neurotransmitter activates the postsynaptic receptor

Figure 4. Acetylcholine is released from the postsynaptic receptors and is immediately destroyed by the enzyme, acetylcholine esterase. This prevents an acetylcholine molecule initiating more than one depolarisation. The wave of depolarisation speeds off down the nerve cell membrane.

situated in the postsynaptic membrane and so recommences the impulse. Some snake toxins are able to interfere with the interaction between the neurotransmitter (e.g. acetylcholine) and the postsynaptic receptor (e.g. acetylcholine receptor — AchR). There are a specific group of snake toxins, the α-neurotoxins, which do exactly this. The best known of these horrifically potent toxins is α-bungarotoxin (α-BG) from the Banded Krait (*Bungarus multicinctus*) and the Thai Krait (*Bungarus candidus*) which mimics the shape of the acetylcholine molecule and therefore fits the AChR so blocking it. So when a person (and presumably an animal) is bitten by a Krait, they feel numbness in the region of the wound because the α-BG has inhibited local sensory (i.e. taking a message to the brain) neurotransmission, this leads to flaccid paralysis and later systemic effects inhibiting respiration and causing death. It would take about 21 mg of α-BG to kill a human.

Not only the Kraits produce α-neurotoxins. The Indian Cobra (*Cobra naja naja*) also utilises one of the most potent of all the snake toxins, cobratoxin, which is also an α-neurotoxin. Cobratoxin is phenomenally toxic; only 4.5 mg is needed to kill a human. In fact, a single Cobra can produce sufficient toxin to kill ten men. The Sea Kraits (e.g. *Laticauda semifasciata* from Malaysia) are the most toxic of all snakes; they produce erabutoxin which is an α-neurotoxin of unbelievable potency, but fortunately, as discussed earlier, they have small jaws, which makes it difficult for them to bite a human.

Cell Poisons

The neurotoxins discussed above do not kill neurones, but rather interfere with their neurotransmission activity. The cell poisons or cytotoxins actually kill cells. They cause regions of cell death (or necrosis), which, if left unattended would lead to gangrene. But, perhaps more importantly, if they reach crucial organs in the body (e.g. the heart) the cell death which would occur might be fatal. There are three types of cytotoxins which are classified according to their mode of action.

Phospholipases

I have mentioned cell membranes many times above in my discussion of the neurotoxins. It is now necessary to outline the structure and function of the membrane in order that the mechanism of action of the cytotoxins can be fully appreciated.

All cells are surrounded by a membrane which is based on a bilayer of phospholipids with their polar head groups (water attracting parts of the molecule) facing out towards the water-based cytoplasm (cell contents) or intracellular space. Floating in the membrane (the membrane is a liquid) are numerous proteins and other molecules all having very specific and important functions. For example, some proteins form channels to allow molecules to enter or leave the cell (e.g. the ion channels of the neuronal membrane) and others have complex carbohydrate chains attached which form the basis of cell-cell recognition (blood groups are expressed in this way). The fluidity of the membrane is crucial to facilitate movement of the membrane's component molecules. If the membrane is damaged in such a way that its fluidity is changed this will severely affect cellular function. The phospholipases remove one of the fatty acid chains of the phospholipid molecule so forming a lysophospholipid.

Lysophospholipids have detergent properties and dissolve a small region of the membrane so forming a hole which allows free passage of water and ions. This results in the cell rupturing and spilling its contents.

Heart poisons

The heart poisons or cardiotoxins are, in reality, general muscle poisons. However their effect on the muscle of the heart is by far the most important in terms of killing the snake's prey or attacker. These incredibly potent toxins bind to specific biochemical sites on the surface of muscle cells causing depolarisation. The muscle cell works in a manner akin to neurones in that it transmits information, to initiate a contraction, by exchanging + and − ions across its cell membrane. In the case of voluntary muscle (e.g. the muscles of the big toe, to continue our analogy) the depolarisation originates from a nervous impulse from the brain transmitted via a series of nerves. In the case of the involuntary muscle (i.e. muscle that is not under

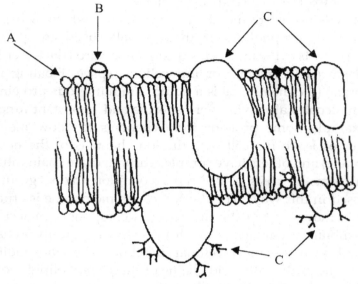

Figure 5. A schematic representation of a cross-section of a biological membrane, showing the phospholipid bilayer (A), and membrane pore (B) and other membrane proteins (C). The whole thing is a liquid with the proteins floating in the sea of phospholipids.

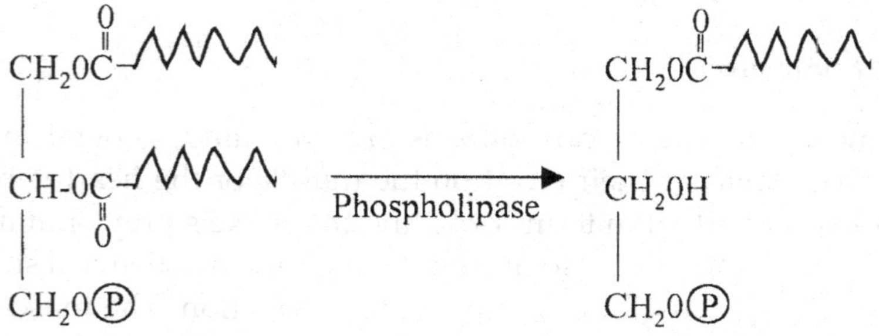

Figure 6. Loss of a fatty acid from a phospholipid to produce a lysophospholipid. Lysophospholipids are detergent-like and denature the biological membrane.

conscious control, e.g. the heart) the depolarisation is initiated within the organ itself.

Whether voluntary or involuntary muscle is affected the outcome is the same, the cardiotoxins prevent muscle contractions. Clearly if this effect is directed at the muscle of the heart, the heart will beat irregularly and perhaps stop beating altogether. Death is inevitable and extremely fast in this latter case.

Only the Cobras (Genus *Naja*) and Ringhals (Genus *Hemachatus*) produce venoms which contain cardiotoxins. These are intensely poisonous. Only 50 mg of the cardiotoxin from the Thai Cobra (*Naja naja siamensis*) would kill an adult human.

There is one other toxin, about which very little is known, which affects the heart and very rapidly causes death. This is sarafotoxin from the Middle Eastern Burrowing Asp (*Atractapsis engaddensis*); it interacts with a specific receptor (the endothelin receptor) which causes blood vessel constriction (vasoconstriction). Sarafotoxin causes massive constriction of the blood vessels with a concomitant rise in blood pressure. This blood pressure rise can be so rapid and great that death results from blood vessel (e.g. aortic) rupture.

Blood cell toxins

The blood cell toxins have a specific effect upon only one blood cell, the red blood cell (RBC) or erythrocyte. They cause the rupture of erythrocytes (haemolysis). This is clearly a very undesirable eventuality, which in its own right would result in death. However it is difficult to decide whether this property of several of the snake toxins is particularly important from the point of view of their fatality potential, because it is slow and it is very likely that one of the other toxins in the venom would have killed the unfortunate recipient before the haemolytic toxin had a chance to exert its effect.

There is very little known about the haemolytic toxins, indeed it is often difficult to distinguish them from the cardiotoxins, because the cardiotoxins often also have haemolytic properties.

Chemistry of the Snake Toxins

All of the snake toxins discussed in this chapter are proteins. Indeed all of the important snake toxins are proteins. There is evidence that some snake venoms contain toxic metals, however their concentrations in the venoms are too low to result in death.

The snake venom toxic proteins are ingenious in their design. They fall into two major categories. Enzymes which destroy important biological molecules (e.g. phospholipases) and proteins whose shapes mimic other biological molecules so disrupting the actions of the latter (e.g. postsynaptic receptor blockers).

The biological molecule mimics are particularly ingenious molecules of death. They are protein molecules which often fit into membrane receptors (which are huge proteins) intended to accept small non-protein biological molecules. They achieve this mimicking effect by the protein folding in such a way as to expose specific amino acids, which in some way resemble the natural receptor binding molecule. For example, α-BG is a small protein molecule with a molecular weight of 8000 daltons comprising 74 amino acids. It mimics acetylcholine which is, by comparison, tiny (molecular weight = 181 daltons). The interaction between α-BG and the AChR is not fully understood, however it is known to involve specific non-covalent interactions, including hydrogen bonds and Van der Waals forces. Surprisingly, ionic interactions are not thought to be involved. In a "normal" chemical sense these interactions would be regarded as weak, but the binding of α-BG to the AChR is amazingly strong; it has a dissociation constant of the order of 10^{-11} M which means that the binding is to all intents and purposes irreversible.

The fact that α-BG is such a potent toxin is illustrated well by its incredible affinity for the AChR, but also because it appears that only one (or perhaps less!) molecule of α-BG is needed to incapacitate a single AChR. In toxicology, the dose of a toxin is generally very much greater than that which actually arrives at the toxin's site of action. This appears not to be the case for the incredibly potent α-BG. To illustrate this affinity further, studies have been carried out from which it has been calculated that 1 mole (i.e. 8000 g) of α-BG will bind to 90,000 g of receptor molecule.

The first of the receptor-binding snake toxins to be studied and its structure elucidated was erabutoxin-b from the Banded Sea Snake (*Laticauda semifasciata*). It has two cysteine amino acid residues which form a covalent (S-S bridge) between two sections of the peptide chain. This forms a loop which is thought to be important in its binding to the AChR. It is possible that the conformation (shape) and electrostatic topography (surface charge) of this loop resembles acetylcholine and fools the receptor into accepting this rogue molecule — α-BG and erabutoxin-b are false keys to the postsynaptic receptor's door.

These examples illustrate well the ingenious way in which the snake toxins exert their deleterious effects upon the snake's prey or attacker and how little is known about the specifics of the interactions between the toxins and the unfortunate recipient's body.

Uses of the Snake Toxins

It might seem impossible that any of these terrible poisons could be put to good use, but many have been surprisingly useful, and no doubt others will have uses in the future when our understanding of their modes of action increases. They have very important uses in experimental biology and biochemistry, in particular in neurobiology where they have been used to investigate the workings of the nervous system. In fact, perhaps surprisingly, they have been the main tool that neurobiochemists have used to elucidate the workings of the postsynaptic receptor and its role in neurotransmission.

As a result of studies with α-BG it has been possible to study myasthenia gravis, a debilitating neuromuscular disease which involves the sufferer's body directing antibodies against the postsynaptic receptor. This is a fortuitous sting in the tail for a molecule that was "designed" to kill. From its use has come knowledge which has helped in the treatment of human disease.

In a more general sense, studies on the modes of action of the snake toxins have significantly increased our understanding of the workings of the nervous system and so, indirectly, helped in the understanding and treatment of neurological diseases.

As yet the snake toxins have not been used directly in medicine perhaps because of their very specific mode of action which is directed (often) at the nervous system. Other plant and animal toxins, however, have been used very successfully indeed in human medicine and it is possible that snake toxins could be used in a similar way. But that is for the future.

Treatment of Snake Bites

Unfortunately many snake bites occur in remote geographical locations and so the sufferer cannot be taken to a hospital or to a doctor. For this reason it is likely that these people will die. Because of the remoteness of many attacks the deaths never find their way into the statistics and we probably underestimate the importance of snake bites as a cause of death.

If, however, the bite occurs in a large city or town and the patient can be taken to the hospital quickly, providing the hospital holds the appropriate antiserum it is likely that the patient will survive. In Australia, for example, hospitals hold stocks of antisera for the snakes that live in the area.

Antisera are made by "milking" the snake. To do this the animal is forced to evenomate into a small pot. This is often achieved by tying a muslim pad over the top of a laboratory beaker and forcing the snake to bite into the pad. By doing this the snake evenomates into the beaker. This is not the safest task in the world! The venom is then injected into rabbits or goats at a sufficiently low dose not to cause toxicity. The animal makes antibodies against the toxin. These can be isolated from the animal's blood, purified and injected into a bitten human. When injected they combine with the toxin in the person's circulatory system so sequestering it and rendering it harmless. In order for antisera to be effective they must be administered very soon after a bite.

Some snake venoms are so toxic that if a bite is sufficiently severe, and the snake evenomates and has not bitten anything or anybody for a day or two, it is extremely unlikely that the patient will survive. Bites from the Indian Cobra (*Cobra naja naja*), Black Mamba (*Dendroapsis polylepis*) and Banded Sea Krait (*Laticauda semifasciala*) fall into this category. Recently a young man swimming in the Mediterranean Sea died very quickly with no history of life threatening illness. He was later found to have been bitten by a Sea Krait. This illustrates very well how quickly snake bites can kill their victims.

In conclusion, snakes have evolved a very specialised system for protecting themselves and immobilising and killing their prey. The molecules of death that they produce are incredible in design and ferocity of action. This approach to defence and feeding has meant that humans (and many other animals) are wary of snakes (it is probably more accurate to describe most people as terrified), which, of course means that the snakes are less likely to be preyed upon than many other animals. This is a successful evolutionary trait. It is interesting, however, that several animals have evolved specifically to prey upon this abundant group of animals and so have a supply of food that is not contested by most of their potential competitors. A good example of such a creature is the mongoose from India. This is a small rodent that preys almost entirely upon snakes and their eggs. It is able to immobilise and kill the lethal cobra by biting it just behind its head. This is an excellent approach, because the snake cannot turn and bite the mongoose in this position.

There is one, as yet, unsolved mystery in the snake world. Surely when snakes evenomate they swallow a small amount of their own poison, and when they eat their prey that has been killed by their own lethal injection they ingest these lethal chemicals. Why then are snakes not killed by their own toxins? This is perfect evolution. The snakes produce some of the most lethal poisons in the world, but appear to be immune to their effects.

Acknowledgements

I thank David Zehms and Emma Burke for proof-reading the manuscript and for making helpful suggestions, and Margaret Tanner for her expert computer skills.

20

Spider Toxins

R. H. Waring

Effects and Occurrence

When a five-year-old girl was rushed to the emergency clinic of an Arizona hospital, the medics on duty recognised the symptoms. She had been getting dressed and had put her foot into her shoe. Immediately she felt a pricking sensation and after about an hour, had severe back, leg and abdominal pain. When her foot was examined, a puncture mark was seen, surrounded by dead white skin, then a red ring of inflamed tissue. She was, in fact, a typical victim of a Black Widow spider.

The child was given antivenin and soon recovered, but the team in the emergency clinic knew that if she had not been treated, she would have had days or even weeks of pain, nausea, vomiting and headache, and might have died.

It is relatively common for spiders to be venomous — the United States has 20,000 species and only two of these are harmless. However, most spiders do not have the capacity to penetrate human skin and even fewer can deliver large quantities of venom. Only a small number of species are dangerous to human beings, the best known being the Black Widow (*Latrodectus mactans*), the brown recluse spider (*Loxosceles reclusa*) in the United States and the redback and funnel-web spiders in Australia. Most spiders possess a pair of venom glands in their prosons (see Figures 1 and 2); these are surrounded by spiral bands of muscle that contract and squeeze the venom into hollow fangs at the tip of the chelicerae. Spiders can be "milked" to collect the venom to produce antibodies. The initial drop of venom is relatively less toxic than the rest, containing much

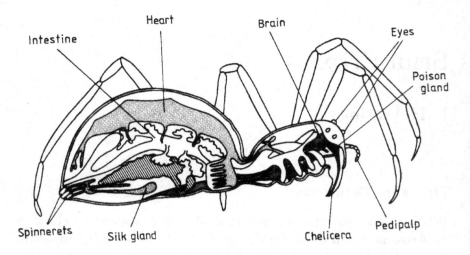

Figure 1. Anatomy of a spider.

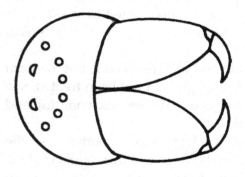

Figure 2. Prosons, the spider's poison glands.

less of the poisonous peptide fraction. When venom is removed, the spider makes more, but as the liquid is slowly absorbed back in the spider the newly regenerated product is less harmful than that found after prolonged storage and concentration. Most spiders produce a range of toxins with different effects, possibly to deal with a variety of potential prey; the primitive hunting spider *Plectneurys tristis*, for instance, produces at least 50 different peptide venoms, while the American funnel-web (*Agelenopsis aperta*) has at least three different types of toxins, which synergise the effects of the others.

Why do spiders produce venoms and how do they cause the reactions which are seen? Generally spiders prey on insects, disabling their victims before eating them often some time later. Venoms, then, are the way in which spiders paralyse their food supply. As the neurotransmission pathways which link brain to muscles via the nervous system are common to both insects and humans, a large dose of poison will affect the muscle systems in both cases. The tissue death (necrosis) which can be seen at the area surrounding the bite occurs because the venom contains enzymes which can dissolve proteins and membranes, enabling the spider to feed on the juices produced. Spiders do not have sets of teeth and they rely on primitive digestion systems with "sucking stomachs" where the sac is squeezed flat by sets of muscles which then release to draw in liquid. This capacity to liquefy their food is a useful evolutionary strategy.

Mechanisms of Toxicity

Biochemists and toxicologists have long been fascinated by the variety of poisonous compounds produced in venom from any species of spider and have used particular toxins to explain the specific mechanisms involved in neurotransmission. An early finding was that the venoms produced by spiders are not necessarily unique to arthropods. Many spiders produce polyamines (Figure 3) and closely related compounds have been found in venom of the solitary digger wasp *Philanthus triangulum*. Biologists often speak of "convergence" when different species appear to have evolved similar complicated chemical structures which are aimed at a common target, in this case the calcium (Ca^{++}) ion channels in a sub-type (NMDA) of the glutamate receptor.

What, then, are these pathways in the central (brain) and peripheral nervous systems, which are common to all species, and on which spider toxins can exert their effects? Nerve cells (neurons) are the functional units of the brain and have membranes which are "excitable" over their entire surface. Neurons communicate with each other by releasing chemical substances across the gap

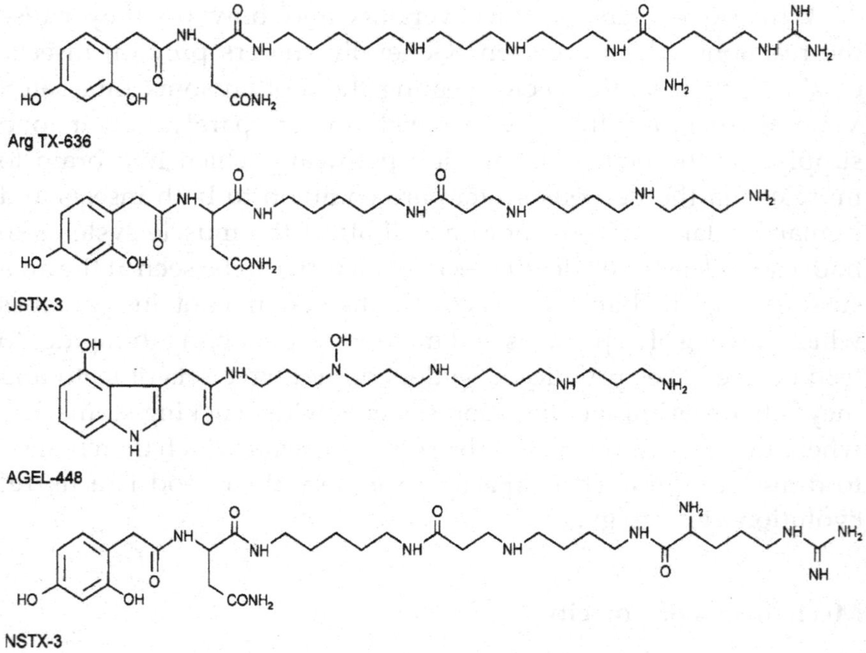

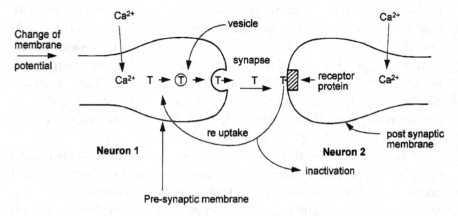

Arg TX-636

JSTX-3

AGEL-448

NSTX-3

Figure 3. Structures of polyamine spider toxins.

Figure 4. Mechanisms of neurotransmission.

(synapse) between them. These neurotransmitters can activate or inhibit the output of the target cell (see Figure 4) and do so by interacting with a protein complex (receptor) on the target neuron. This interaction is dependent on calcium (Ca^{++}) ion entry into the initiating neurone and is linked with transportation through specific channels in the target neuron and generation of a small electric potential. After this, the neurotransmitter is released from the target cell and either enters a re-uptake system which restores it to the initiating neuron or alternatively is metabolised and inactivated. Anything which alters this complex sequence of events will also alter the process of neurotransmission and hence brain function. In nerve and striated muscle cells, messages can also be passed on by electrical excitability. This is the ability of a cell to generate a short-lasting depolarisation or reversal of the membrane potential ("action potential") in response to electrical stimulation. If no message is being passed along the cells, there is an excess of negatively changed ions on the inner membrane surface. This is the "resting potential" and is between −60 and −70 mV, although as the membrane is very thin this value would be equivalent to about 100,000 volts across a 1 cm thick membrane. Rather like seawater, the ions outside the cell membrane are rich in sodium (Na^+) and chloride (Cl^-), while inside the cell there is a high level of potassium (K^+) ions. There are channels for both sodium and potassium ions which are "voltage activated", opening transiently when the membrane change changes. Initially, as the action potential is propagated, the Na^+ ion channels open briefly, so that sodium ions move inwards across the membrane. As more Na^+ ions move in, these channels close (inactivation). Potassium ions then move across the membrane outwards; the situation is finally restored by an energy-dependent process (Figure 5). These ion channels are complexes of proteins and carbohydrates which cross the membrane and appear to have a hollow core down which the ions can travel. This channel is lined with polar groups while the outside of the protein, which is embedded in the lipid membrane, is largely non-polar. In evolutionary terms, these structures are ancient, and are thought to have existed over 500 million years

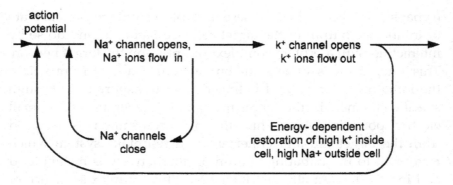

Figure 5.

ago. Most channels are "gated", opening or closing as a response to electrical or chemical signals. All types of voltage-gated ion channels have rather similar structures. Each is an aggregate of four similar or even identical peptide units, existing either as part of a single very long peptide chain (Na^+ channels) or as independent subunits (K^+ channels). Each of the four units has six helical sections which cross the membrane and are arranged rather like bundles of firewood. The four units are arranged symmetrically around the ion channel, which has a central aqueous pore (Figure 6). Spider toxins appear to primarily affect this ion-channel linked form of neurotransmission and at least two main classes of compound have become apparent, the "polyamine" toxins and the "cystine-knot" toxins; both these bind to a site inside the ion channel, blocking the movement of ions. The environmental prey is first left paralysed and then finally dead.

Types of Toxin

Polyamine toxins have been isolated from the venoms of Agelenids, orb-web spiders (argiotoxins from Argiope and Araneus species), trap-door spiders, Joro spiders (JSTX) from *Nephila clavata* (NSTX), and from *Dolomedes okefinokensis*, "a fishing spider" which is capable of immobilising vertebrate prey including fish. These compounds usually have an aromatic, often phenolic ring at one end of the

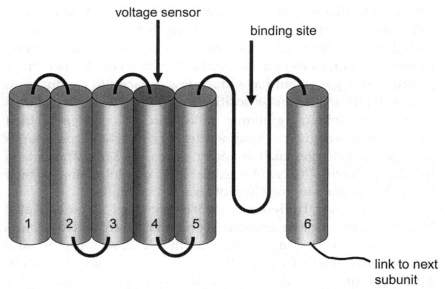

One of 4 domains of the Na⁺ channel

Figure 6.

molecule, then a polyamide or polyamine backbone, sometimes with one or more amino acid residues. The backbone finally ends in an amino ($-NH_2$) group or a guanidino residue (-NH-CH(=NH)-NH$_2$) (Figure 3). At physiological pH ($\sim$ 6.8), these structures acquire a positive change on the end of a long chain. As such, they have the capacity to mimic cations such as Na^+ and K^+ and are a similar size, fitting into their ion channels and blocking their function. Some of the polyamine toxins potentiate or increase the cell's response to glutamic acid, a natural neurotransmitter, although others are known to block the action of glutamate non-competitively while also inhibiting the uptake of glutamate into the initiating neuron. One of the toxins from *Dolomedes okefinokensis* blocks calcium (rather than sodium or potassium) channels so that relatively small changes in chemical structures can alter the ion specificity.

Recent studies have suggested that polyamine toxins may bind at the same or similar receptor ion channel sites as endogenous compounds like spermine, which is a diamine. A number of different forms of K^+-selective ion channels have been identified, with different pharmacological properties. Kir (K^+ inward rectifier) channels conduct ions in the inward direction when the membrane potential is negative but have much less capacity to conduct ions outward if the membrane is positively charged. These Kir channels have acidic residues (glutamic or aspartic acids) deep in their hollow cores, which act as binding sites for cations and compounds with positive charges such as spider toxins. As Kir channels tend to "damp down" the cellular excitability, any inhibition of their function will magnify the effect of an action potential.

There are several different types of receptors which are activated by the neurotransmitter amino acid glutamate and involve calcium ion fluxes. The one most studied, the NMDA receptor, also responds to N-methyl-D-aspartic acid, hence the name, but others exist, including the AMPA and kainate-responsive sub-types. These glutamate receptors mediate fast neurotransmission in the brain and spinal cord, while also being involved in the induction of neuronal cell death and in memory processes. AMPA/kainate subtypes can be blocked by diamines such as spermine, preventing the flux of Na^+ and Ca^+ ions across the membrane, and some of the spider toxins also act at this spermine site. Most work, however, has been carried out with the NMDA subtype of receptors. Prolonged activation of all glutamate receptors, but particularly this subtype, leads to neuronal degeneration and cell death, as seen, for example, in motor neuron disease. NMDA receptors require glycine as a co-agonist with glutamate and are partly blocked by extracellular Mg^{2+}. Like the Kir channels, these receptors bind both diamines and polyamine spider toxins and again the polyamine toxins act to plug the ion channels by binding to sites deep within their hollow centres. α-Agatoxins (from *Agelenopsis aperta*, the American funnel-web spider) specifically target glutamate receptors by this mechanism, although some polyamines also bind to the Mg^{2+} sites on the receptor. Argiotoxin 636 (from *Argiope lobata*) is known to complete with Mg^{2+} and MK801, a specific glutamate-binding antagonist, so that compounds of this type may have more than one mechanism of action. Interestingly, some spider toxins show great selectivity for particular areas of the brain; one of the components of venom from *Nephila clavata* specifically blocks a glutamate sub-type receptor in the hippocampal neurons. As the hippocampus is an area of the brain involved in memory, bites from this spider may lead to problems in recalling information.

Other polyamines also affect specific points in the neurotransmission sequence. A toxin (DTX9-2) from the weaving spider (*Diguetra canities*) causes paralysis. Similarly funnel-web spider venom contains components (μ-agatoxins) which are similar to those found in scorpion and sea-anemone toxins; these polyamines all act at the Na^+ channels, slowing inactivation and leading to repetitive firing of action potentials. Heteropodatoxin (from *Heteropoda Venu* toxin) is a polypeptide with 29–32 amino acids which blocks the outward K^+ current and is closely related to tarantula toxins. However these toxins always work on the same principle, being effective because they have a long flexible molecular structure which can adopt an extended linear configuration interacting deep inside ion channel pores. They behave, in fact, rather like a finger plugging a hole.

Not all toxins which inhibit ion channel currents have the polyamine structure. A venom constituent, HF-7 which has been isolated from the funnel-web spider *Hololena curta* also blocks non-NMDA glutamate-sensitive Ca^{2+} channels. However, unlike the cation-forming polyamines, this compound can exist as a mono- or di-anion, with a net negative charge. Structurally, it is quite different from the polyamines, as it is an acetylated bis(sulphate ester) of a glucose derivative of guanosine (Figure 7). Possibly the electrostatic repulsion of the two sulphate groups ensures that the molecule again has an extended configuration. The specificity for Ca^{2+} channels suggests that this sub-type has basic amino acid residues at critical sites. As *H. curta* venom also contains the neurotransmitter noradrenalin, bites from this species lead to vasoconstriction and restlessness in the victim as well as the usual symptoms.

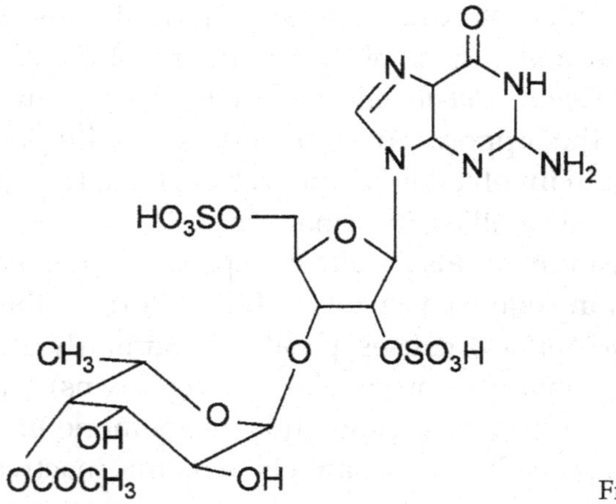

Figure 7.

The American funnel-web spider (*Agelenopsis aperta*) is usually found in the deserts of southwestern United States where it builds large sheet webs with retreats under rocks and logs. As well as α- and μ-agatoxins it produces ω-agatoxins which have a complex disulphide bridged peptide structure that block glutamate-sensitive receptor channels in insect muscle. The Australian Blue Mountains funnel web (*Hadronyche versuta*) similarly produces polypeptides which are neurotoxic to a variety of insect species. One of these compounds, atracotoxin-HV1 (ACTX-HV1) is a 37-amino acid peptide which is harmless to newborn mice and man but lethal to the larvae of *Helicoverpa armigera*, the cotton bollworm. As this has commercial implications the structure of ACTX-HV1 has been investigated in the hope that it may lead to synthesis of more effective pesticides which have a low mammalian toxicity and are not environmentally harmful. As might be expected, the structure is very complex for such a small molecule (Figure 8). There is a hydrophobic globular core and three sulphur-sulphur bonds formed by cross-linkage of cysteine residues. Two of these disulphide bonds are packed inside the core, while an amino-acid loop protrudes in a "hair-pin" or "finger"-type structure. Cysteine 1 binds with cysteine 4, cysteine 2 with cysteine 5 and cysteine 3 with cysteine 6. In ACTX-HV1, the central disulphide bridge passes

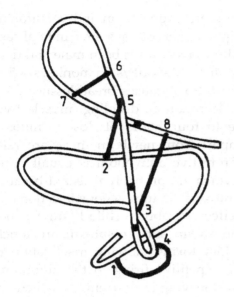

— disulphide bridges
● conserved arginine residues   Figure 8.

through a loop formed by the other two disulphide bonds and the connecting pieces of polypeptide backbone. Huwentoxin (HWTX-1) is a neurotoxic peptide purified from the venom of the Chinese bird spider Selenocosmia huwena. It has the same cysteine-binding scheme as ACTX-HV1 and affects the postsynaptic nicotinic acetylcholine receptor. Other venom components of this spider have haemagglutination capacity, causing blood clotting, but the main toxic effects are due to HWTX-1 and its homologues which all have the "cystine-knot" type of structure.

Although some cystine-knot type compounds, such as ACTX-HV1, do not affect human beings, this is not the case for robustoxin, a 42-amino acid peptide from the Sydney funnel-web spider, *Atrax robusta* or for its analogue versutoxin from *Hadromyche versuta*. Both toxins cause hypertension, then hypotension and circulatory failure, breathlessness, muscle twitching, salivation, tears, sweating, nausea, vomiting, diarrhoea, lung oedema and pain. As can be seen from this list of symptoms, robustoxin and versutoxin affect both the central and peripheral nervous systems. In fatal cases, the positively identified spiders have always been male and death has occurred between 15 minutes and six days. In monkeys, a 5 mg/kg intravenous dose of robustotoxin causes breathlessness, a severe drop in blood pressure, lacrimation, drooling, muscle twitching and death within three to four hours of dose administration. Non-primate animals, apart from newborn mice, are relatively little affected and the protective antivenom is usually prepared by immunising rabbits which are probably resistant because they produce endogenous antitoxins; an intravenous injection of 15 mg of crude venom has no effect on rabbits while 12 mg has no effect on cane toads. Both robustoxin and versutoxin are ionchannel inhibitors, like other spider toxins, slowing inactivation of Na^+ channels. This results in repetitive firing in the autonomic and motor nerve fibres, in effect leaving the current "switched-on".

Death from *A. robustus* envenomation is now uncommon (13 cases were reported between 1927 and 1984) since the development of a specific antivenom in 1980.

The "cystine-knot" configuration is found in a number of polypeptides from biological sources and may make the toxin more resistant to degradation in the victim's body. In fact, this arrangement is not only found in spider toxins (Figure 8). Venomous toxins from the cone snail (*Conus geographus*) have a very similar three-dimensional structure to the ω-agatoxins from the funnel-web spider. These have four cysteine-cysteine bonds (Cys1-cys4, cys2-cys5, cys3-cys8 and cys6cys7) and a tightly-folded structure (Figure 9) preventing Ca^{++} entry through the voltage-sensitive Ca-channels. As a number of neurotransmitters such as acetylcholine, serotonin, dopamine, γ-aminobutyric acid (GABA) and glutamate are released *via* this mechanism, ω-agatoxins have a wide range of toxic effects. However, since neuronal damage after head trauma or stroke is partly due to accumulation of calcium ions inside cells, it has been suggested that derivatives of the ω-agatoxins may have some therapeutic uses.

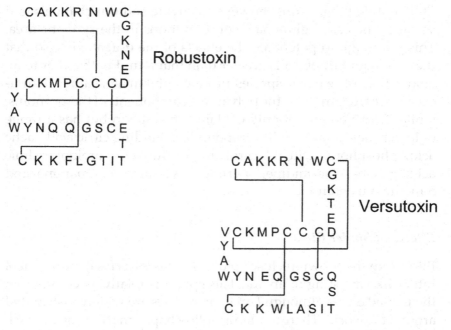

Figure 9.

Interestingly, a synthetic derivative with the same three-dimensional shape as ω-agatoxins but without di-sulphide bonds is also able to raise antibodies to the native toxin. Recent work has shown that there are three arginine residues which are "conserved", i.e. found in most toxins of the type. They have strong positive changes and appear to lie almost in a line on the curved surface of the model of the toxins. It seems probable that they interact with a similar-shaped pattern of negative ions on the inner surface of the calcium channel, forming strong bonds and effectively sealing the pore against ion fluxes.

Usually, spiders synthesise a range of metabolic poisons which often act in concert to synergise each other's effects. For instance the polyamine α-agatoxins block glutamate receptors, but only when glutamate levels are high. Apparently, these toxins require channel opening before they can gain access to their binding site.

The peptide μ-agatoxins, however, activate sodium channels and so cause increased glutamate concentrations in the synaptic area. This is enough to potentiate the effects of the α-agatoxins, so that the final paralytic effect is greater than that found with either toxin alone. The orb weaver species produce glutamate, rather than μ-class toxins, to synergise the polyamine components of their venoms, while *Atrax* venom not only contains atraxatoxin but has a range of lower molecular weight compounds, including citric acid, lactic acid, phosphoric acid, glycerol, urea, glucose, γ-aminolevulinic acid, glycine, spermidine, spermine, tyramine, octopamine and 5-methoxytryptamine.

Effects of Spider Bites

Bites from the brown recluse spider (*Loxosceles reclusa*) initially feel rather like the sting of an ant. This spider is relatively common in the United States although *Loxosceles* species are widely distributed around the world. There is often a violin-shaped marking on the back (hence the name "fiddleback spider") and the spider has three pairs of eyes. It is nocturnal and not normally aggressive. Both females (1 to 1.5 cm long) and the slightly smaller males are venomous. These brown spiders travel widely in search of new web sites and mates and are most active at night during spring and summer months, emerging from woodpiles and under stones. Some species prefer homes with warm undisturbed environments; in Chile *L. laeta* is called "arana de detras de los cuadros" ("spider behind the pictures"). Brown spiders are not normally aggressive to humans but will bite if trapped against skin. After the initial bite, the skin often itches, then becomes red with a blanched area surrounding the lesion and an outer red ring. The reddened area enlarges over the subsequent six to eight hours after developing haemorrhages while the bite site forms a blister which subsequently ruptures. The whole area becomes swollen and painful, and as time goes on, the central bite becomes an area of dead tissue. This necrosis can invade large areas of muscle tissue, serious bites giving lesions as large as 80 cm^2 which require plastic surgery for reconstruction. Patients can have a range of systemic symptoms, including fever, nausea, vomiting, joint pains, delirium and stomach cramps, with jaundice and lysis of red blood cells. This destruction of tissue occurs because the spider venom contains enzymes which can break down tissue components such as collagen and hyaluronic is carried out by acid and split sphingomyelin, an integral part of cell membranes. Sphingomyelinase D is a protein traction of 32,000 dalton and is probably the most important component, although other lipases are present, with proteases and ribonucleases. The venom also has vaso-constrictive and coagulation properties. Breakdown of the membranes of the red blood cells releases haemoglobin, the oxygen-carrying protein in blood. This can clog the kidneys, leading to renal failure. The excess haemoglobin is broken down in the liver to give a series of pigments, one of which, bilirubin, is yellow. High blood levels of bilirubin give the skin the yellowish tinge of jaundice while the excretion of excess pigment via the kidney leads to production of dark urine. Victims of brown recluse spider bites often also have "petechiae", a rash of tiny red spots caused when the sphingomyelinase D breaks down the underlying tissue and the red blood cell membranes so that blood leaks from capillaries into tissue spaces. Corticosteroid treatment is often helpful in reducing haemolysis and inflammation, while antivenin can be used if it is available. Hyperbaric oxygen therapy within 48 hours of the bite decreases the size of the necrotic ulcer. Dapsone, a drug used in tuberculosis therapy, has also been found useful although its mechanism of action is not clear. In all cases, treatment with rest, ice compresses and antibiotics is recommended, but healing may take several weeks to several months. Hobo spiders (*Tegenaria agrestis*) are native to Europe (although they are now found in the Pacific northwest of the USA) and fairly large (14 mm in length). They are found in houses, wood stores and yards, building sheet webs that narrow to a funnel shape; the initial bite is painless but the area becomes swollen and hard within 30 minutes, then developes as a blister which ruptures to leave a scar. Nausea, vomiting and diarrhoea often occur in victims rarely, headache and hallucinations have been reported; the toxins are similar to those of the brown recluse spiders.

Widow spiders (*Latrodectus species*) are found over most of the world in continents with temperate or tropical climates. Although both male and female are venomous, only the female has fangs large and strong enough to pierce human skin. They are larger (1–1.8 cm long) than the males (0.3–0.5 cm) and usually found in wood piles, outhouses, barns and garages although they also come into houses. Black widows (*Latrodectus mactans*) are the best known, with a red hour-glass pattern or red spots on shiny black. The Australian equivalent, the Perth red-back spider (*L. hasselti*), also causes vomiting with sweating and generalised pain which is particularly severe in the abdomen. About 2000 people per year have confirmed bites although only about 20% of these have serious poisoning. The venom contains neurotoxins, as well as proteases and hyaluronidases. After the initial bite, victims usually have dull pain at the site and pain or cramps in the muscles, with "fasciculations" or twitchings. Sweating is common, as is rigidity and pain in the abdominal muscles and the patient is usually very agitated and restless with a rapid pulse, high blood pressure, headache and dizziness. Calcium gluconate injected intravenously sometimes reduces muscle pain and muscle relaxants can be helpful, but antivenin is the most appropriate therapy in general, with oral corticosteroids to avoid reactions from administration of antivenin.

The neuroactive agents from widow spiders are called α-latrotoxins; they appear to form trans-membrane channels and usually have high molecular weights (~130,000). They cause massive release of neurotransmitters as they stimulate fusion of the synaptic vesicles (which contain neurotransmitters) with the pre-synaptic membrane. The Australian red back spider often bites people who are in bed or asleep and the venom has a long-term effect, with patients recovering partially and then collapsing with further pain, numbness of an affected limb and even paralysis, still being affected two months later. Studies have shown that antivenom is fully effective for up to three months after the initial bite and probably for longer; long-term sequelae to red-back bites is avoided if larger doses of antivenom are given at the start of treatment. In a study of 68 red-back spider (*L. hasselti*) bites in Australia, there was severe pain in 62% of cases, prolonged (>24 hours) pain in 66% of cases and systemic effects in 35% of victims while an analysis of cases with *L. curacaviensis* bites in Brazil found that limb pain was present in 29% , tremor and rigidity in 29%, generalised sweating in 28% and abdominal pain in 17% so that there are different features of latrodectism in different countries. In an American review of 163 black widow bites in Arizona, the most common initial symptoms included generalised abdominal, back and leg pain. Antivenom therapy was successful as 52% of patients not receiving antivenom required hospital treatment compared with only 12% of those who did have antivenom, all of whom recovered within one hour of therapy.

Funnel-web spiders (*Atrax robustus*) are very selective about their habitats, preferring rain-forests although they will live in eucalyptus forests and even suburban gardens if the soil is damp and undisturbed. They build silk-lined tunnels in the ground and are found over most of Australia and Tasmania. These spiders bite their victims repeatedly, usually clinging on firmly, but death is unlikely if the antivenom is available. There are two phases of envenoming following funnel web bites; initially there is massive autonomic nervous system excitation with simultaneous nicotinic, cholinergic and adrenergic components, so that the local piloerection and muscle twitching (fasciculation) start within a few minutes of being bitten. This then extends and is generalised within 20 minutes, accompanied by adrenergic components (hypertension, rapid heart beat) and cholinergic components (diarrhoea, salivation, lacrimation, coma). Fatal respiratory arrest may occur at this stage but usually this "autonomic storm", due to a slowing of voltage-gated sodium channel inactivation, will subside within one to two hours and the patient becomes more conscious but suffers from progressive hypotension, lung oedema and respiratory depression.

As can be seen, spiders put a great deal of metabolic energy and effort into synthesising toxins which will paralyse and kill their prey. They are succesful because they can exploit pathways of neurotransmission which are common to most living creatures. In the process, their toxins can tell us how our brains work and can shed light on the still-mysterious processes of neural function.

Antidotes

Antivenoms (usually made by injecting a tolerant animal with the venom) are the best therapy with adrenaline required in cases of anaphylactic shock. Antibiotics, corticosteroids, ice-packs and rest can be used as supportive therapy while oxygen, mechanical ventilation and intravenous fluid support may be needed in severe cases and atropine may be useful in reducing salivation. Severe hypertension and tachycardia may respond to β-blockers; most reports suggest that benzodiazepines are useful sedatives.

Case Histories

1. Loxosceles

A 35-year-old woman who was seven months pregnant was dressing one morning when she felt a sharp sting under her armpit. She hastily took her T-shirt off and shook it as a brown spider with darker brown markings on its back fell out onto the floor. After two hours, the site had become red and inflamed with a blister at the centre of the bite. A day later, the site looked rather like a "bull's eye", with a blue blister at the centre, then a white hard ridge and a reddened peripheral region. Two days later, the wound had become necrotic and itching and the woman finally went to the hospital where the wound was cleaned, cold compresses were applied and antihistamines with tetanus prophylaxis and antivenom were administered. However, this was too late to avoid the extended necrosis, which extended down her body as the venom had flowed down with gravity. Her doctors explained that because the venom had stayed on the original site, she had avoided the systemic effects of muscle aches, chills, fever, nausea and vomiting. Nevertheless, the wound ulcerated and then slowly healed just before her baby was born; the child was unharmed but the mother was left with a large area of deep scarring.

2. *Phoneutria*

A 23-year-old market worker was moving bunches of bananas early one morning in Sao Paulo, Brazil when he felt a sharp pain in his hand and realised he had been bitten by a large grey aggressive spider with a 3.5 cm body, 6 cm long legs and 5 mm long fangs. He recognised it as a female "aranha armadeira" or "warrior spider" (*Phoneutria nigriventer*) with the characteristic red brush of hairs around the fangs. The bite became extremely painful, and the victim noticed localised sweating at the bite site, where the hairs on his skin were standing on end. The pain radiated up to his chest and he felt his heart start to race then became dizzy and nauseous. He felt chilled all over and began to drool and vomit; at the same time he had an erection. Frightened by the fact that he could no longer focus his eyes, he begged to be taken to the hospital, where the wound was cleaned and his arm was immobilised and elevated with warm compresses to achieve local peripheral vasodilation. Local anaesthetics, tetanus prophylaxis and antivenom were then given and the victim recovered within 36 hours. The doctors told him that while they knew that toxins in *Phoneutria* venom included aspartic acid, glutamic acid, histamine, hyaluronidase and serotonin, there were unidentified protein/peptide fractions (which activated kallikrenin-kinin) and sodium channel poisons (that potentiated action potentials along axons) for which they had no antivenom.

3. *Tegenaria*

A 47-year-old man was clearing away rubbish in his basement in Seattle when he felt a slight stinging pain on the palm of his hand. A large brown spider with a 1 cm body and 4 cm leg span ran away very fast to hide behind some logs and the man realised that he had disturbed the horizontal silk mat spun by a hobo spider (*Tegenaria agrestis*) which had then attacked him. After 30 minutes, the bite area was sunken and very painful, with a surrounding ring of inflammation that increased to a 5 cm diameter. Within ten hours of the bite, the victim had severe headache, with nausea and vomiting, weakness and impaired vision. He found that he was unable to remember the circumstances of the bite and knew that he had memory loss, which fortunately was only temporary. Blisters developed within the wound crater within a day and then ruptured and coalesced to give a larger crater. The victim was treated with cool compresses, wound cleansing, tetanus prophylaxis and analgesics, while the arm was immobilised and elevated. The surgeons considered that scar excision with split-thickness skin grafting might be necessary when the bite had finally healed after about six weeks.

4. Atrax/Hadronyche

The boy was looking for a lost ball in his garden one summer in Sydney, Australia, when he disturbed the spider that had built a silk-lined funnel-shaped web between the tree roots. It was large with a shiny black body 4 cm long and a 6 cm leg span, and it promptly sank its prominent fangs into his hand. The boy knew enough to call for help as almost immediately he felt sick, then started to vomit. When his father found him, he was breathless, covered in sweat, frothing at the mouth and his muscles were twitching. He complained that his mouth felt "tingly all round" and his father realised that it must have been a bite from a Sydney funnel–web spider (*Atrax robustus*), one of the most poisonous spiders in the world and that there was very little time before his son died from lung oedema and circulatory failure. Hastily he splinted the boy's arm and wrapped the wrist and hand with a pressure bandage to slow the lymphatic spread of the neurotoxic venom to the central circulation. Then, telling his son to stay as still as possible, he rushed him to the local hospital that he knew kept supplies of antivenom. Once there, two ampoules of antivenom were administered intravenously every 15 minutes until the neurotoxic symptoms resolved. Then the bandage was slowly released in an intensive care unit with more supplies of antivenom on hand in case the neurotoxiciy re-started. Fortunately, the antivenom was also active against *Hadronyche* funnel web spiders and against *Missulena* mouse spiders so that precise identification of the spider was not essential. The boy made a good recovery with no reactions to the antivenom preparation ("serum sickness").

5. *Latrodectus*

The woman was just shifting the table in the garden before the barbecue when she felt a sudden sharp pain in her hand. Startled, she jumped back as a black spider with a red band on its back and a 4 cm leg span scuttled away. A tiny puncture site was visible on her hand. She realised it must have been a "red-back" or "widow spider" (*Latrodectus hasselti*) and remembered the stories her grandfather had told her about the pioneer days in the Outback, when red-backs lurked in every "dunny" or outdoor privy and were liable to bite the unsuspecting user on the buttocks or genitalia. Within 30 minutes, her limbs and lower back had become rigid and painful, with involuntary spasms and cramps while her abdomen felt as rigid and painful as when she had appendicitis. She started to sweat and realised that her face had become contorted and tears were streaming down. She was rushed to the hospital where the bite site was cleaned and ice packs were applied with analgesics and tetanus prophylaxis. She was in great pain and as the muscle cramps had not subsided and her blood pressure was rising, the medical team administered opioids intravenously with sedative benzodiazepines, then a purified Fab-fragment *Latrodectus* antivenom to avoid the serum sickness (anaphylaxis) and death sometimes seen in the past when unrefined antivenoms were produced in horses. She was able to leave the hospital 24 hours later, although the pain continued for the next week.

Suggested Further Reading

Ellenhorn MJ and Barceloux DG (1988). In: *Medical Toxicology: Diagnosis and Treatment of Human Poisoning*. Elsevier, New York, pp. 1148–1152.
Klaasen CD (ed). (1996). *Casarett and Doull's Toxicology: The Basic Science of Poisons*, 5th ed. McGraw-Hill, New York, pp. 831–835.

21

STRYCHNINE

R. M. Harris

Description

Strychnine is a member of the alkaloid group of chemical compounds which are widely distributed amongst plants and some animals. They have a bewildering array of structures, usually comprising ring systems and almost always containing at least one basic nitrogen atom. It is the latter which is responsible for the alkaline nature of these molecules and is believed to be responsible for their bitter taste. The different structures endow the alkaloids with a vast range of pharmacological properties, many of which have been used since ancient times for both medicinal and magical/religious purposes. Due to the complexity of these chemicals, most alkaloids were isolated from their natural source before their structures were determined. Consequently, their names tend to reflect their pharmacological action, the name of their discoverer or, in the case of strychnine (isolated from *Strychnos* species) their botanical source.

Although strychnine was first isolated in the early 19th century by Pelletier and Caventou, it has a particularly complex structure (Figure 1) which was not elucidated until the mid-1940s, largely as a result of a multinational effort by Robinson's group at Oxford, Woodward's group at Harvard and Prelog's group at Zurich. Their brilliant achievement is still regarded as one of the triumphs of classical degradative chemistry. Woodward's group continued with the work and eventually devised a method for synthesizing strychnine in the laboratory which was published in 1954. The method was made especially complicated by the fact that strychnine,

(A)

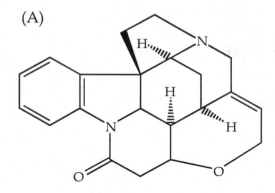

(B)

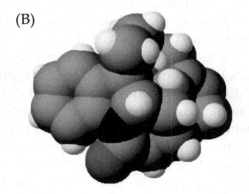

Figure 1. The structure of strychnine. A strychnine molecule shown (A) in line form and (B) as a space-filling representation. In this line diagram and others in this chapter, wedged bonds project above the plane of the paper and dashed bonds project into it.

in common with many other organic molecules, can exist in both left and right-handed (i.e. mirror image) forms. This property, which is a consequence of the tetrahedral shape of carbon atoms, is referred to as chirality and occurs when each of a carbon atom's four bonds is connected to a different chemical group. Under these circumstances, it is possible to arrange the groups in two unique sequences and this produces molecules which are mirror images of each other. Although the molecules are chemically and physically identical, their chiral nature means that they are optically active and can rotate rays of plane-polarised light either clockwise (+) or anticlockwise (–). In nature, because these molecules are synthesised by other molecules which are themselves chiral, only one form is made (as an analogy, a mould designed to make rubber gloves for the left hand cannot make gloves to fit the right hand). However, when chemicals are synthesised in the laboratory, unless the starting materials are already chiral, the groups bind in a random fashion producing a racemic (i.e. one-to-one) mixture of the clockwise and anticlockwise forms which must then be separated. Hence, the original method for strychnine synthesis required no less than 15 steps to achieve the correct, optically active conformation. Since then, many other methods have been devised and strychnine still remains an object of fascination for many organic chemists.

In the pure state, strychnine consists of a white, light-sensitive, crystalline powder with a melting point of 286°C. It is only sparingly soluble in water, but has an extremely bitter taste which can be detected at concentrations of about 8 ppm. It is, however, highly soluble in alcohol and its optical activity is such that a solution of 0.5 g/ml in absolute alcohol in a 10 cm long chamber at 20°C will rotate polarised light at a wavelength of 589.3 nm anticlockwise by 104°.

Strychnine is extremely toxic with the lethal dose for an adult being about 50 to 100 mg when taken orally. When administered intravenously the lethal quantity is much smaller and the onset of symptoms is hastened considerably. In children, the fatal dose can be extrapolated *pro-rata* with body weight, but adverse reactions have been documented from doses as small as 2 mg. At the opposite extreme, prompt medical attention has enabled some people to survive doses many times larger than the normally lethal amount, the maximum so far recorded being about 3750 mg. As well as ingestion and injection, strychnine can also be absorbed across the skin and, if it is inhaled, across the lining of the lungs.

Strychnine may be identified chemically by the production of a deep violet-blue colour when 0.1 g is added to 3 ml of a 1%

solution of ammonium vanadate in sulphuric acid. The colour gradually changes to purple and then, on dilution with water, to cherry-red. If a small crystal of strychnine is dissolved in two to three drops of sulphuric acid, an intense violet colour which changes through red to yellow, is produced when a crystal of potassium dichromate is passed through the solution. By contrast, today's modern chromatographic techniques allow strychnine to be detected at concentrations below 1 ng/ml of blood, but although the early chemical methods were comparatively crude, they were still sufficiently sensitive and precise to provide evidence for the conviction of several murderers in the 19th and early 20th centuries.

Source

Strychnine occurs naturally in plants of the genus *Strychnos* which encompasses about 200 species of trees, shrubs and lianas. The plants are found world-wide in both sub-tropical and tropical zones and are sometimes grown in gardens for interest rather than ornamental value. Strychnine is actually found in relatively few members of the genus *Strychnos*, but these plants are known to produce a wide range of alkaloids which may be tetanising (cause convulsions), curarising (cause paralysis) or some cases both depending on the dose. The concentration of toxic alkaloids varies between species to the extent that some of the fruits can be consumed by humans. The Zulu people use the fruits of the monkey orange (*S. spinosa*) to make an alcoholic beverage and as a cure for snake-bite, while the fruits of the black monkey orange (*S. madagascariensis*) are used to treat dysentery and to prepare a type of jam. However, in general, all members of the genus should be treated with caution and particular care taken when handling the bark, roots and seeds in which the alkaloids tend to be most highly concentrated.

   Strychnine was first isolated from St. Ignatius beans (*S. ignatii*), a vine found in the Philippines and introduced to Cochin China where it was highly regarded as a medicine. However, *S. nux-vomica* (Figure 2), a medium-sized tree native to India, is now used as the commercial source of this compound. The trees can grow up to 75 feet tall and bear clusters of small white, green-tinged flowers with an unpleasant smell. When fertilised, these develop into large orange berries containing up to five disc-shaped seeds. When dried, the seeds usually contain between 1.1% and 1.4% strychnine plus a similar quantity of the related alkaloid brucine.

Figure 2. Leaves and flowers of *Strychnos nux-vomica*.

Strychnine is prepared from the seeds by grinding them to a paste with slaked lime and water. The paste is dried and mixed with a suitable organic solvent (formerly benzene was used, but now less harmful solvents are substituted, e.g. chloroform). The slaked lime makes the mixture alkaline and under these conditions the strychnine dissolves readily in organic (i.e. oily) but not water-based solvents. The particulate debris is then removed by filtration and the liquid shaken with dilute sulphuric acid. This neutralises the effects of the lime and when the organic solvent and acid layers separate (cf. oil and vinegar salad dressing) the strychnine is dissolved in the latter. The organic solvent is decanted from the strychnine solution which is rendered alkaline by the addition of either ammonia or sodium hydroxide. This drastically reduces the solubility of the strychnine molecules in water and, since there is no organic solvent present to dissolve them, they precipitate and are removed by filtration. The crude strychnine can then be purified by further combinations of acids, alkalis and organic solvents until the product complies with the necessary analytical criteria.

Uses

Current orthodox medical opinion is that strychnine has no therapeutic value. However, its selective neurotoxic properties (see next section) are still widely exploited by researchers studying nerve functions in a variety of neurological conditions including Parkinson's disease, schizophrenia and stroke.

Until fairly recently, strychnine was used as an ingredient in "over-the-counter" tonics — a practice which has caused many accidental poisonings, particularly of young children. The seeds of S. nux-vomica, both in powdered form and as a tincture, were used to treat indigestion, chronic constipation and, by virtue of their bitter taste, to stimulate the salivary glands and thereby increase appetite. One of the more infamous recipients of this type of therapy was Adolf Hitler who, amongst many other medications, was prescribed pills containing extracts of belladonna (deadly nightshade) and S. nux-vomica as a remedy for indigestion and flatulence. Strychnine extracted from the seeds was also used as a bitter and as a tonic for the circulatory system. Since both heroin and cocaine are also white, bitter tasting powders, strychnine has been added to both of these drugs in order to make users believe they are being sold a purer product. Several addicts have received fatal doses as a result of this sinister practice.

In 19th century Russia, injections of strychnine were prescribed for the treatment of alcoholism where it was considered to be beneficial for curing not only the acute symptoms and delirium tremens but also for abolishing the desire to drink. A more obscure side-effect of strychnine is that small doses enhance the perception of the colour blue. It was therefore used to treat ambylopia (dull vision) before it was realised that this condition was usually due to a poor diet and/or mild cyanide poisoning. Strychnine, almost infinitesimally diluted, is still used by homeopaths for treating a variety of disorders including asthma, haemorrhoids, jaundice, migrane smoking, snoring and the effects of alcohol. *S. nux-vomica* is also still found as a constituent of some Chinese herbal medicines usually after some form of processing to reduce the toxicity — this is not always successful and accidental poisonings have been recorded. In larger quantities, strychnine is used as a pesticide to control rodents and other small mammals. However, due to the agonising nature of the death which it inflicts, its use in the UK is strictly limited by the Cruel Poisons Act. Unfortunately, this property meant that it was also widely favoured as an "ordeal poison"; if the subject survived then they were innocent but if not then they were guilty and a ghastly example had been made. Extracts from *Strychnos* species are still used in the manufacture of curare and similar concoctions which are smeared on the tips of darts and arrows by native tribesmen in some parts of Africa and South America. Here, advantage is taken of the fact that strychnine and other alkaloids are very much more potent when administered intravenously than orally. As a consequence, prey killed by a curare-tipped dart is perfectly safe for human consumption. (As a historical note, it is from the Greek *toxikon pharmakon* (poison/drug for arrows) that the words "toxicology" and "pharmacology" are derived.)

Scientifically, strychnine is a valuable research tool for studying the nervous system. Another constructive use of strychnine employs both its chiral and alkaline properties to separate racemic mixtures of organic acids into their (+) and (−) forms. When (−) strychnine molecules are allowed to react chemically with the acid molecules, the resulting strychnine-acid hybrids are either (− +) or (− −). Since these molecules are no longer mirror images, which would be (+ −) or (+ +) respectively, they no longer have identical physical properties and may be separated. It is then relatively straightforward to remove the strychnine moiety from the hybrid molecules leaving the original racemic acid mixture resolved into its (+) and (−) constituents.

Mechanism of Toxicity

Although it is much less potent than the so-called "nerve gases", the main target for strychnine is the nervous system. In particular, it primarily affects the motor nerves in the spinal cord which control muscle contraction. A detailed account of how nerves transmit signals can be found in the following chapter on tetrodotoxin. An impulse is triggered at one end of a nerve by the binding of neuroexcitatory chemicals (also called neurotransmitters) to receptors present on the surface of the cell. However, in addition to excitatory receptors, nerve cells are also studded with receptors for neuroinhibitory chemicals. In the presence of neuroinhibitors, a far greater proportion of the excitatory receptors must bind neurotransmitters before an impulse is initiated. One type of neuroinhibitory receptor binds the amino acid glycine (Figure 3). Strychnine will also bind to this receptor at a slightly different, but partially overlapping, site. Consequently, although strychnine will bind to the glycine receptors, it does not have an inhibitory effect and prevents the natural ligand, glycine, from binding (rather like a van parked in front of a driveway preventing the returning householder from parking their car in their garage). This means that nerve impulses are triggered in response to lower than normal levels of neurotransmitters (Figure 4). Since the motor neurones are involved, the slightest stimulus causes the victim to undergo the violent, agonising convulsions which are so characteristic of strychnine poisoning. Death usually occurs from asphyxiation caused by spasms of the diaphragm, thoracic and abdominal muscles which prevent the victim from breathing.

The violent muscular activity has other side-effects which, if the victim survives the convulsions, can hinder recovery and cause serious long-term damage. These effects, which result both from the biochemistry underlying muscular contraction and the mechanical damage to which the straining muscles are subjected, include:

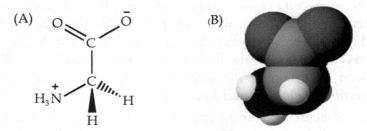

Figure 3. The structure of glycine. A molecule of the amino acid glycine shown (A) in line form and (B) as a space-filling representation.

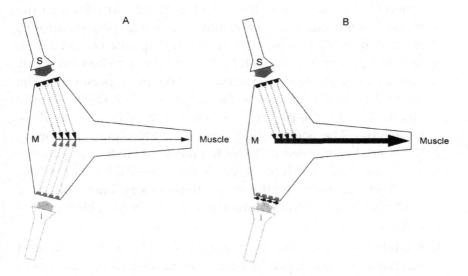

Figure 4. Schematic diagram showing the blockade of the inhibitory control of motor neurones by strychnine.

Under normal conditions (A), nerve impulses from the motor neurones (M) to the muscles are under the control of neurotransmitters (➤) released by the stimulatory (S) and inhibitory (I) nerves. These bind to specific receptor molecules (▼) on the surface of the neurone which trigger the release of intracellular messenger molecules (┄┄┄➤). The equilibrium between the different intracellular messengers controls the rate of contractile impulses sent to the muscle (————➤). In the case of strychnine poisoning (B), strychnine molecules (➤) bind to the inhibitory receptors but do not trigger the release of inhibitory intracellular messengers. This results in a massive imbalance in favour of stimulation which leads to uncontrollable spasms of the muscles.

(1) Hyperthermia (raised body temperature). As anyone who has removed their coat while chopping logs on a cold winter's day will testify, physical exertion produces heat. In the case of strychnine poisoning, body temperatures of 43°C (109°F) have been documented in patients who have survived but come perilously close to multi-organ failure.

(2) Lactic acidosis. During sudden, strenuous activity, the heart is unable to pump blood fast enough to supply sufficient oxygen to the muscles. As a consequence, the muscles switch from aerobic (oxygen using) to anaerobic metabolism which, although far less efficient, does not require oxygen. The by-product of mammalian anaerobic metabolism is lactic acid which accumulates in the bloodstream until it can be removed when aerobic conditions are restored. While someone who has just sprinted for a bus can rest and pant to re-establish their normal body chemistry, this option is not available to victims of strychnine poisoning. The pH of the blood is usually in the range of 7.35–7.45 (i.e. slightly alkaline) and if the pH drops below 6.8 the patient is unlikely to survive. The acidification caused by strychnine poisoning has been documented to lower the blood pH to 6.61 in patients who later made a full recovery. In one extreme case, however, a patient who survived the acute effects of strychnine, but later died from multi-organ failure, was found to have a blood pH of 6.46.

(3) Rhabdomyolysis. This is a condition where the heat and mechanical stresses produced in the muscles causes the fibres to break down. This releases large quantities of myoglobin (a simpler form of haemoglobin found in muscles) into the blood stream which can overload the kidneys and lead to renal failure.

However, all of these side-effects are correctable (see section on antidotes). Provided that the involvement of strychnine is recognised quickly, all but the most severely poisoned patients can be expected to make a full recovery without any lasting physical damage.

Metabolism and Detoxification

The alkaloids form part of a complex array of chemicals which plants use to deter animals, including man, from eating them. Unable to run from predators, many plants rely on overt toxicity and/or a nasty taste as a first or second line of defence. Obviously an animal whose body can destroy these poisons before toxic concentrations can accumulate is able to utilise a far greater range of foods and is at a particular advantage during times of famine. As a consequence, there has been an evolutionary "arms race" between plants producing gradually more potent toxins and animals gradually improving the mechanisms by which they can metabolise and excrete them. In mammals, although there are many inter-species variations, a basic repertoire of detoxification enzymes is always present. However, if a change in the diet overloads a particular pathway, additional quantities of the appropriate enzymes can be rapidly synthesised.

Prior to the discontinuation of the use of strychnine as a drug, analysis of patients' urine samples showed that usually less than 15% of a dose was excreted unchanged. However, this can vary between 1%–30% depending on the individual and the size of the dose. Most of the details of strychnine metabolism have been worked out in Japan using rats and rabbits. In the rat, strychnine has been found to be a potent inducer of cytochromes P450 2B1 and 2B2 suggesting that these enzymes play a major role in its detoxification. This confirms an observation in 1962 when it was noted that rats pre-treated with phenobarbitone (which also induces the cytochrome 2B family) were far more resistant to strychnine poisoning than untreated rats. Enzyme induction explains why, that provided the patient does not die, smaller proportions of larger doses tend to be excreted unchanged because the body has more time to respond to the xenobiotic challenge. By the late 1980s at least seven metabolites had been identified (Figure 5) with strychnine N-oxide being the most prevalent. Strychnine has also been found to induce members of the glutathione-S-transferase (GST) and UDP-glucuronosyl transferase (UDPGT) families of enzymes. Many products of cytochrome

P450 metabolism are prone to form free radicals which can bind to DNA, potentially leading to genetic mutation and, under the right circumstances, triggering carcinogenesis. Glutathione conjugation, mediated by GST, is a protective mechanism which neutralises the free radicals before they can cause damage. Conjugation with both glutathione and glucuronic acid (*via* UDPGT) increases the solubility of the xenobiotic and facilitates excretion. However, despite the observation that strychnine induces GST and UDPGT, the relevant metabolites have yet to be identified. Overall, the toxicokinetic half-life of strychnine in humans appears to be around ten to 16 hours, which means that the poison will be virtually cleared from the body in three to five days.

Suicides and Homicides

Although it is a highly effective poison, strychnine has never been widely used for either suicide or murder. In the case of the former, the ghastly death which it inflicts means that it has tended to be used either in ignorance or when no other alternative is available. However, it has been intentionally self-administered by some members of the Free Pentecostal Holiness Church as one of several methods which they use to provide a divine test of their faith. There have been several cases where drug addicts, mistaking the white, bitter powder for cocaine, have snorted strychnine with lethal results. It is also likely that the inadequate removal of strychnine used to separate the optical isomers of some drugs has contributed to the deaths of some addicts. With regard to murder, the main problem with strychnine is overcoming its intensely bitter taste. During the American prohibition years, cocktails of strychnine and whiskey were sometimes used by criminals both to remove, and to make an example of, undesirable associates.

There were, however, several high-profile 18th and 19th century murder cases in which strychnine was involved. The most famous was probably that of the Lambeth Poisoner, Dr. Thomas Neill Cream. Born in Scotland, his family moved to Canada where he graduated in medicine from McGill College, Montreal on 31 March 1876. Shortly

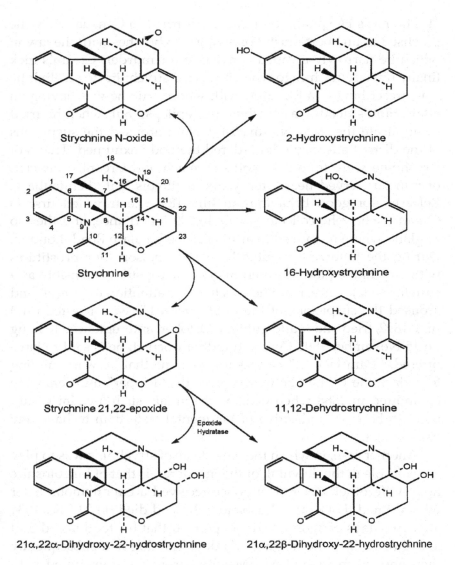

Figure 5. Known metabolites of strychnine.

afterward he met a Flora Elizabeth Brooks whom he married after nearly killing her when he performed his own abortion on their unwanted child. He left for England, underwent further training at

St. Thomas's Hospital, London and returned to Canada where he practised as an abortionist. Following two brushes with the law in which he narrowly escaped conviction for murder, Cream's luck finally ran out in Chicago when he added strychnine to an elixir he was prescribing to a Mr. Stott with whose wife he was having an affair. Stott's death was attributed to epilepsy but, when he tried to implicate the local pharmacist, Cream aroused the suspicions of the district attorney who ordered the body exhumed. Traces of strychnine were found in Stott's stomach and Cream, convicted of murder, served nearly ten years in Illinois State Penitentiary. Released for good behaviour in July 1891, he travelled first to Canada, where he collected a US$16,000 inheritance, and then to England where he moved in at 103 Lambeth Palace Road, London. During the following months he was to poison four prostitutes with strychnine (administered in gelatine capsules ostensibly as a cure for acne). Once again he was to draw attention to himself and aroused the suspicions of the local police. He was arrested on 3 June 1892, tried and found guilty in October, and sentenced to hang on 15 November 1892. As the trapdoor opened, he is said to have uttered, "I am Jack ..." suggesting to some that he was claiming to be Jack the Ripper. However, since the "Ripper" murders were committed in 1888 when Cream was in jail, such a claim would seem to be more indicative of his mental state than to have any basis in fact.

Another case, in which the bitter taste of strychnine was to play a major role in the downfall of the murderer, is that of Lincolnshire housewife Ethel Major. The police received an anonymous letter which stated that Ethel's husband, who had died on 24 May 1934 after two days of agony, had complained that his food tasted foul and thrown it to the neighbour's dog. The dog had subsequently died and when its and Mr. Major's corpse were exhumed both were found to contain strychnine. During the police investigation, Ethel unwittingly implicated herself when she stated that she was unaware that her husband had "died of strychnine poisoning", before mention had been made of strychnine. She was tried in November 1934, found guilty and, despite a recommendation of mercy by the jury, was hanged at Hull prison on 19 December 1934.

Antidotes

There is no true antidote to strychnine poisoning — once it has entered the body, there is no known substance which can be administered to prevent it blocking the glycine receptors. However, although the effects of this can easily be lethal, if the severity of the convulsions can be controlled, the patient will naturally clear the poison without physically suffering from any long-term effects.

There are five main points to consider when treating victims of strychnine poisoning, and these may essentially be applied to the treatment of any form of poisoning for which no specific antidote is available.

1. Respiration. This must be maintained and monitored, particularly during the first few hours of treatment.
2. Convulsions. Not only does the control of these facilitate further treatment, but most of the long-term damage is due to the side-effects of the extreme physical activity involved.
3. Blood chemistry. Correction of the acidosis produced by muscular activity and replacement of lost electrolytes due to the hyperthermia it induces greatly improves the patient's prognosis.
4. Prevention of further absorption. Although free strychnine is absorbed rapidly from the gut, it is often taken in a form (e.g. rodenticide pellets) which retards absorption. By the time that the symptoms have started to appear, often only a relatively small portion of the total dose has been absorbed. The destruction or removal of the strychnine in the gut prevents further absorption and shortens the period of exposure for the patient.
5. Enhancement of excretion. Once strychnine has entered the body, it is only cleared relatively slowly. If the rate can be increased then the blockade of the inhibitory nerves will be dissipated that much sooner.

A typical scenario of a successful treatment might be as follows. The patient is rushed into casualty in a distressed condition having already had a series of convulsions in the ambulance. The ambulance staff have already ascertained the cause and the patient is given an intravenous injection of diazepam (a skeletal muscle relaxant which also acts as an anti-anxiety agent) immediately on arrival. After several minutes, during which the patient is encouraged to breathe deeply, he/she is placed under general anaesthesia coupled with a stronger muscle relaxant. The stomach is washed out with saline solution, after which a slurry of dilute (1:5000) potassium permanganate and activated charcoal is administered *via* a nasogastric tube. Potassium permanganate is a strong oxidising agent which reacts with many organic molecules and renders them less harmful — unfortunately it is also highly irritating to the mucous membranes and can only be used in dilute form. Activated charcoal binds many different types of molecule and physically prevents them from being absorbed by the gut. In order to minimise stimulation, the patient is allowed to awaken in a quiet, darkened room where the treatment is continued. To facilitate excretion of strychnine *via* the kidneys and to moderate the effects of rhabdomyolysis, the patient is given a saline drip to supply an excess of fluids which increase the flow of urine. Meanwhile, the blood chemistry is slowly brought back to normal by intravenously administering sodium bicarbonate to raise the pH and potassium chloride to correct the potassium deficiency. The latter is partly due to the effects of hyperthermia and partly due to the diuresis. Control of the convulsions is maintained by a combination of diazepam and phenytoin (used in the treatment of epilepsy). During this period, the patient must be carefully monitored to ensure that the diazepam does not exacerbate the acidosis which is one of its potential side-effects. This regimen is continued until there is no sign of increased reflex spasms in the neck and limbs following mild stimulation. Provided that the patient reaches hospital before too many convulsions have occurred, a complete physical recovery may be expected within eight hours of admission although the psychological trauma may take longer to resolve.

Case History

With the greatly reduced availability of strychnine as both a pesticide and constituent of over-the-counter tonics, strychnine poisoning has become increasingly rare. Fatalities are rarer still because, provided that the poison is recognised quickly, prompt treatment can prevent the victim from succumbing to even a very large dose. The following case history is based on the Coroner's case report of a suicide in Allegheny County, Pittsburgh, USA.

At about noon, on a May day, a middle-aged man concluded that life was no longer worth living. He managed to swallow just over half a tin of mole poison which contained over 250 mg of strychnine. The alarm was raised shortly afterwards by his neighbours who heard loud moans of pain coming from his apartment. The police were called and broke into the flat at about 12:49. The victim was found to be fully conscious, but was lying face down on the living room floor in a state of obvious agony. He was moved to a sofa where, complaining of excruciating abdominal pain, he underwent a series of convulsions. During the spasms, his body became rigid and adopted the classic opisthotonos posture associated with strychnine poisoning – his spine arching so that he was resting on his head and heels. Unable to breathe, the patient turned blue but after about half a minute his body relaxed. Seconds later he had a second attack and fell unconscious. The emergency team began cardiopulmonary resuscitation but were unable to revive him although an electrocardiogram showed that his heart was attempting to beat for a further 20 minutes.

At the autopsy (commenced three hours post-mortem) the body was that of a well-developed middle-aged man 5'8" tall and weighing 12½ stone. The body was extremely rigid as if in rigor mortis although it had barely begun to cool down. When the stomach was opened it was found to contain a thick, green sludge which matched the colour of the rodenticide pellets. On analysis, the stomach contents were found to contain 175 ppm strychnine (totalling 213.5 mg from an estimated dose of 266 mg) and a much lower concentration (4.1 ppm) was found in the small intestine. The poison had not had enough time to reach the large intestine and no traces were found. The stomach wall contained 14.9 ppm strychnine which was the highest concentration found in the body tissues and suggests that this was the primary site of absorption. Despite this, the mucosa lining the stomach was normal in appearance and the only internal organs showing signs of gross damage were the lungs which were swollen and congested. Strychnine was found in the blood (whole blood 3.3 ppm, plasma 2.6 ppm) and was apparently being concentrated and excreted by the liver (6.2 ppm) into the bile which had already attained a concentration of 9.2 ppm. By contrast, the concentration in the kidneys (3.2 ppm) was similar to the blood, while that found in the urine was much lower (1.4 ppm). Ironically, the lowest quantifiable concentration occurred in the cerebrospinal fluid (0.08 ppm). Traces were just detected in the cerebrum itself, but could not be found in other parts of the brain which was normal in appearance. The upper part of the spinal cord showed signs of congestion and, when examined microscopically, it was found to be riddled with minute haemorrhages. Some of the nerve cells had undergone chromatolysis, a form of cell death which occurs when neurones have become energetically exhausted from repeated firing.

The symptoms presented by the patient were typical of strychnine poisoning which is characterised by the presence of violent convulsions which are triggered by the slightest stimulus and may also include a rapid side-to-side movement of the eyes (nystagmus). The fact that the patient usually remains conscious until immediately prior to death distinguishes strychnine poisoning from other convulsive disorders such as epilepsy and overdoses of some antipsychotic drugs based on phenothiazine. The sudden onset of the convulsions coupled with the absence of fever rules out meningitis and meningoencephalitis. It should be noted, however, that other people have survived doses of strychnine far greater than that taken by this particular man, and this underlines the necessity of finding the victim quickly in such cases.

Acknowledgements

I would like to thank Ms. Pauline Hill of the School of Biological Sciences (now Biosciences) at the University of Birmingham for drawing the picture of the leaves and flowers of *Strychnos nux vomica* in Figure 2.

Suggested Further Reading

Philippe G, Angenot L, Tits M and Frédérich M (2004). About the toxicity of some Strychnos species and their alkaloids. *Toxicon* **44**, 405–416.

For information on famous cases of poisoning and the poisons involved

Farrell M (1992). *Poisons and Poisoners. An Encyclopedia of Homicidal Poisonings*. Robert Hale Limited, London.

Case Reports

Lambert JR, Byrick RJ and Hammeke MD (1981). Management of acute strychnine poisoning. *Canadian Medical Association Journal* **124**, 1268–1270.
Perper JA (1985). Fatal strychnine poisoning — a case report and review of the literature. *Journal of Forensic Sciences* **30**, 1248–1255.

22

TETRODOTOXIN

M. Wheatley

Introduction

It is well-acknowledged that in the distant past life began in the sea, and now for some, life ends there. The charts of ancient mariners often noted at their edges that *"There be sea monsters here"*. The sea can still pose a threat to life today, not from sea monsters, but at the molecular level from natural marine toxins. The term ichthyotoxicity (from *ikhthus*, the Greek word for fish) is used to describe poisoning by eating fish and any other marine animals. Where poisoning leads to an impairment of nerve function, the term pelagic paralysis has been coined. This does not refer to a specific source of toxin ingestion but rather to a set of symptoms. The name derives from the main common feature of such an affliction, namely paralysis and the classical Greek word for the open sea — *Pelagos*.

The most common form of fish poisoning is *Ciguatera* which was first reported by sailors in the 18th century. The name was originally given by Don Antonio Parra, whilst in the Caribbean in 1787, to describe the toxic effects of eating the *"cigua"*, or turban shellfish but is now applied to a set of symptoms arising from eating many species of fish. Outbreaks of poisoning are sporadic and exhibit a dispersed distribution throughout tropical and sub-tropical coastal areas but are particularly common in the Pacific and Caribbean regions. Poisoning is unpredictable, as fish species that are usually perfectly edible suddenly become ciguatoxic in a given area even though they are freshly caught and appear healthy. Furthermore, many different species can be affected in the same outbreak. Over 2000 cases are reported every year often centred on coral reefs. Symptoms appear over a period of hours after ingestion of an infected fish and include tingling, itching, cramps, nausea and most unusually, a reversal in the sensation of hot and cold! Ciguatera only rarely results in death, usually by respiratory or cardiac failure. It has recently been discovered that ciguatoxin is produced by algae and passed onto reef-dwelling fish, especially large carnivorous fish. It appears that disturbances in the coral reef ecosystem can result in ciguatoxic episodes. The toxic algae thrive in dead coral, so undergo a population explosion if the coral is disrupted. Such disruptions can be due to natural forces, such as monsoons and seismic activity, or caused by man during construction projects or even nuclear tests.

Moray eels (*Gymnothorax javanicus*) have also been reported to be occasionally toxic. The symptoms resemble ciguatera but additionally often include excessive salivation. As moray eel toxicity seems most common in eels caught on coral reefs when other fish are ciguatoxic, it seems most likely that moray eel poisoning is actually ciguatoxin accumulated by this formidable predator.

There have been several reports that some species of fish are psychotogenic in that ingestion can cause delirium, mental disorder or hallucinations. Indeed one of the sea chub family, *Kyphosus fuscus*, reputedly causes nightmares and is subsequently referred to in certain regions of the Pacific as the "dream fish".

Obviously not all marine toxins arise from adventitious contamination. Marine species also utilise endogenously synthesized toxins for both defensive and offensive strategies. The Red Sea flatfish, *Pardachirus marmoratus*, secretes a slime which contains the toxin pardaxin. This is a linear 33-residue polypeptide which, upon binding to cells, spontaneously forms pores in the plasma membrane leading to cytolysis and cell death. Interestingly, it has been found that pardaxin is an effective shark repellent.

Examples of marine species using venoms for hunting are too numerous to list here but one which caught my imagination was the piscivorous sea snail *Conus purpurascens* or purple cone. How can a snail survive by catching fish given the relative speed (or lack of it) of the two animals? The answer is that the snail possesses a harpoon-like weapon comprising of a modified tooth, which is hollow and barbed, on the end of an extendible proboscis. This proboscis extends rapidly to strike at passing fish (see Figure 1). Once pierced, venom is immediately injected via the hollow tooth. A good strike generates almost instantaneous rigid paralysis and tetanus of the major fins. If fish escape at this stage they suffer from flaccid (floppy) paralysis. The normal fate however, is that the fish is effectively tethered by the "harpoon" and can be drawn back towards the predator whereupon it is eaten at snail-pace! The snail's venom has been analysed biochemically and found to be a potent cocktail of diverse "conotoxins" rather than a single component venom. Several of these conotoxins have been shown to specifically disrupt the performance of proteins required for normal neuronal functioning. Consequently, when injected in combination by the snail, neurotoxic effects are exhibited by the fish extremely rapidly and are referred to by researchers as the sudden tetanus of prey (STOP) syndrome.

Although the seas have provided an abundance of toxins, restrictions in space allow only two interesting examples to be presented here in depth.

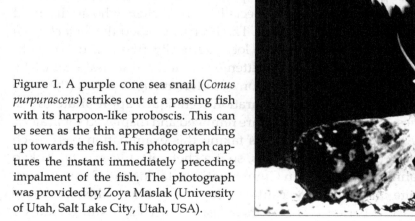

Figure 1. A purple cone sea snail (Conus purpurascens) strikes out at a passing fish with its harpoon-like proboscis. This can be seen as the thin appendage extending up towards the fish. This photograph captures the instant immediately preceding impalment of the fish. The photograph was provided by Zoya Maslak (University of Utah, Salt Lake City, Utah, USA).

Marine Toxins

An interesting feature of the puffer fish, and indeed the origin of its name, is its tendency when alarmed to inflate itself into a spiky sphere by swallowing water (Figure 2). Impressive as this defence strategy is, it does not deter the Japanese who view puffer fish fillet, or *Fugu*, as a gastronomic delicacy. Puffer fish is represented in the Japanese by two characters *"kawa"* and *"buta"*, which have the literal meaning of "river pig" (Figure 3). The taste is apparently typical of white fish and it is liked for the texture of the flesh. The puffer fish has a second line of defence however, in that the flesh contains a potent neurotoxin. The toxin in *fugu* is tetrodotoxin which is named after the order Tetraodontiformes (meaning four-toothed) to which puffer fish belong. Only traces of tetrodotoxin exist in the muscle and occasionally this can generate a tingling sensation. Obviously, the generation of tingling does not rate highly as a form of defence against predation. However, although only trace amounts of tetrodotoxin exist in the muscle, the toxin is concentrated in the liver, ovaries, intestine and skin. In addition, female puffer fish are more toxic than males due to the high level of tetrodotoxin in the ovaries. Tetrodotoxin levels are also seasonal being highest when the fish are spawning (March to June). As the toxin is not destroyed by freezing or cooking, it is vital that the *fugu* musculature is not contaminated by viscera when it is prepared for eating. Consequently, in Japanese restaurants, *fugu* flesh has to be cleaned and dressed by specially trained chefs who are licensed by the Japanese Government. The license is called the *fugu chirishii menkyo* and is issued by the local authority (the *Ken* or *To*). The exam has two sections, a written exam and a practical exam. The practical includes identification of different *Fugu* species (there are over 30) plus different preparation techniques appropriate to the species. The poisonous parts are removed and displayed separately. Only one in three candidates taking this examination pass. Each prefecture has its own exams, so if a chef gets a license for Tokyo he cannot set up shop in Gunma without taking the Gunma exam. Despite all of these precautions, for some diners, this expensive culinary experience is also their last. In fact *Fugu* still remains a

Figure 2. A fully-inflated puffer fish.

Figure 3. "Puffer fish" in Japanese.

common cause of food poisoning in Japan with approximately 50 deaths annually. *Sabu fugu* has no toxin at all and *Tora fugu* is the most poisonous. Perversely perhaps, it is the *Tora fugu* which is the most popular and the most expensive.

Puffer fish have a widespread distribution in warm and temperate waters and puffer fish poisoning has been recognised by man for a long time. A puffer fish which has been identified as *Tetraodon lineatus* is shown in a fifth dynasty Egyptian tomb dating from c. 2500 BC. Puffer fish and *Tetraodon* roe are also reported in traditional Chinese herbal medicine, being cited in the *Pen-T'so Chin* (Herbal) written in the second or first century BC and again in the *Pen-T'so Kang Mu* (Great Herbal) dating from 1596. Although well-recognised in the Orient, particularly in coastal areas, puffer fish poisoning was an unknown phenomenon to Europeans until the 18th century. From 1690 until 1693, the German physician Engelbert Kaempfer was at the Dutch Embassy in Nagasaki and had what at that time was a rare opportunity to study all aspects of Japanese life. In 1727 his observations on Japanese government, buildings, minerals, flora and fauna were published in London as the *History of Japan*. In this book, puffer fish are well described and indeed illustrated. Moreover, it is recorded by Kaempfer that should a whole fish be eaten it would "unavoidably to occasion death". It would have served the great English explorer and cartographer Captain James Cook well if he had managed to procure a copy of Kaempfer's book before embarking on his second journey around the world in 1774. On the evening of 7 September 1774, *HMS Resolution* was off Cape Colnet, New Caledonia. The captain dined with Mr. Johann R. Forster, the naturalist on the expedition, and Mr. Forster's son. Their repast consisted of liver and roe from a fish which was identified as the puffer fish *Tetraodon scleratus*. Captain Cook wrote in his journal that he ate a piece of liver "about the size of a crown piece" (this coin has a diameter of 38 mm). The effects of puffer fish poisoning followed soon after. "About three or four o'clock in the morning (of 8 September 1774) we found ourselves seized with an extraordinary weakness and numbness over all our limbs attended with numbness of sensation like to that caused by exposing one's hands and feet to a fire after having been pinched much by frost. I had almost lost the sense of feeling; nor could I distinguish between light and heavy bodies, of such that I had strength to move, a quart pot full of water (1.136 litres) and feather being the same in my hand". It was then reported by Mr. Forster that "The Captain threw up very little, I a great deal and my son less than I, but a great deal more than the Captain". The vomiting probably prevented further intoxication and happily all three eventually made a full recovery but it took four days. On 4 September 1845, two members of the ship's company aboard the Dutch Brig-of-War *Postillion* were not so lucky after eating puffer fish in Simon's Bay on the Cape of Good Hope. Mr. Hellmuth the ship's surgeon reported them dead after just 17 minutes! Indeed the lethal effects of tetrodotoxin are so dramatic that it was used on more than one occasion against the British Secret Service agent 007. The Russian spy Rosa Klebb kicked James Bond in the right calf with a toxin-coated steel needle protruding from the toe of her shoe in *From Russia, With Love*. The resulting symptoms presented were of classical tetrodotoxication...... "Numbness was creeping up Bond's body......There was no feeling in his fingers. They seemed as big as cucumbers. His hand fell heavily to his side......Breathing became difficult......Now he had to gasp for breath......Bond felt his knees begin to buckle......". Later, in *Dr. No*, this poison was indeed identified as originating from *Fugu*. Sometimes the puffer fish poisoning is a complete surprise. In 1977 in Italy, three people died of tetrodotoxication after eating frozen puffer fish from Taiwan which had been incorrectly labelled as angler fish.

From a biochemist's point of view, it was always going to be important to purify the toxin, determine its chemical structure and elucidate its biological mode of action at the molecular level. It was already recognised that tetrodotoxin was present in high concentrations in puffer fish ovaries as this had been determined empirically by bioassay, namely when people ate the ovaries they died! Consequently, this was a good rich source from which to purify the toxin. The first purification was reported in 1911 by Tahara and generated a product which was about 4% pure. However, it was not until the early 1950s that pure crystalline tetrodotoxin was obtained. Although the toxin was called "spheroidine" when first purified, the original name of tetrodotoxin was later universally adopted. After many years of chemical analysis the structure of the toxin was finally solved in 1964 and is shown in Figure 4.

The tetrodotoxin molecule has a molecular weight of 319.28 ($C_{11}H_{17}N_3O_8$) and possesses a very unusual structure which is based around a rigid central cage-like scaffold. Its full systematic chemical name is octahydro-12-(hydroxymethyl)-2-imino-5,9:7, 10a-dimethano-10aH-[1,3]dioxocino[6,5-d]pyrimidine-4,7,10,11, 12-pentol, so it is no wonder scientists have retained the name

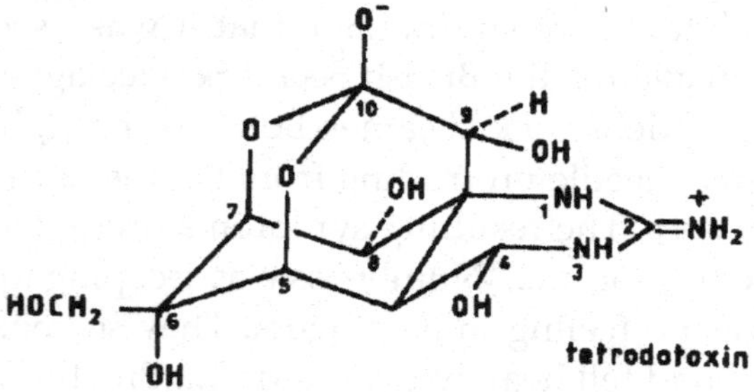

tetrodotoxin

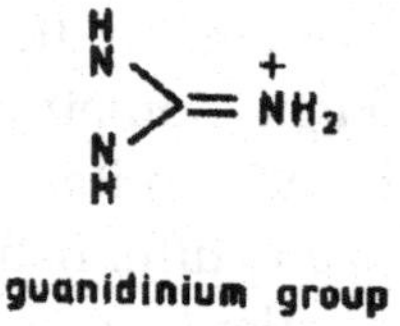

guanidinium group

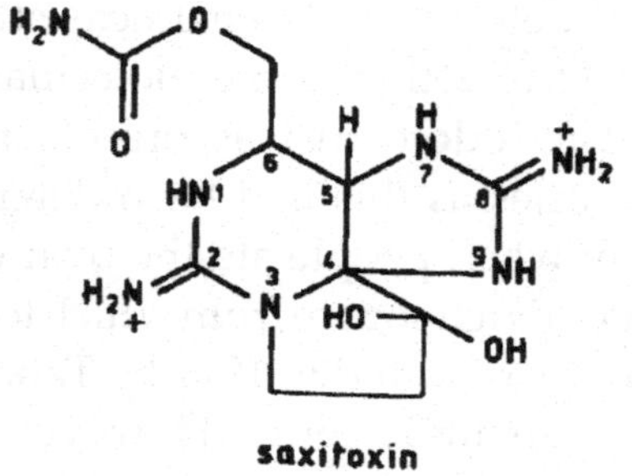

saxitoxin

Figure 4. The chemical structures of tetrodotoxin and saxitoxin. The numbering refers to the specific position of atoms within the molecule and is cited in the text. The guanidinium group is critical for the toxicity of these compounds and its formula is also shown.

tetrodotoxin. It has also been established that the absolute configuration of crystalline tetrodotoxin is the zwitterionic form, i.e. possessing both a negative charge (attached to carbon-10) and a positive charge (contributed by the guanidinium moiety). The structure of the toxin is generally stable and it is not inactivated by heating. It has been shown however, that it is destroyed in alkaline conditions. When toxicological tests were performed on tetrodotoxin, it was found to be one of the most toxic substances known. It is 275 times deadlier than cyanide and 50-fold more lethal than strychnine (also described in this book), so Captain Cook and the Forsters had lucky escapes all those years ago. Because in certain parts of the world the local populace have persisted in eating puffer fish species, there is a small but steady admission of people into intensive care clinics suffering from puffer fish poisoning. As a result, the progressive physiological ramifications of ingesting tetrodotoxin are comprehensively reported in medical literature. The onset of symptoms following puffer fish poisoning usually occurs between two and 30 minutes after ingestion. Indeed the first symptoms are often perceived by the person as they are still eating the fish. The time of onset, together with the range and severity of the toxic effects, is determined by the amount of tetrodotoxin consumed and by individual variation. The following symptoms were recorded after an otherwise fit 25-year-old man ate contaminated *fugu*. After five to ten minutes he noticed a tingling in his lips, his tongue and on the right side of his mouth. This was followed by dizziness, headache, a constricting feeling in his throat, nausea and vomiting (as with Captain Cook and the Forsters above). He found speaking progressively difficult, had facial flushing and complained of tightness in his upper chest. There was a gradual ascending paralysis, his legs weakened and he collapsed. Poor muscular co-ordination, respiratory distress, an inability to speak (aphonia) and difficulty in swallowing (dysphagia) were also observed. He developed blue discolouration of the skin (cyanosis) due to the presence of deoxygenated blood caused by the defective respiration and had low blood pressure (hypotension). In a normal individual the blood vessels are slightly contracted due to the action of the sympathetic nervous system. In this patient, the toxin damaged certain nerves, thereby reducing this sympathetic tone of the vasculature. The effect of this was that the blood vessels dilated. This dilation was responsible for generating the observed hypotension and furthermore, explains why the face of the individual looked flushed. Finally, he experienced impaired consciousness and convulsions before suffering respiratory failure resulting from paralysis of the diaphragm, severe hypotension and cardiac arrhythmias. The respiratory paralysis and cardiovascular collapse culminated in his death. All of these symptoms are caused by tetrodotoxin affecting nerves in the brain as well as inhibiting motor, sensory and autonomic nerves in the periphery. It is generally the case, that if a patient is admitted to a clinic with puffer fish poisoning then the treatment administered is merely supportive. There is no specific antidote to tetrodotoxin. Vomiting is induced by administering an emetic to remove undigested toxin. However, violent vomiting may have already occurred naturally (as mentioned above for Mr. Forster on Cook's *HMS Resolution*) due to tetrodotoxin acting on the brainstem. In addition, the stomach is washed out with 2% (weight/volume) sodium bicarbonate solution. The rationale for this is, as noted above, that the toxin is unstable in the alkaline conditions existing in the presence of bicarbonate. Activated charcoal is then given to absorb any remaining toxin and thereby render it unavailable to the patient's system. Oxygen therapy and artificial ventilation would also be supplied to support the impaired respiratory system following tetrodotoxication. Support would also be provided to the circulation in the form of volume expansion infusions to combat the detrimental effects of the severe vasodilation/hypotension that the toxin generates. Experience has shown that if a victim survives 24 hours then they will probably make a full recovery.

Although there is no antitoxin available currently, there have been reports in the scientific literature that antibodies have been successfully raised against tetrodotoxin. Antibodies are generated by injecting small amounts of the chemical of interest (called the antigen) into an animal. This challenges the animal's immune system as the antigen is recognised as "foreign". Antibodies which specifically bind and neutralise the antigen are duly produced by the animal. These can then be harvested, purified and used subsequently as a reagent for therapeutic or scientific applications. The obvious question to this claim is probably, how is it possible raise an antibody to something as lethal as tetrodotoxin when injecting even small amounts will kill the mouse? The answer lies in the fact that prior to immunisation, the toxin was chemically coupled to haemocyanin, a protein from the Keyhole Limpet (often abbreviated to KLH). This is a routine procedure in immunology and ensures that a good immunological response is obtained to what would otherwise be a small molecule. With tetrodotoxin, this formation of a chemical couple had the useful ramification of rendering the molecule non-toxic. Consequently, antibodies which recognised normal tetrodotoxin were obtained by using the non-lethal toxin − KLH conjugate to inject the mouse. With reference to Figure 4, the chemical coupling was via the guanidinyl group of the toxin whereas the antibody bound to the OH-groups on C-4 and/or C-9 of the tetrodotoxin. Interestingly, it has also been shown that prior injection of this antibody was able to protect mice from a dose of tetrodotoxin which was lethal to normal mice. This raises the possibility that at some point in the future, it might be possible to administer antibodies to a patient which would bind circulating toxin and thereby neutralise it.

The creeping helplessness experienced by those suffering from tetrodotoxin poisoning has led to the intriguing suggestion that puffer fish is used in voodoo practices in areas of the Caribbean, notably Haiti, to produce the "living dead" or zombies. Analysis of such "zombie potions" revealed that they did indeed contain dried puffer fish together with arachnids, reptiles, annelids, human bones and flesh plus some psychoactive plants such as *Mucuna* and *Datura* species. However, when five such powders were investigated further, tetrodotoxin was absent despite the presence of puffer fish remnants. This observation is explained by the discovery by the investigators that the potions were highly alkaline. As noted above, tetrodotoxin is destroyed under such conditions. Nevertheless, one could perhaps envisage that occasionally a potion is obtained that is not quite so alkaline.

The natural presence of tetrodotoxin undoubtedly provides the puffer fish with an efficient deterrent to any prospective predators and as such endows the fish with a survival advantage. This is obviously in addition to the more benign, but nevertheless effective, co-existing defence strategy exhibited by these fish of inflating into a spherical shape when threatened. Furthermore, the existence of the toxin in puffer fish roe ensures that the protective cloak of the tetrodotoxin is present throughout the life-cycle of the fish. Although possession of the toxin is used only passively as a defence mechanism by most puffer fish, this is not universally the case. Several species of puffer fish belonging to the genus *Takifugu* possess specialised exocrine glands in their skin which contain the tetrodotoxin. When alarmed, for example by the approach of a predator, fish of this genus expel the contents of these glands into the water where it is thought that the toxin may act as a repellent to any attacker. In addition to defence, another action for tetrodotoxin has been suggested by studies using the puffer fish *Fugu niphobles*. It was established that in this species the toxin acted as a chemical attractant, or pheromone, to spermating males but not to sexually active females. Furthermore, the sensitivity of the males to responding to the toxin was exquisite, such that they could detect as little as 5 ng (0.000000005 g) per litre. Studying females caught during the spawning season (June), it was found that tetrodotoxin was released through the vitelline coat of the oocytes (a stage of egg maturation) during ovulation. During the breeding season, *Fugu niphobles* form schools and rush to beaches for spawning. It has been postulated that the tetrodotoxin released into the ovarian cavity by ovulated oocytes is discharged from the cloaca prior to spawning and has a role in forming these breeding schools of puffer fish.

From what has been cited above, sound scientific reasons could be put forward as to why the puffer fish species have developed biochemical pathways for the synthesis of such a complicated molecule. The molecular structure of the tetrodotoxin is quite unlike any of the end-products or intermediates of the metabolic pathways which have been described in fish. Given this complex nature of the molecule, it was long thought by biologists that the toxin was a peculiarity of the puffer fish. Consequently, it was envisaged that the puffer fish species had evolved discrete metabolic pathways to synthesize this elaborate structure. This would imply that the necessary chemical reactions were not present in the biochemical repertoire of other animals. In the 1930s, developmental biologists studying embryology and the control of the differentiation of tissues often employed the experimental strategy of transplanting rudimentary organs and limbs between the embryos of different amphibian species. Dr. Twitty used egg clusters of the Californian newt *Taricha torosa* for such experiments, largely because this species of newt was to be found on the campus of Stanford University where he undertook his investigations. Surprisingly, when he transplanted the rudimentary limbs from *T. torosa* onto other newt species, the host newt developed a paralysis before eventually recovering. Obviously, the newt *T. torosa* contained a powerful neurotoxin. Furthermore, a similar neurotoxin was found in other species of newt which belong to the family *Salamandridae*. These included the Japanese newt *Cynops pyrrhogaster* and the East American newt *Notophthalmus viridescens*. Originally this newt toxin was called tarichatoxin after the genus of the Californian newt. However, after extensive neuropharmacological and structural analysis it was concluded that tarichatoxin was actually tetrodotoxin. Interestingly, it had been reported independently that both puffer fish and certain newts were resistant to tetrodotoxin poisoning. However, these two animals are not related in zoological terms. Although the two species belong to the same phylum, *Chordata*, they belong to different classes, namely *Osteichthyes* and *Amphibia*, respectively. Does this mean that the metabolic pathways for synthesizing tetrodotoxin were developed during evolution not once, but twice? Although this is obviously not impossible it does not seem very likely given the complex nature of the molecule. Nevertheless, both puffer fish and newts, as well as their eggs, are protected by the toxin.

As has already been cited, causes of tetrodotoxin poisoning and subsequent fatalities have been reported for centuries. Up to a point this is understandable because the inhabitants of coastal villages have relied heavily on fish as the mainstay of their diet. It might be reasonable to suppose that tetrodotoxin poisoning resulting from ingestion of newts would not occur. However, the human spirit of adventure knows no bounds and reports have appeared in the medical literature of humans being poisoned by newts. The Oregon rough-skinned newt (*Taricha granulosa*) has caused two fatalities after being swallowed. Following a drinking session in 1971 a 26-year-old man ate five newts for a bet! Following the onset of symptoms similar to puffer fish poisoning he received treatment and survived. In the subsequent medical report, the advice given to physicians treating newt ingestion in the future was similar to that recommended for treating puffer fish poisoning, with the addition of removing the animals from the stomach by gastroscopy! The average 10 g newt contains c. 250 micrograms of toxin which is enough to kill 1500 mice.

Given that tetrodotoxin had been shown to exist in the *Salamandridae* family as well as in the *Tetraodontidae* fish, scientists began looking for the toxin in other families within the *Amphibia*. It has long been observed that when nature wants to indicate that something is dangerous, it is often brightly coloured as a warning. There are many examples of such "Keep Off!" signs. One example is the highly coloured frogs found in Central and South America. These small frogs are indeed both marked with vivid colours and are highly poisonous. In fact, potions prepared from the skin of such frogs have been used since ancient times as a poison for coating arrows. Studying a range of brilliantly coloured frogs from the families *Atelopidae* and *Dendrobatidae*, Drs. Fuhrman and Mosher discovered that several of these species contained tetrodotoxin in their skin and sometimes in their eggs. These species included the green and black *Atelopus varius ambulatorius*, the mottled blue-grey/rusty brown *Atelopus chiriquiensis* and the gaudy *Atelopus varius varius* with its black, red and yellow markings. All of these frogs are taxonomically related and are native to Costa Rica. The Panamanian black and orange frog *Atelopus varius zeteki* also contains tetrodotoxin. Subsequently it has become well-established that tetrodotoxin is present in a wide range of animals and is not restricted to the phylum *Chordata*. For example, the toxin has been reported in a variety of gastropod molluscs including the Japanese shellfish *Babylonia japonica* (the Ivory Shell) and *Charonia sauliae* (the Trumpet Shell). Tetrodotoxin has also been found in starfish (phylum *Echinodermata*), certain crabs (phylum *Arthropoda*) such as the Taiwanese crab *Lophozozymus pictor* and is also found in *Pseudopotamia occelata* (phylum *Annelida*). Recent food-poisoning incidents in South Taiwan were reported to be due to tetrodotoxin in the shellfish *Niotha clathrata* and *Zeuxis scalaris*. Although outbreaks of human tetrodotoxin poisoning following the consumption of marine organisms, other than puffer fish, are rare they do occur occasionally. In 1995, 17 people in Taiwan were treated for the typical clinical symptoms of tetrodotoxification after eating gastropod molluscs. The marine molluscs were later identified as *Nassarius castus* and *Nassarius conoidalis*. Specimens of both of these species were collected from the area. When they were chemically analysed the presence of the toxin was confirmed. Interestingly, as with the puffer fish species, it has been found that the level of the toxin present in the shellfish is highly seasonal. The Xanthid crab *Zosimus aeneus* has also been implicated in poisoning. When specimens of the crab were collected from various locations including the Negros Islands, Fiji, the Philippines and Okinawa, all were shown to contain the tetrodotoxin. All of these various species cited above use tetrodotoxin in a passive way, in that its presence in their bodies or eggs will deter predators. However, the cephalopod molluscs *Hapalochlaena lunulata* and *H. maculosa* use the toxin in an altogether more aggressive manner as it constitutes part of the offensive, rather than the defensive strategy. *Hapalochlaena lunulata* and *H. maculosa* are commonly referred to as blue-ringed octopi (see Figure 5). The blue-ringed octopus is common in the southern oceans and particularly in the waters around Australia. Its toxin, originally termed maculotoxin but unambiguously identified as tetrodotoxin in 1978 by Dr. Sheumach and co-workers, is present in the posterior salivary glands rather than the more usual (for puffer fish) distribution in the gonads and/or skin. As such, the toxin constitutes a major component of this octopus' venom. Although it is only 10 cm in length, the bite of an adult blue-ringed octopus presents a real danger to humans. Death can occur as quickly as 90 minutes following a bite. Secret agent 007, James Bond, was confronted by tetrodotoxin from this source when he was threatened with a blue-ringed octopus in *Octopussy*.

Although originally named after the tetraodontidae fish, it is now very apparent that the occurrence of tetrodotoxin is not restricted to this family. In fact it is not even the unique preserve of one phylum, as it has been identified in a series of taxonomically unrelated animals representing at least five different phyla, as noted above. The toxin has no obvious similarity to any known animal or plant natural product, so its synthesis does not merely result from the simple modification of a related (non-toxic) precursor. Whilst it was thought that tetrodotoxin only occurred in fish and newts, it was just possible that the toxin had been "invented" twice

Figure 5. The blue-ringed octopus.

during evolution. However, it is totally unfeasible that it has been independently developed on five separate occasions. An alternative scenario is that the tetrodotoxin is not actually synthesised by the puffer fish, newts, etc. but that it is merely acquired and accumulated by them from an exogenous source.

A common experimental protocol for establishing metabolic pathways is to utilise a precursor molecule which incorporates a radioactive label. The radioactivity is easily identified and quantified. This approach enables the biochemist to follow the fate of the radiolabelled precursor as it fluxes through a biochemical pathway, generating radiolabelled intermediates as it does so. In this way the newts *Taricha torosa* and *Taricha granulosa* were fed [^{14}C]acetate, which is a radiolabelled form of a metabolic building block. It was found that although guanidinium-containing compounds such as carnosine and creatine were efficiently labelled by the [^{14}C]acetate, the tetrodotoxin (which also contains a guanidinium group; see Figure 4) remained unlabelled. These data are perhaps indicative that the toxin does not have an endogenous origin. This conclusion was supported by the observation that if puffer fish, or newts, are bred in captivity (sometimes referred to as cultured puffer fish), then they do not contain the poison. If however, these cultured puffer fish (or newts) are then allowed close contact with wild puffer fish, they subsequently acquire tetrodotoxin with the usual concentration and tissue distribution. In addition, tetrodotoxin could be detected in the cultured puffer fish some time after they had been fed the livers of toxic fish. All of these data suggest that the actual source of the tetrodotoxin is an exogenous organism and that the toxin is either accumulated through the normal food-chain or via some symbiotic relationship between the exogenous organism and the fish. A number of bacteria have now been identified which produce tetrodotoxin. Of particular interest in this regard are a series of experiments by Dr. Yasumoto and his colleagues who established that calcareous red algae of the *Jania* species collected from several locations, including the seas off the Gambier Islands, French Polynesia and Okinawa, contained tetrodotoxin. Bacteria were then isolated from the red algae and cultured. *Alteromonas* species present in this culture were shown to produce the toxin. These experiments demonstrate quite clearly how the toxin can move through the food-chain in this coral reef environment. The *Alteromonas* bacteria produce the tetrodotoxin and are transferred to other animals, including herbivorous fish, which feed on the red algae. *Vibrio* species of bacteria isolated from the gut of Xanthid crabs have also been shown to generate tetrodotoxin. Furthermore, the toxin has been identified in cultures of *Pseudoalteromonas haloplanktis tetraadonis* species obtained from the skin of the puffer fish *Fugu poecilonotus*. This probably explains the initiation of tetrodotoxin production observed in the cultured puffer fish following exposure to wild puffer fish. When the gut of the newt *Taricha granulosa* was washed and the isolates grown on agar plates, bacterial colonies which produced tetrodotoxin were identified. Consequently, as for the puffer fish species, the primary source of tetrodotoxin in newts appears to be specific bacteria in the food-chain. This concept of tetrodotoxin moving up through the food-chain has been strengthened by the recent observation that tetrodotoxin-producing bacteria in marine sediments can transfer the toxin to bacterivorous organisms such as protozoa.

Although it has long been known that tetrodotoxin ingestion affects nerve function, the precise molecular basis for the observed neurotoxic actions has only been elucidated relatively recently. In order to understand how the toxin works, it is necessary to understand how nerves transmit their signals under normal circumstances. Nerve cells (neurones) are "excitable" in that they are capable of generating self-propagating electrical signals known as action potentials. The resting neurone has a potential difference of approximately 70 mV across the cell membrane with the inside negative relative to the outside. This is generated by differences in ionic concentration within and without the cell. If the neurone is electrically stimulated, then this resting potential of −70 mV becomes less negative. If this depolarisation reaches a critical level, referred to as the threshold potential, the neurone develops a series of changes in potential which move in a wave of depolarisation down the nerve. This is known as an action potential, during which the potential difference of the membrane transiently reverses to about +50 mV in a few tenths of a millisecond. This is followed by a phase of repolarisation back to the resting potential, all within approximately two milliseconds. During an action potential there are periods when it is impossible, or more difficult, to initiate a second action potential. These are referred to as absolute, and relative, refractory periods, respectively (Figure 6). Changes in the membrane permeability to specific ions underlies the generation of an action potential. The depolarising stimulation causes a selective and specific increase in the cell membrane's permeability to sodium ions. This increase in sodium ion conductance leads to Na$^+$ entering the neurone and it is this inward sodium current which is responsible for the depolarisation. This increased sodium conductance is transient. As it falls back to a low level, the potassium conductance increases thereby generating an outward potassium current which re-polarises the neurone membrane. The depolarisation-dependent increase in sodium conductance is

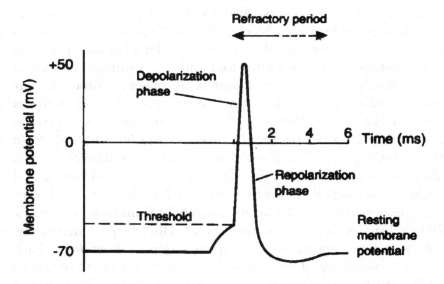

Figure 6. An action potential. The diagram illustrates the sequential changes in membrane potential that occur during an action potential. The refractory period is divided into absolute (solid line) and relative (dashed line). See text for details.

mediated by specialised transmembrane proteins which form a pore, or channel, through the cell membrane. When the neurone is resting, and therefore polarised, these channels exist in a closed conformation which prevents ion flux. Depolarisation of the neurone induces the sodium channel to adopt an open conformation. It is these characteristic properties which has lead to these proteins being referred to as "voltage-gated ion channels". Voltage-gated sodium channels become inactivated following membrane depolarisation. This is a slower process than the voltage-induced activation, so during an action potential the sodium conductance rises then falls. The phenomenon of channel inactivation underlies the refractory period mentioned earlier. A similar voltage-gated situation exists for potassium channels except that the K^+ channels activate more slowly than Na^+ channels thereby generating a potassium current after the sodium current. Unlike Na^+ channels, K^+ channels do not inactivate so that they adopt only open and closed conformations and not an inactivated conformation.

Electrophysiological studies on squid nerve fibres by Moore, Narahashi and co-workers, lead to the discovery in 1967 that tetrodotoxin selectively blocks the voltage-sensitive sodium permeability of nerve membranes. The specificity of this action has made tetrodotoxin a powerful tool and it is routinely employed by neuroscientists in their research. A major advance in our understanding of neuronal functioning was achieved between 1984 and 1988 when Professor Shosaku Numa and his colleagues successfully cloned a number of voltage-gated Na^+ channels from the electric organ of the eel *Electrophorus electricus* and rat brain. This allowed the architecture of the channel protein to be addressed and related to its function for the first time. The Na^+ channels are all very large proteins composed of approximately 2000 amino acid residues. Detailed analysis of the sequences revealed that the protein contains within it four repeats which display sequence homology. Each repeat is composed of six α-helical membrane-spanning segments labelled S1 to S6 (see Figure 7). The loop between S5 and S6 forms a hairpin composed of two short segments which are also membrane-associated and have been designated as SS1 and SS2. The amino-

and carboxy-terminals are on the cytoplasmic side of the membrane. The four repeats are arranged in a pseudosymmetrical array around a central pore (Figure 8). The SS1 and SS2 domains form the lining of this central pore. Therefore each of the four repeats contributes one quarter of the wall of this channel (Figure 8). The segments S1, S2, S3, S5 and S6 are all hydrophobic but S4 is positively charged. In fact, S4 has a positively charged residue at every third position and it is this domain which provides the voltage sensor of the Na^+ channel. Consequently, S4 initiates the change in conformation of the channel from the closed to open state in response to membrane depolarisation. In support of this it has been found that mutations within the S4 segment can alter the voltage at which this gating of the channel occurs by as much as 20 mV. Specific residues near to the mouth of the pore have a key role in determining the Na^+ ion

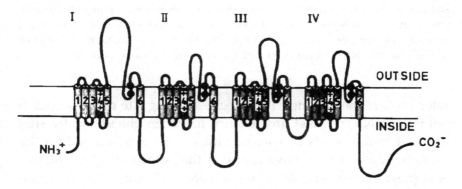

Figure 7. A schematic representation of the voltage-sensitive sodium channel. The single polypeptide chain is composed of four repeats (I, II, III, IV) each comprising six transmembrane domains represented by cylinders numbered 1–6. Transmembrane domain 4 is the voltage sensor and contains positively charged residues. The hairpin loop between 5 and 6 in each repeat lines the pore of the sodium channel protein and is divided into the SS1 domain (entering the membrane) and the SS2 domain (leaving the membrane). The two black circles on the SS2 domain of each repeat represent the position of individual amino acid residues important for tetrodotoxin binding. The arrangement of the repeats in three-dimensional space is shown in Figure 7. In the folded conformation, these important residues form two rings at different depths in the pore of the channel. See text for details.

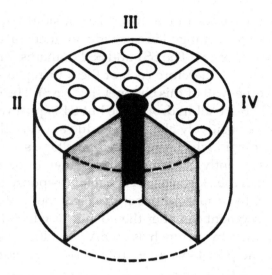

Figure 8. Schematic representation of the architecture of the voltage-sensitive sodium channel. The orientation in the plasma membrane of the polypeptide chain shown in Figure 6 is presented. The four repeats (Roman numerals) are arranged around a central pore. The top of the six transmembrane domains within each repeat is shown. Repeat I has been omitted for clarity. See text for details.

selectivity of the channel. The precise binding site of tetrodotoxin within the Na^+ channel architecture has been elucidated by site-directed mutagenesis of specific amino acid residues within the protein. These studies have revealed that substituting a glutamic acid (negatively charged) at position 387 with glutamine (neutral) decreased tetrodotoxin binding 10,000-fold. This residue is in the SS2 domain of the first repeat. Interestingly, acidic (negatively charged) amino acids occur at the analogous position in the SS2 domain of the other three repeats and it was found that they were all required for toxin binding. As the SS2 domains of the four repeats line the wall of the pore, these four residues effectively cluster to form a ring in the mouth of the channel. A second ring of SS2 residues only slightly deeper into the channel is also necessary for tetrodotoxin binding (Figure 7). If any of the negatively charged residues in these two ring clusters is neutralised, toxin binding is reduced and the ability of the channel to conduct Na^+ ions is also impaired. Therefore, the negative charges are fundamentally important to both the conductance of Na^+ ions and to binding the tetrodotoxin. This perhaps is not surprising as both the sodium ion and the toxin carry a positive charge which would interact with the negative charges in the mouth of the channel. The absolute necessity for the guanidinium group in tetrodotoxin (see Figure 4) is entirely consistent with this scenario. Consequently, the molecular basis for the effects of tetrodotoxin is that it binds to a site located near to the extracellular side of the pore of the Na^+ channel thereby impeding the movement of Na^+ ions through the channel. The interaction of the toxin with the channel is not entirely electrostatic however. Cardiac sodium channels possess all of these eight SS2 "ring" residues but they bind tetrodotoxin with 200-fold lower affinity and, unlike brain Na^+ channels, are blocked by external cadmium or zinc. The cardiac Na^+ channel differs from that of brain and skeletal muscle by having a cysteine at position 374 rather than an aromatic amino acid such as tyrosine or phenylalanine. This position is in the SS2 domain and is juxtaposed to the "ring clusters". Substitution of this cysteine in the cardiac channel by an aromatic residue increased toxin binding and decreased sensitivity to cadmium and zinc. The converse mutation in the brain channel made it more "heart-like" by reducing tetrodotoxin binding and increasing sensitivity to Cd^{2+} and Zn^{2+}. Therefore, this aromatic residue also has a key role in toxin binding. Interestingly, many local anaesthetics used in the clinic also block Na^+ channels.

Tetrodotoxin is not the only marine toxin to block voltage-gated Na^+ channels. Saxitoxin has a very different structure (see Figure 4) but has a virtually identical biological action. Although saxitoxin has two guanidinium groups, only one of these (at position 7, 8, 9) is analogous to the guanidinium group in tetrodotoxin and is directly involved in blocking Na^+ channels. Saxitoxin is responsible for paralytic shellfish poisoning (PSP) which results from eating contaminated mussels, cockles and clams. Indeed the toxin is so named after being extracted from the Alaskan butterclam (*Saxidomus giganteus*). An early report of the effect of this toxin is given in Captain George Vancouver's account of exploring the British Columbian coastline in 1793. On 15 June, some seamen breakfasted on roasted mussels, felt ill within the hour and one of them, John Carter, was dead within five hours. Outbreaks of PSP are associated with a massive proliferation of micro-organisms floating in the surface waters of the sea, particularly photosynthetic algae, or dinoflagellates, of the genus *Gonyaulax*. Although present throughout the year, given the right conditions such as warming sea temperature, sunlight for photosynthesis, nutrients and a calm sea, there can be a population explosion of *Gonyaulax*. This can be of such intensity that the sea actually changes colour to brown, cream or red and is strangely luminescent at night. For this reason these blooms of algae have been referred to as "red tides". Photosynthetic micro-organisms (including *Gonyaulax*) have been described as "the grass of the sea" and as such are vital to marine food-chains. *Gonyaulax* species contain saxitoxin and large amounts of this toxin enter the food chains subsequent to a red tide. Bivalve molluscs feed continuously by filtering small particles from the sea water. Consequently, following a red tide these filter-feeders concentrate the saxitoxin. Unfortunately, the shellfish harvested commercially such as mussels (*Mytilus edulis*), cockles (*Cardium edule*), scallops (*Pecten maximus*) and Queen scallops (*Chlamys opercularis*) are remarkably resistant to saxitoxin, so appear healthy to the local fishermen who collect them as normal. Although red tides are usually associated with the warm waters of the tropics, a major outbreak of shellfish poisoning occurred in Northumberland, England. I can personally testify that the seas off Northumberland are not warm and bathing there is like swimming in ice-cream but nevertheless, a bloom of *Gonyaulax tamarensis* occurred in mid-May 1968 giving rise to the first recorded PSP epidemic in England. On 14 May unusual luminescence was observed at night in the waters around the Farne Islands and Holy Island (Lindisfarne). Between 23–31 May deep brown discolouration of the sea was reported by fishermen four miles off Blyth and ten miles ENE of Tynemouth. Plankton in the North Sea is regularly monitored by recorders which are routinely towed by merchant ships. These revealed a very high phytoplankton-count (measured as chlorophyll A) and an increasing concentration of *Gonyaulax tamarensis* throughout May with a peak between 11–19 May. Mussels collected from Holy Island were sold in Newcastle-upon-Tyne on 30 May and the first of 78 people developed saxitoxin poisoning 90 minutes after eating cooked muscles. Symptoms included vomiting, clumsy movement, limb weakness, inability to talk and tingling in the mouth "like a dentist's local anaesthetic". Warning signs such as that in Figure 9 were erected along the Northumbrian coast. Saxitoxin contamination is measured in mouse units where one unit is the amount of toxin to kill a 20 g mouse. Symptoms in humans appear after ingestion of 5000 units, are severe at 20,000 units and usually lethal at 30,000 units. At the start of June 1968, mussels from Holy Island equalled the world record for saxitoxin contamination with 20,800 units per 100 g flesh but remained healthy. Fortunately all 78 people poisoned recovered. This was partly due to the British habit of cooking mussels by boiling thoroughly for about 20 minutes. The water which would contain 30%– 40% of the saxitoxin is then discarded. Interestingly, two men who both ate 30,000 units of toxin after drinking ten pints of beer (5.7 litres) each, suffered no ill effects. Either the beer neutralised/diluted the toxin or the symptoms of saxitoxin intoxication were masked by ethanol intoxication!

Saxitoxin also entered other food chains. The sand-eel (*Ammodytes tobianus*) forms a major part of the diet of many sea birds. In late May 1968, large numbers of dead sand-eels were reported washed ashore and floating in the fishing grounds around the Farne Islands. Obviously they had received saxitoxin from the dinoflagellates via the fish lavae and small crustaceans on which sand-eels feed. The effect on the local birds which feed on these eels, particularly shags (*Phalacrocorax aristotelis*), was catastrophic. Approximately 80% of the 1200 shags in the Farne Islands area died. Sea bird colonies are closely studied on the Farnes and of the 298 nests present on the 18 May only nine were still occupied by the 23 May 1968. Deaths of Arctic tern (*Sterna macrura*) and common tern (*S. hirundo*) occurred a week after the peak of shag fatalities. The symptoms of dying birds were similar to saxitoxification in humans

Figure 9. Contaminated shellfish warning notice. These notices were posted by the Environmental Health Department at various positions around the coastline of Northumberland following contamination of shellfish by algae. This notice was at Chare Ends on Holy Island (Lindisfarne). Hitherto, shellfish had been harvested from the mussel beds here continually from at least the seventh century, when land was granted to Saint Aiden by King Oswald of Northumbria to found Lindisfarne Priory.

viz. vomiting, paralysis and respiratory failure. Birds feeding off-shore such as puffins (*Fratercula arctica*) and kittiwakes (*Rissa tridactyla*) were largely unaffected as the highest toxin levels were inshore. Surprisingly, eider ducks (*Somateria mollissima*), which feed predominately on mussels, were unaffected. To date no scientific basis for the preservation of the eider ducks has been proposed. In Northumberland itself, there are those who attribute the eiders' survival to their protection by the Northumbrian saint, Cuthbert (634–687). Cuthbert was Bishop of Lindisfarne, lived many years on the Farne Islands and was known to have a particular affection for eider ducks. Indeed, in the area of the saxitoxin poisoning outbreak, eiders are still referred to locally as "St. Cuthbert's birds". By the end of August saxitoxin contamination of shellfish had fallen to safe levels. At no time was saxitoxin detected in edible crabs (*Cancer pagurus*), lobsters (*Homarus vulgaris*) or commercial fish.

From what has been presented on tetrodotoxin and saxitoxin, it can be appreciated what a two-edged sword these toxins are. On the one hand they have proven to be invaluable probes for the biochemist in the quest for understanding the workings of the nervous system and on the other, they are impressively efficient "molecules of death" with no antidote. Better in the test tube than in the stomach!

Acknowledgements

I am grateful to Liz Kamai (Faculty of Engineering, Gunma University, Japan) and to her family and friends for invaluable information on Japanese "*Fugu* culture". The excellent photographs of the inflated puffer fish and the blue-ringed octopus were taken by Dr. Roy L Caldwell (Department of Integrative Biology, University of California at Berkeley, USA) to whom I was introduced via the cephalopod website run by Dr. James B. Wood (Department of Biology, Dalhousie University, Halifax, Nova Scotia, Canada). The traditional Japanese calligraphy was by Ichita Kamei. The algae contamination notice on Lindisfarne (Holy Island), Northumberland, was photographed by Barbara A. Hanson.

Suggested Further Reading

Catterall WA (1993). Structure and function of voltage-gated ion channels. *Trends in Neuroscience* **16**, 500–506.
Cephalopod Web Site: http://is.dal.ca/ceph/wood.html.
James Bond novels by Ian Fleming were originally published by Jonathan Cape Ltd.

Puffer Fish Web Site: http://fugu.hgmp.mrc.ac.uk/fugu/pffp/pf.html.
Ritchie KB *et al.* (2000). A tetrodotoxin-producing marine pathogen. *Nature* **404**, 354.
Kao CY and Levinson SR (eds.) (1986). Tetrodotoxin, saxitoxin and the molecular biology of the sodium channel. In: *Annals of the New York Academy of Sciences*, Vol. 479.

23

THALLIUM

D. B. Ramsden

Introduction

The discovery and isolation of thallium was the consequence of deliberate research. William Crookes extracted selenium from the seleniferous mud deposits obtained from the lead chambers used to produce sulphuric acid at the Tilkerode works in Harz. The residue remaining after distillation was supposed to contain tellurium, identified some 60 years before, but when examined spectroscopically a new bright green line was observed. This was attributed to the presence of a previously unknown element which was named thallium, as Crookes explained in a paper, "*On the existence of a new element, probably of the sulphur group*" accompanying the announcement of his discovery on 30 March 1861; "from the Greek θαλλός, or Latin *thallus*, a budding twig — a word which is frequently employed to express the beautiful green tint of young vegetation; and which I have chosen because the green line which it communicates to the spectrum recalls with peculiar vividness the fresh colour of vegetation in spring". The word *thallus* had also entered usage in botany a few years earlier to describe, "a vegetable structure without vascular tissue in which there is no differentiation into stem and leaves and from which true roots are absent" as often seen in algae and fungi.

Some difficulty in classifying thallium was at first experienced, for it resembles the alkali metals (e.g. sodium, potassium, etc.) in some particulars and lead in others. Thallium shows little resemblance, if any, to aluminium, but the Periodic table could not find any space for it other than in this group. It is now appreciated that no great likeness should be expected between elements in rows so far apart, even if they do occupy the same group (cf. lead and silicon, mercury and magnesium).

Thallium is obtained from the dust deposited in the flues of the pyrites burners used in the manufacture of sulphuric acid. The separation of thallium from other metals depends on the fact that it is the only metal with a soluble carbonate and an insoluble chloride. It is a bluish-white metal, easily deformed and softer than lead, being present at concentrations of 0.6–0.7 ppm in the earth's crust, and although it can be found in pure metallic form it is more commonly associated in mineral combination.

Thallium salts have been utilized as a depilatory, being marketed to treat scalp ringworm as they rapidly caused alopecia (baldness) enabling ointments to control the fungal infection to be applied more effectively. Thallium-containing creams could be purchased over-the-counter until the 1950s and were frequently used to remove unwanted facial hair. Following fatalities from these procedures, the treatment fell out of fashion and the sale of thallium and its salts became strictly controlled. Thallium salts have also been employed therapeutically to treat the night-sweats of tuberculosis, venereal infections such as syphilis, dysentery and gout, all with doubtful success. The only current medical use is that of thallium isotopes (^{201}Tl) being administered intravenously as the dipyridamole complex to permit myocardial imaging in patients with coronary artery disease. The results of the test are used to predict which patients have a good long-term prognosis.

Other uses of thallium and its salts are in the manufacture of imitation jewellery, low-temperature thermometers, ceramic semiconductor material, scintillation counters for radioactivity quantitation and in optical lenses to which it confers a high refractive index. Its widespread use as an ant and rat poison, although extremely effective as the rodents cannot recognise its insiduous action and therefore return to eat more of the bait, has been banned for the past 30 years because of its severe human toxicity.

Human exposure to thallium occurs by oral, dermal and inhalation routes. Cases of accidental toxicity usually occur through industrial exposure of workers in coal-fired power plants, lead and cadmium smelting operations and glass factories. Environmental toxicity from thallium can occur in communities living near cement plants that emit thallium dust. This settles in the neighbourhood and contaminates soil and plants. As it is chemically similar to potassium, thallium can be readily assimilated into plants and so reaches the food chain; consumption of home-grown vegetables and fruit in contaminated areas can give urinary thallium levels of greater than 75 µg/litre (normal levels less than 2 µg/litre) indicating significant environmental exposure although the symptoms of toxicity may not be evident. Concerns have also been expressed over rainwater draining through spoil-banks from lead and zinc mines and infiltrating drinking supplies.

Deliberate intake, in an attempt to commit suicide, is now relatively rare owing to the limited availability of thallium compounds. In the past, rodenticides were the main source of thallium-related suicides; the rat poisons, usually as pastes, were taken in water or a beverage or as a sandwich spread. Apart from accidental or deliberate intake, thallium has long been recognised as an ideal murder weapon. The victim is usually unaware, especially in the early days, of the presence of the toxin and as such it approximates the "perfect poison", appearing fictionally in many detective novels including one penned by the crime-writer, Agatha Christie ("The Pale Horse"). In reality, it has been used as a means of political assassination and also in domestic killings and attempts to cause malicious criminal damage. The reader is referred to the list at the end of the chapter for further references and details of its various nefarious applications.

Thallium is extremely toxic, the lethal oral dose ranging from 0.5 to 1.0 g of soluble thallium salts. These salts can be readily absorbed across the skin, even when protective rubber gloves are worn. Gastrointestinal absorption is also rapid and cell membranes pose no barriers to thallium ions, which are known to cross the placenta into the developing foetus. Thallium appears in the urine within one hour of exposure, but the elimination half-life varies between two and 30 days, depending upon how much material has been absorbed and over how long a period, so that both acute and chronic toxicity can occur.

Toxicity Profile

Acute exposure

With acute intoxication, the exposure to a large single dose, there is usually an initial hypotension and bradycardia owing to direct effects of the sinus node and cardiac muscle, followed by hypertension and tachycardia thought to be due to vagal nerve degeneration. Myocardial necrosis may occur with dysrhythmias and heart failure. The nervous system problems range from peripheral neuropathies to an ascending paralysis, via coma to death from respiratory failure. If the thallium intake was via the gastrointestinal tract the expected symptoms of colic, nausea, vomiting and diarrhoea will be present followed by centrilobular necrosis of the liver. If the patient survives they will show, in part, the symptoms described for chronic toxicity, although central and peripheral nervous system abnormalities may persist including ataxia, tremor, foot drop and memory loss.

Chronic exposure

With repeated exposure to small doses the symptoms of toxicity can be variable and rather diffuse, requiring experience in diagnosis in many instances if permanent damage is to be avoided. If the intake is oral, within a few hours abdominal pain and diarrhoea will occur as may be associated with food poisoning. Within hours to days, a painful tingling in the hands and feet ("glove and stocking" polyneuropathy) develops and progresses to "burning" sensations that may be accompanied by partial paralysis. However, the most startling finding in thallium poisoning is that of alopecia. if death is delayed for two to four weeks the hair falls out almost completely. This triad of gastroenteritis, polyneuropathy and alopecia is peculiar to thallium poisoning and the hair loss is the single most specific diagnostic clue.

White streaks (Mee's lines) may be seen on finger-nails and toenails if the intake of thallium has occurred over a long period. This may be followed by slowly progressive deterioration in visual function due to lesions in the retinal nerve fibres, and then by confusion and convulsions, sometimes with hallucinations. When death finally comes it is after degeneration of the heart, liver and kidney; cardiac and renal failure being the ultimate causes.

Mechanism Of Action

The exact mechanism of thallium toxicity is unclear. Like other heavy metals, thallium binds to sulphydryl groups of proteins and mitochondrial membranes thereby inhibiting a range of enzyme reactions and leading to a generalised "poisoning". Hair proteins form a tightly cross-linked structure with many bonds between adjacent sulphur atoms to give disulphide linkages (these linkages are broken and reformed when hair is "permed"). In the presence of thallium, this cross linking is disrupted and thallium-sulphur structures are formed instead, leading to a very weak network. If the hair shafts are examined under polarised light, dark bands can be seen caused by empty spaces in the disrupted hair proteins. These weaken the structure enabling it to break easily when combed or brushed and fall out in tufts almost "overnight".

Throughout the body, thallium ions (Tl^+ 1.54Å) exchange with potassium ions (K^+ 1.44Å) and interfere with oxidative phosphorylation by inhibiting Na^+/K^+ ATPase. Thallium easily crosses the blood-brain barrier to concentrate in the central nervous system with highest levels in the grey matter. Potassium ions, together with sodium, play an essential role in neurotransmission and the dilution with and substitution by thallium ions causes gross disturbance. Axonal degeneration and swelling with distended mitochondria and secondary myelin loss all point to disruption of membrane integrity, possibly due to damage to the

Na^+/K^+ ATPase motivated membrane pumps which maintain the essential ionic balance within these electrically active tissues. Both the lens and optic nerve are normally rich in potassium ions and the introduction of thallium ions enables binding to melanin-type pigments within the structures of the eye and generally affects function and precipitates damage resulting in swelling of the optic disc and visual deterioration.

Treatment

The usual gastrointestinal decontamination procedures are followed, emesis, lavage and activated charcoal, within the first few hours of ingestion. Repeated doses of activated charcoal may be useful as this substance has been shown to effectively bind thallium ions.

The drug D-penicillamine has been used in therapeutic approaches to thallium poisoning and as this normally binds to sulphydryl groups on proteins it may be beneficial in that it can displace thallium ions and so allow them to be excreted. This line of therapy must, however, be used cautiously. Early use of N-acetylcysteine and dithiothreitol, which would work on the same principle, actually increased symptoms and damage as high levels of thallium were displaced from tissue storage sites into the general blood circulation. Dithiocarb, a similar chelating agent, caused a redistribution of thallium to the central nervous system precipitating a worsening of symptoms. Generally, chelating agents are not effective. Potassium chloride diuresis replaces thallium intracellularly and increases thallium excretion but, similarly, may enhance toxicity by elevating thallium blood levels.

The most effective antidote is the dye Prussian (Berlin) blue (potassium ferrichexacyanoferrate) $(K_4[Fe(CN)_6]_3)$ which is able to exchange a potassium ion for a thallium and the resulting stable chemical complex, which is not absorbed, is excreted from the body. This is only of use in increasing faecal excretion and dosages are given orally. However, there is evidence that thallium is excreted against a concentration gradient (active secretion) along the entire length of the gastrointestinal tract as well as in the digestive juices (and bile) and thus can be trapped whilst passing through the gut before potential reabsorption.

Case Histories

Attempted assassination

Four members of a political organisation consumed a snack prepared by their seemingly amiable host. Within two days they all experienced abdominal pain and within a week this pain was followed by burning sensations and a loss of feeling in their hands and feet. After three weeks most of their hair had fallen out and, following this diagnostic sign, they were found to have high levels of thallium in their blood and urine, shed hair and nail clippings. All patients recovered completely after treatment.

Attempted murder

A middle-aged woman came to the hospital with a history of repeated attacks of complete hairloss over a period of ten months. She had diffuse pain in both legs and had experienced some gastrointestinal disturbance, with alternate diarrhoea and constipation. She had noticed a lack of sensation in her fingertips and also tingling and numbness in her feet. She also reported a slow progressive loss of vision which had started about six months after the first attack of alopecia. Traces of thallium were found in her urine and blood. It was later discovered that her husband had several times attempted to poison her with rat poison which contained thallium salts. After a follow-up examination six years later she still retained partial blindness.

Malicious contamination

Four teenagers came to hospital three days after one of them had unexpectedly received marzipan sweets expensively packaged in a box from a well-known supplier. Two of the teenagers had each eaten a piece of marzipan, whilst the other two had shared a piece between them. The next day they had experienced gastrointestinal pain then burning sensations and tingling in their hands and feet. A tentative diagnosis of thallium poisoning was made and atomic absorption spectroscopy showed that each piece of marzipan contained a potentially fatal level of thallium. Antidote therapy with Prussian blue was started. One of the teenagers who had eaten a whole piece of marzipan almost died. The others developed hypertension and tachycardia and then alopecia after about three weeks. All finally recovered without any obvious long-term clinical problems.

Suggested Further Reading

Prick JJG, Smith WGS and Muller L (1955). *Thallium Poisoning*. Elsevier, Amsterdam.

Polson CJ, Green MA and Lee MR (1983). *Clinical Toxicology*, 3rd ed. Pitman Books Ltd, London, Chap. 23, pp. 447–457.

EPILOGUE

So, what can be learnt from a tour through the chapters of this book? What makes a good poison? Are the most potent molecules made by man or fashioned by nature? Is enough known about the mechanisms of poisoning to construct effective antidotes, or even a universal antidote (a true mithridate), a panacea to all toxic ills? Engaging aspects to ponder over by oneself and perhaps discuss with colleagues.

Before despondency and doom descends, let the issue be considered in perspective. Remember that with our food we take into our bodies every day many thousands of essentially unknown chemicals. These foreign entities appear to cause no immediate or short-term, and probably no serious long-term, harm. Those particular chemicals that do show measurable deleterious effects are notorious because they are so few in number. Not millions of them — just a few that are wise to avoid. Nevertheless, this relative handful of molecular species may prove fatal if introduced into the human organism. Although, even here, perhaps surprisingly, differing degrees of responses to the same toxic insult are often observed. The legion influences of nature and nurture will throw up endless variants within populations producing cohorts that will fare better than others when exposed to a toxicological challenge. Mankind, overall, will undoubtedly survive but within the species the individual units, us, have a finite span. We all fall prey to toxic events, a few percent of damage here and a few more percent of damage there and, if one manages to avoid other pitfalls, these injuries will accumulate to assist in our eventual undoing. Unfortunately, living is probably bad for you.

> "Out - out are the lights - out all!
> And, over each quivering form,
> The curtain, a funeral pall,
> Comes down with the rush of a storm,
> While the angels, all pallid and wan,
> Uprising, unveiling, affirm.
> That the play is the tragedy, 'Man',
> And its hero the Conqueror Worm".

The Conqueror Worm
(to become the verses of Ligeia)
Edgar Allan Poe 1809–1849

INDEX